VOLUME FIFTY SIX

ADVANCES IN CHEMICAL ENGINEERING

Advances in Polymer Reaction Engineering

ADVANCES IN CHEMICAL ENGINEERING

Serial Editor

DAVIDE MOSCATELLI

Department of Chemistry,
Materials and Chemical Engineering "Giulio Natta"
Politecnico di Milano, Milano, Italy

VOLUME FIFTY SIX

ADVANCES IN CHEMICAL ENGINEERING

Advances in Polymer Reaction Engineering

Edited by

DAVIDE MOSCATELLI
Department of Chemistry,
Materials and Chemical Engineering "Giulio Natta"
Politecnico di Milano, Milano, Italy

MATTIA SPONCHIONI
Department of Chemistry,
Materials and Chemical Engineering "Giulio Natta"
Politecnico di Milano, Milano, Italy

ELSEVIER

ACADEMIC PRESS
An imprint of Elsevier

Academic Press is an imprint of Elsevier
50 Hampshire Street, 5th Floor, Cambridge, MA 02139, United States
525 B Street, Suite 1650, San Diego, CA 92101, United States
The Boulevard, Langford Lane, Kidlington, Oxford OX5 1GB, United Kingdom
125 London Wall, London, EC2Y 5AS, United Kingdom

First edition 2020

ISBN: 978-0-12-820645-4
ISSN: 0065-2377

For information on all Academic Press publications
visit our website at https://www.elsevier.com/books-and-journals

Publisher: Zoe Kruze
Acquisitions Editor: Sam Mahfoudh
Editorial Project Manager: Leticia M. Lima
Production Project Manager: Denny Mansingh
Cover Designer: Alan Studholme

Typeset by SPi Global, India

Contents

Contributors

Miren Aguirre
POLYMAT and Kimika Aplikatua Saila, Kimika Fakultatea, University of the Basque Country UPV-EHU, Joxe Mari Korta Zentroa, Donostia-San Sebastian, Spain

José Maria Asua
POLYMAT and Kimika Aplikatua Saila, Kimika Fakultatea, University of the Basque Country UPV-EHU, Joxe Mari Korta Zentroa, Donostia-San Sebastian, Spain

Andreas Daiss
Digitalization in Materials Science R&D, BASF SE, Ludwigshafen am Rhein, Germany

Peter Deglmann
Digitalization in Materials Science R&D, BASF SE, Ludwigshafen am Rhein, Germany

Dagmar R. D'hooge
Laboratory for Chemical Technology (LCT); Centre for Textile Science and Engineering (CTSE), Department of Materials, Textiles and Chemical Engineering, Ghent University, Ghent, Belgium

Mariya Edeleva
Laboratory for Chemical Technology (LCT), Department of Materials, Textiles and Chemical Engineering, Ghent University, Ghent, Belgium

Marcel Fassbender
Institute of Technical and Macromolecular Chemistry, University of Hamburg, Hamburg, Germany

Edurne González
POLYMAT and Kimika Aplikatua Saila, Kimika Fakultatea, University of the Basque Country UPV-EHU, Joxe Mari Korta Zentroa, Donostia-San Sebastian, Spain

Shaghayegh Hamzehlou
POLYMAT and Kimika Aplikatua Saila, Kimika Fakultatea, University of the Basque Country UPV-EHU, Joxe Mari Korta Zentroa, Donostia-San Sebastian, Spain

Juri Ilare
Department of Chemistry, Materials and Chemical Engineering, Politecnico di Milano, Milano, Italy

Stefano Lazzari
Digitalization in Materials Science R&D, BASF SE, Ludwigshafen am Rhein, Germany

Jose R. Leiza
POLYMAT and Kimika Aplikatua Saila, Kimika Fakultatea, University of the Basque Country UPV-EHU, Joxe Mari Korta Zentroa, Donostia-San Sebastian, Spain

Andree Lischewski
Digitalization in Materials Science R&D, BASF SE, Ludwigshafen am Rhein, Germany

Gerrit A. Luinstra
Institute of Technical and Macromolecular Chemistry, University of Hamburg, Hamburg, Germany

Yoshi W. Marien
Laboratory for Chemical Technology (LCT), Department of Materials, Textiles and Chemical Engineering, Ghent University, Ghent, Belgium

Yury Orlov
Digitalization in Materials Science R&D, BASF SE, Ludwigshafen am Rhein, Germany

Werner Pauer
Institute of Technical and Macromolecular Chemistry, University of Hamburg, Hamburg, Germany

Gianmarco Polotti
Department of Chemistry, Materials and Chemical Engineering "Giulio Natta", Politecnico di Milano, Milano, Italy

Eduard Schreiner
Digitalization in Materials Science R&D, BASF SE, Ludwigshafen am Rhein, Germany

Mattia Sponchioni
Department of Chemistry, Materials and Chemical Engineering, Politecnico di Milano, Milano, Italy

Amit K. Tripathi
Department of Chemistry, University of New Hampshire, Durham, NH, United States

John G. Tsavalas
Department of Chemistry, University of New Hampshire, Durham, NH, United States

Hugo Vale
Digitalization in Materials Science R&D, BASF SE, Ludwigshafen am Rhein, Germany

Paul H.M. Van Steenberge
Laboratory for Chemical Technology (LCT), Department of Materials, Textiles and Chemical Engineering, Ghent University, Ghent, Belgium

Kristina M. Zentel
Institute of Technical and Macromolecular Chemistry, University of Hamburg, Hamburg, Germany

CHAPTER ONE

Perspectives, pillars, and examples of polymer reaction engineering in the 21st century

John G. Tsavalas* and Amit K. Tripathi
Department of Chemistry, University of New Hampshire, Durham, NH, United States
*Corresponding author: e-mail address: john.tsavalas@unh.edu

Contents

Abstract

In this chapter, we reflect on the very definition of polymer reaction engineering (PRE) now versus from its origin paying homage to original pillars of the field and the trajectory it has evolved along to date. We open with a collection of perspectives from direct conversations with former chairs of the Polymer Reaction Engineering conference series as well as a selection of major contributors that we had the opportunity to ask. We highlight the common threads and themes both on the origins of the field, birthed from chemical reaction engineering, and the noted challenges we are facing today. We project examples of the future direction of PRE with commentary and describe our confidence in the continued needs for our expertise with new polymer classes and markets. Now nearly a century from the birth of the synthetic macromolecular market, we also remark on the critical role of the determination of reaction rate coefficients by our PRE community as well as a perspective on a basis for the continued challenge to directly determine those for certain monomers that have been in use for many decades. Here, we contrast direct analytical determination by pulsed-laser

Advances in Chemical Engineering, Volume 56
ISSN 0065-2377
https://doi.org/10.1016/bs.ache.2020.07.006

polymerization methods, coupled with molecular weight distribution analysis, with modeling based approaches extrapolating from kinetics output. We are encouraged by the impact PRE has made through the 20th century and we are certain the field and society will feel the same as we continue to contribute through the 21st century.

1. The evolution of polymer reaction engineering

We can acknowledge that polymer reaction engineering (PRE) was born from chemical reaction engineering (CRE). I enjoyed a rendition of this statement from a close colleague who phrased it as *PRE having started as a "costola" of CRE, under the impulse of people like Harmon Ray and Archie Hamielec.* These foreparents of ours, among others, continue to receive our thanks and admiration for their pioneering work. It is useful to note that even as recently as this past year Harmon Ray continues to receive acknowledgment for this, embodied in the 2019 Neal R. Amundson Award for Excellence in Chemical Reaction Engineering. Indeed, the definition of our parent field of CRE is equivalently a discipline focusing on the flow phenomena of mass transfer, heat transfer and reaction kinetics. It is the latter that will become a specific focus presented in this chapter, yet it behooves us to start with a reflection on the very description of PRE itself. Some have recently characterized PRE as a mature field and with that descriptor come different potential interpretations. On one hand, one might connect maturity with a connotation of reduced rate of further development, while instead we can surely agree that our own sophisticated contributions to science and engineering have come after we achieved some level of "maturity" post-academic matriculation. Indeed, this maturity is what has allowed us the vision to see new paths and needs and to bring our expanding toolboxes to them toward progress. We would like to open with a statement that we firmly believe that maturity does not bring a negative connotation to our minds, but instead an opportunity to expand our boundary conditions while simultaneously revealing ever more clarity within our current workspace. It is the specific topic of rate coefficients, and how to approach those not yet directly measured, that we will discuss later in the context of predictive modeling in this chapter. We are players in this field that are coming after a long history of respected contributors and we humbly leverage their vast body of work and learnings. The opportunity for us to enter these paths now we acknowledge is also largely due to timing and the advent of new computational tools, methods and hardware that were not available in a practical sense to our foreparents.

So, how does the determination of rate coefficients directly relate to the reactor that influences the type of polymer product one obtains? We are all familiar with the phrase "a product of process" so often professed within our PRE community. The link is, of course, kinetics. Rate coefficients are a function of temperature and in concert with monomer concentration(s) are directly impacted by heat transfer and mass transfer aspects that correlate with the reactor and process conditions. That contribution to kinetics is then further influenced by the diffusion of species and the viscosity of the reaction environment (critical considerations with macromolecular reactions); once again directly influenced by heat and mass transfer. These are knowingly simplified descriptors put forth to emphasize this core message: PRE serves the purpose of understanding how to leverage the chain of direct correlations between the smallest of molecules in the reaction, to the larger macromolecules born from them, and the reactor itself.

Our field has enjoyed a long tradition of PRE conferences (since 1980) as a forum to discuss our technical progress, meet in person to network and engage in debate openly outside of the projector room, and to forge new questions and targets for continued progress. These have traditionally been held every 3 years with offset cycles in the North American region, Europe, and Asia-Pacific. Participation and engagement have been from a worldwide composition. In the context of two of the most recent PRE conferences I had the opportunity to participate in (as Chair of Polymer Reaction Engineering X (PRE 10) in the Dominican Republic in 2018, and as a keynote speaker at the DECHEMA 13th International Workshop on Polymer Reaction Engineering in 2019 in Hamburg, Germany), we return here to the definition of PRE. 2020 has been a year for reflection from multiple global perspectives and it thus feels relevant to report on some level of reflection here as well; in this case, not regarding the human element, but instead specific to the technical aspects of our topic. I asked a simple question to many of our colleagues at those meetings, including past PRE conference chairs and major contributors to the current era of the field (as in, they were present at such events affording me to speak directly with them). The following is by no means meant to purposely exclude anyone—these are simply collections of perspectives from close colleagues that I had the opportunity to ask. There are many common threads in these and some truly exceptional insights that we believe important to share here as we consider our current and future eras ahead in PRE. To respect the request for anonymity by some of the contributors, we will elect to maintain all as such, but in an attempt to maintain voices the responses captured below will be left essentially verbatim to the source and **italicized**.

2. The question and request for perspectives

"Can you please describe how you define polymer reaction engineering today—and versus how you would have defined it a decade or so ago? What are the exciting new developments, evolution of how PRE is utilized, and/or concerns?"

3. Excerpts from the responses

Essentially, and without playing with words, it is reaction engineering applied to polymerizations. In a traditional sense, this means understanding the relationship between the result of the reaction(s) one is interested in and the design and operation of the chemical reactor in which the reaction(s) take place. From my point of view, the special thing about PRE has always been the very strong relationship between the product one obtains and the reactor used to make it. To make styrene, one alkylates benzene, then dehydrogenates the resulting ethyl benzene to obtain styrene. From a chemical reaction engineering point of view we need to build the kinetic model and understand how the residence time distribution and operating conditions of our plant impact the yield (purity) of the final product. But regardless of the level of performance of the process we design, the styrene that results is always styrene with a single, unique set of properties. However, if we take that styrene and polymerize it via a bulk process, a suspension process or an emulsion process, in batch, semi-batch or continuous processes, then we will often end up with very different products—chemically they are the same, but their physical properties will be different because we are making, in essence, an almost infinite number of different molecules (as opposed to the styrene process where we make one!). How the molecules are assembled in the reactor—controlling this, modeling this, monitoring this—is the specificity of PRE. Of course we have the additional constraints linked to the exothermicity (often) and viscosity, the non-ideal kinetics, and the large number of "known unknowns" (gel effect, penultimate effect, side reactions in polycondensing systems, actions of cocatalysts and kinetic mechanisms in coordination polymerization, particle nucleation and coagulation in emulsion polymerization, etc.) that we need to deal with. So, at least from a traditional sense, one could say that understanding the kinetic and physical mechanisms that are essential for us to develop quantitative descriptions of the polymerization is the traditional basis for PRE. Earlier in the field, modeling the kinetics, chain growth and chain architecture were the basic building blocks of PRE. These were of course associated with particle generation and stabilization in heterophase systems—I am thinking here of things like DLVO models for particle stability for instance—or particle growth and morphology in heterogeneously catalyzed olefin polymerizations. Of course, one cannot forget the need to include basic, fundamental science like thermodynamics (just like in small molecule chemical reaction engineering)! All-in-all the PRE that people of my generation grew up with, and that was developed by pioneers like Archie Hamielec, Harmon Ray, and A.R. Berens (just to name a few), was centered on this approach. Of course, as time moved on, the tools that we used to develop and solve these problems evolved, so fundamental PRE now uses tools like advanced population balance modeling, Predici, and Monte Carlo approaches to arrive at more complete, accurate answers to PRE problems.

However, I find that as time goes by a lot of these fundamental PRE problems are giving way to more trial-and-error approaches as we attempt to impose PRE methodologies on a wide range of newer materials, for example, hybrids and composites. But really, in my view it is the applications that we are funded by that seem to change from the academic point of view, and in particular in North America; not the actual nature of PRE. PRE remains PRE—it is what we do with it that is evolving, and to a large extent what our public research funding sources ask of us that also evolves. Of course, we can extend our approach to CFRP and PISA and these newer topics, but from my point of view this is once again the application that evolves, not the toolbox!

We started with the above perspective as it truly captures many salient points and reflects well with our own. In fact, we are encouraged by the consistent message thread spanning the bulk of the comments shared here. In the next perspective, the topics of focus and the challenges addressed by our PRE community past vs present are contrasted.

In the past, the focus was predominantly on understanding the polymerization kinetics. At the time there was still not a thorough understanding of kinetics. This included:

- *measuring kinetic rate constants*
- *accounting for heat transfer effects, and the coupling of heat transfer with reaction kinetics*
- *accounting for mass transfer/mixing/diffusion effects, in homogenous systems, heterogeneous systems and catalysts*
- *mathematical modeling of processes was of high interest*
- *generally, this period involved a lot of "analysis"*

In the present, as well as for my thoughts on the future, I would say that we understand kinetics pretty well, for most systems. We are also now quite advanced on most of the other bullets above. Now, greater emphasis is on designing and being able to predict material structure and morphology and designing new materials with enabling properties: stimuli-responsiveness, smart materials, memory retaining materials (e.g., shape). It appears there is a general shift away from focus on the reactor to the materials themselves. Now, and in the future, renewably sourced/green materials also offer lots of opportunities and demand (not at the current price of oil, but this movement is young). In general, this period now and ahead involves less "analysis" and more "discovery."

This shift to "discovery" brings the engineer closer to the scientific bench. In addition to environmental and societal push toward renewable and sustainable sourcing of raw materials, there is also a well warranted surge of public pressure to ban long-life disposable polymer products. Impressively making an international statement, France was one of the first to ban single-use plastics (polyethylene bags) in 2015 (Solomon, 2020) with cutlery, plates, and cups in 2015 and now after January 1, 2020 fully banning single-use plastic straws, coffee stirrers and cotton swabs (Woods, 2020). France is followed closely by China banning non-degradable bags in major cities by the end of 2020 (with all cities and towns by 2022) and restaurants banning non-degradable straws by end of 2020 (BBC News, 2020). In fact, the EU has legislation proposed to ban single-use plastics throughout the EU by 2021 (the same is being proposed in Canada) and with all plastic packaging produced forward to be recyclable by 2030. The U.K. projects a slower timeline targeting 2043, and the United States is also progressing with eight states thus far with a full ban on single-use plastic bags (started by California in 2015 and now including Connecticut, Delaware, Hawaii, Maine, New York, Oregon and Vermont). Slowly this awakening of responsibility for our products is propagating from recommendation to law Worldwide. Elimination is one approach, but certainly not a universal one. Here the field opens to the younger concepts and creative solutions to insert orthogonal triggers into the polymer structure suited to compostability and biodegradability. Stability over a specified timeframe, or susceptibility to controlled environmental conditions suited to deconstruct the polymer, bring new chemical structural species not only with regard to sidechain functionality but more importantly in the polymer backbone itself. These new structures may demand adjustments or even new approaches to polymerization methodology and likewise the suited conditions of manufacture. Clearly the polymer reaction engineer will be in high demand to address these challenges and opportunities. World War II sparked a need and concomitant surge in the variety and tonnage of polymers produced which fueled PRE. Here, a need arises again; both for the polymer chemist to produce the novel design and the polymer reaction engineer to bring that through to the end-product for the market.

Returning to the assessment above, the differentiation between focus past vs present is echoed in the next perspective. Here we are building a consistent message that PRE is a tool, or toolbox, but in generations it becomes applied differently. This is almost described explicitly as so below, which contains a merging of two contributions who had very similar messages on certain points.

The core of PRE has not changed substantially over the years; PRE provides the tools to design and carry out the polymerization process to produce high performance polymers in a safe, cost-effective, consistent, scalable, sustainable and environmentally friendly way. As often these desired characteristics are in conflict, choices should be made. However, there are things that have substantially changed with time. Some of them are listed below:

- *Taking Professors Hamielec and Ray as examples of the pioneers in this field, they basically applied Chemical Reactor Engineering to polymerization, focusing mainly in large scale processes. It is difficult to find in their publications ideas such as sustainability and environmentally friendliness that currently weight heavily in the decisions.*
    - o *People like Harmon Ray and Archie Hamielec initiated the transformation of a largely empirical discipline into a solid coherent piece of science where accurate experimentation and rigorous mathematical modeling integrated to understand the fundamentals and subsequently also the applicative aspects of the discipline.*
    - o *Initially the large industrial polymerization processes were considered, starting from the enormous field of polyolefins, and moving to all the others covering practically all what we currently have. The paradigm was to connect the reaction conditions to the polymer molecular characteristics, and from these then proceed to predict the application properties so as to achieve the dream of "macromolecules by design." We are still struggling with this last step.*
    - o *In my view the future of PRE, like for most mature sciences, is to expand the boundaries and enter specific sectors and try to use our consolidated approach to solve their problems: bioplastics, controlled delivery, gels, as a few examples.*
- *Knowledge of the polymerization process: The impressive development of characterization techniques combined with a wider use of computational techniques including Quantum Chemistry has led to a much deeper knowledge of the polymerization mechanisms and this is leading to better process strategies.*
- *Structure-property relationships: Another consequence of the improvement in characterization is the development of better structure-polymer relationships that help to define better the aimed polymer structure.*

- *Increasing use in small scale processes. There are many specialty polymers that are produced on small scale (sometimes in kilograms) but the performance requirements need intensive use of PRE concepts.*
- *An extension of the previous point is that the concept of the reactor has changed. Now it is the place where the polymerization occurs (e.g., a tooth or a surface). This is not often recognized by the PRE community.*
- *PRE work is now conducted in close contact with many other disciplines: Materials Science, Polymer Physics and Chemistry, Computational Chemistry, Biology, Medicine, to name a few.*

PRE has plenty of challenges and therefore opportunities. A challenge is to incorporate in our work, tools and concepts that are well developed in other fields. Another big challenge is to identify non-traditional fields where PRE can play an important role. These challenges are also opportunities. A concern is how to maintain the feeling of belonging to the PRE community when working in these non-traditional fields.

To this last point, we have purposely opened our doors to non-traditional fields to PRE in the past decade or so of PRE conferences around the globe. Invitations for talks from chemists for whom the term scale up would lead to a quite different definition from that of our traditional community has been a deliberate move on our part. This is on one hand a strategy to find these new fields "in need of" PRE, but also for the expanded community to observe the power of the discipline in PRE that has been developed thus far. Indeed, the world's problems require teams to generate creative and efficient solutions. We are encouraged by the dynamic of mutual welcome and respect that we have witnessed thus far in the meetings of such "representatives" at recent events.

The next several perspectives shift from general assessments on the definition and evolution of PRE to some of the drivers and motivation from Industry.

Over the past two decades or so, the products and processes of emulsion polymerization have seen significant changes. Engineers, working with bench scientists, have had to scale up higher and higher solids latex formulations without violating the heat removal capabilities of the available reactors and still be able to move, store and sometimes ship the more viscous materials without destroying the colloidal stability of the latex. Polymerization processes have moved to higher temperatures to accommodate increased heat transfer and, at the same time, safety concerns have moved processes to lower and lower instantaneous monomer levels within the reactors. At such low monomer levels, additional reaction mechanisms have come into play and have required re-formulation to overcome the sometimes-deleterious effects. At the end of the polymerization process, the removal of residual monomers to levels at or below

100 ppm (based on latex) has added an additional step to latex processing. Laboratory and pilot-scale reactors began to more reasonably resemble plant scale reactors so that the scale up times and success rates could be improved. Additionally, in that same period of time, a new class of "hybrid" emulsion polymers has been developed, even including inorganic materials. These have required new types of processing steps to be commercialized so as to be able to produce the new composite latexes economically.

The above highlights certain scenarios that are presented with industrial needs in mind that otherwise can go unnoticed at the bench scale. We often are concerned with developing the proper monomer choices, process conditions, and even reactor geometry and inner surface material (glass vs steel, for example) so as to generate manageable kinetics that generate macromolecular architectures of target properties. Yet, even the finishing of reaction for the final several 1000 ppm of monomer(s), which would seem trivial compared to the bulk of the reaction, can have dramatic impact on final properties (for example, influencing the multi-phase morphology of composite latex systems, as described above). Here, a true PRE project surfaces in determination of the influence of proper reagent choices, even from the start of the polymerization, so as to produce all aspects of the product within target specifications and in concert with process conditions and reactor configurations available at hand.

Continuing with industry and consumer in mind, polymer manufacture in the future is likely to be divided into two main categories: high value-added specialty products and low-cost commodities. A new criterion for PRE to consider here is not only the economics of scale, but the lifetime of such products post-production.

High value-added, low-volume specialty products will continue to be developed with little chance of remanufacture due to the complexity of the polymer and/or composite nature of the product. These are likely to be designed for long use life. They are likely to be produced in smaller, batch or semi-batch reactors in order to accommodate the complex nature of the molecular and supramolecular structure, and to facilitate rapid market response.

High-volume, low-cost commodity polymers of the future are likely to be produced in very large, world-class continuous reactor systems to take advantage of the economies of scale. As awareness of the problems of plastics disposal becomes greater, these, mostly single-use, polymers are likely to be designed for ease of reuse, recycle or remanufacture.

Hidden in plain sight in the two paragraphs above is a return to the concept of the necessary intertwine of product and process in PRE; one informs the other in a dependent fashion.

Process and product are closely linked, as reactor choice and operating condition directly influence the end-use material properties through their effect on polymer molecular weight, branching and composition distributions. It is the distributed nature of the polymer product that necessitates both specialized experimental techniques to characterize its molecular structure and reaction kinetics, and specialized simulation techniques to develop models to represent its production. Thus, PRE practitioners apply their knowledge to guide the selection of reactor configuration and operating conditions for existing and new polymer grades and to discriminate between kinetic and physical effects, with the ultimate goal of establishing first-principles modeling as a robust tool for process and product design and optimization.

Modeling has been inherently embodied in PRE from the onset. Yet the phrase "*distributed nature of the polymer product*" above that highlights advantages the recent generation of PRE has had the fortune to enjoy/develop. This is captured well in the next perspective.

With the advent of more advanced experimental and modeling techniques, the PRE field is now capable to address each relevant length scale to a greater level of detail. In many of the older studies, both in industry and academia, the polymerization kinetics were designed at the level of microstructural averages with a specific focus on the number and mass average molar mass evolution as a function of monomer conversion. Thanks to pioneering contributions by several groups, more toward the end of the previous century, the focus could be shifted to the design of the complete molar mass distribution with for instance a much clearer understanding of the role of diffusional limitations. Today, we are ready to tackle the kinetic variations of individual chains accounting for many structural parameter variations such as monomer sequences, end- and mid-group functionalities, and branch locations and composition. A transition can also be noticed that the number of reactions considered is increasing, acknowledging that often side reactions are determining the actual polymer product performance. These positive evolutions both in skills and perception will strongly support the development of next-generation polymer synthesis routes but also enable the introduction of the concept of "circularity" for polymers, as chemical recycling is linked to the prior synthesis.

However, we still are facing many challenges in the PRE community. We specifically need to further and systematically understand phenomena at the molecular scale but at the same time we should try to bridge the gap between lab- and industrial-scale PRE research and valorization. Bridging the expertise of engineers, physicists and chemists is here indispensable. The most complex case here is probably dispersed phased polymerization with nanometer scale reaction loci, e.g., emulsion polymerization. Here we encounter competition between reactions and transport in different media and the distributed feature of polymerization processes is maximized with the strong desire to fully address the coupling between chain length, particle size and morphological changes. I am convinced that by joining forces in the next decades we will be again closer to the ultimate goal of a full product spectrum characterization and design from lab up through industrial scale.

As a final perspective to include here, and one that foreshadows a chapter later in this volume, speaks again to expanding boundaries of the reaction space PRE is now attempting to address.

PRE research is a well-established field, with significant scientific contributions dating back to the 1930s. Tremendous advances followed in the subsequent decades, mostly focused on the physico-chemical understanding of polymers and polymerization processes. Since the 2000s our community pushed our knowledge into topics such as charged monomers, RAFT polymerization, secondary reactions in free-radical polymerization, and inverse emulsions.

While these topics are still of great relevance, and many questions remain open, we believe that in the next decades digitalization will be the key driver, expanding the spectrum of PRE, providing new research directions and questions. This will happen through the establishment of one crucial concept, that elegantly connects research and digitalization: the digital twin.

Digital twins combine the strength of physical models with the predictive ability of statistics and machine learning approaches to provide a virtual, dynamic representation of a system, enabling its understanding, control, and optimization.

While ordinary differential equations served our community well for many decades, the advent of individual chain monitoring through advanced population balances and Monte Carlo approaches opened our observation to new levels of scrutiny and resolution on the reactions and developing chains themselves. This did not come without a cost, however, as this detail requires expensive computational resource and time. Whether we consider graph theory or digital twins and machine learning AI approaches inferred above, our pursuit for detail must be balanced with capture under reasonable timeframes and affordable expense. These new tools are certainly exciting and promise great new strides.

Even with increased modeling demands, we must still set assumptions for unknowns or at least our best interpolations from data at hand. This leads us to our own group's current work in this realm specifically on developing an approach to rate coefficients relevant to polymerizations involving divinyl monomers, still to date not explicitly measured, and the resulting topological development of such polymerizations. Indeed, we still use actual experimental data from our forefather Archie Hamielec as a source of validation for not only our modeling but even our experimental kinetic work derived from high performance differential scanning calorimetric analyses of the polymerizations themselves. We stand on very tall shoulders.

4. Kinetic rate coefficients as a core element of PRE

With the future of PRE in mind, it is again useful to summarize one of the core contributions of PRE with the determination of rate coefficients. This is by no means meant to be a review; akin to above, it is simply a perspective (that of our own). We simply summarize elements of this timeline to serve as the basis for why we remain in pursuit of such kinetic parameters even for monomers that have been in wide use for multiple decades.

While products of Nature (starch, cellulose, and natural rubber) were processed commercially in the 19th century, including the vulcanization of natural rubber by Goodyear (1839), synthetic polymers truly arrived in the 20th century. Staudinger birthed the description of the macromolecule in the 1920s (Staudinger, 1920), where Kuhn, Flory, and many others started the first models for molecular weight distributions within reactor processes in the 1930s (Chalmers, 1934; Dostal and Mark, 1936; Flory, 1936; Kuhn, 1930; Schulz, 1935). It was this same timeframe that Nylon (a polyamide) was developed at DuPont by Carothers (1938) and polyethylene at ICI by Fawcett et al. (1939). World War II generated the necessity for invention in the 1940s for synthetic latex, a heterogeneous reaction environment in the form of an emulsion, and eventually this and other reaction environments would bring us to materials formed from poly((meth)acrylates), poly(styrene) and poly(butadiene). If we simply focus our attention to these monomer classes (amides, ethylene, (meth)acrylates, styrene, and dienes) we can describe the majority of commercial polymers still in production today. A century since Staudinger now, one would assume the relevant rate parameters to all those monomer classes would be firmly established and understood. Certainly, values do exist for most perturbations of those structure classes, yet we remain in debate over certain aspects. Part of this challenge resides in the ability to directly measure such reactivity isolated from interdependence to other aspects of the polymerization reaction and process. This is exacerbated for divinyl monomers, such as ethylene glycol dimethacrylate and divinyl benzene widely used for crosslinking, as the pendent (second) vinyl does not experience probability to react until the polymer chain length is greater than that typically suited to analytical determination of reactivity (for example, by pulsed-laser polymerization; to be described shortly). With unknowns at hand, the engineer must develop an extrapolation or interpolation from analogous systems or trajectory of trends within existing data sets. Here, structural classes (i.e., those founding monomer classes, and others, described above) often serve to set initial values to optimize further from.

5. Why are the propagation rate coefficients for the homologous series of methacrylates different?

Seeing this statement here will certainly elicit confusion as there are indeed many contributors to this very question. We don't pose this as a new question by any means, but rather to highlight that the debate on the core elements to the answer remains active. There is excellent recent work reported on this very question (Haehnel et al., 2013, 2014), yet we have found in personal discussions at the three most recent PRE conferences that while there are certainly useful arguments at play, there are other equally consistent approaches. It is deliberate that we open this question via focus on methacrylates as the target, as well. This will play into our own group's approach to rate coefficients for divinyl monomers, so we believe it is useful to remove ourselves from the numeric data for the moment and to review the conceptual elements to this equation.

From a purely structural point of view, the immediate vicinity surrounding the radical on a methacrylate chain end and the vinyl with alpha methyl substitution are equivalent for all methacrylates. From a strictly electronic point of view, the alkyl length on the sidechain has minimal effect on the polarity and electron delocalization within the vinyl bond. So, we must turn toward factors such as the steric impact of the sidechain bulk on the pre-exponential and activation energy Arrhenius contributions to K_p specific to probability of proper orientation for the reaction to occur. We must also consider the entrance to that reaction volume by the penultimate adjacent monomer to the radical chain end (Burke et al., 1997). Concepts such as interaction between monomers, or between monomer and macroradical alike, may also influence their local residence time together thus allowing higher probability for that pair reaction to occur. Indeed, one extreme of this case is found in carboxylic acid based monomers with orders of magnitude difference in K_p in aqueous vs organic solvents as a function of dimerization (Beuermann et al., 2007; Kuchta et al., 2000).

If we invoke these concepts of pairwise interactions (in fact, we should include other nearby monomers, penultimate and further back, also on the "dancefloor"), we start to veer from pure chemical reactivity and into the realm of diffusion, where the time for molecules to translate toward a collision frequency is impacted by their size (individual monomer, dimerized monomers, or macromolecules themselves when we consider termination). This diffusion of reacting species of course is further influenced by concentration and average distance between constitutive pairs. This is not to forget the role of the continuous phase viscosity as we move away from dilute conditions. Glass transition temperature is markedly influenced by the alkyl sidechain length in the homologous methacrylate series, influencing reptation in entangled environments and even the ability for (macro)molecules to rotate again for orientation in the free-radical addition reaction between a monomer and macroradical chain end. This highlights the separate contributions of chemical reactivity and diffusional aspects captured by the various models for total and apparent rate coefficients (D'hooge et al., 2013). We will not expand here on the relative applicability of each of these model approaches to different monomer systems (Arzamendi and Leiza, 2008; Marien et al., 2019; Tobita, 2002), but we leverage this challenge of properly capturing diffusional aspects so critical to macromolecular synthesis as relates to the inherently higher viscosity reaction environment. This is a contribution surely authored by the PRE community.

With assessment of pure reactivity as the goal, we thus attempt to deconstruct and minimize these interdependencies by attempting to measure the reactivity under very dilute conditions and for only very short oligomeric chain length growth. Indeed, the oligomer, shorter than a Kuhn length, is characteristically extended and offers a maximized accessibility for reaction (reactive volume overlap of the species pair) with the radical chain end. Once further propagation occurs such that coiling of the chain develops, the reactive site has opportunity to be buried and inaccessible at times; this generates the onset of contributions separate from the pure reactivity of the pair. This brings us to what emerged and remains as one of the most powerful analytical techniques for measurement of rate coefficients, specifically designed to isolate purely the first few propagating steps toward oligomers in highly dilute conditions; pulsed-laser polymerization.

6. Pulsed-laser polymerization (PLP) and coupled with size exclusion chromatography (PLP-SEC)

Polymerization reactions are parametrized via combinations of reaction rate coefficients representing each potential interaction of relevant species; propagation, chain transfer, termination. A challenge exists in the fact that the literature houses a wide range of values reported for any given rate parameter which was identified as a major concern by the IUPAC Working Party on "Modeling of Kinetics and Process of Polymerization" in the early 1990s (Buback et al., 1988, 1992). The simplest models are restricted to initiation, propagation and termination reactions (where other side reactions are either suppressed or ignored). Theoretical expressions obtained for the rate of polymerization (which can be easily quantified and measured) thus contain a combination of K_p and K_t terms coupled together as K_p^2/K_t, where K_p and K_t are the rate coefficients for radical propagation and termination, respectively. In order to decouple both parameters, the reaction can be run in combination with techniques like electron spin resonance (ESR) spectroscopy to actively measure the radical concentration (Ballard et al., 1986; Lau et al., 1987). The ESR technique is, however, too insensitive for accurate measurement of radical concentrations under realistic polymerization conditions and requires careful calibrations. Another of the early techniques for estimating reaction rates was the rotating sector method (Burnett and Melville, 1945, 1947), in which radicals are formed periodically via a photochemical reaction by switching the illumination source on and off. This and its variant, the spatial intermittent polymerization (SIP) reactor (O'Driscoll and Mahabadi, 1976), do provide some success in reporting propagation rate coefficients, however, any corrections for inclusion of side reactions can be extremely challenging in such techniques (Olaj et al., 1985).

One of the most popular techniques, pioneered by Olaj and colleagues in the late 1980s, is pulsed-laser polymerization (PLP) and remains as the recommended approach by IUPAC (Olaj and Schnöll-Bitai, 1989; Olaj et al., 1985, 1987). In this technique, with some similarity in concept to those mentioned above, primary radicals are generated by short laser flashes in a monomer solution containing a photoinitiator. Radical generation occurs in a very short time period, at least compared to the mean lifetime of polymer chains and even relative to the average duration of a single propagation step. The radical generation can thus be considered rather instantaneous and polymerization occurs between periodic laser flashes. With each laser pulse, new radicals are generated and start propagating. In between the pulses, a fraction of radicals terminate, resulting in a decrease of radical concentration. The laser pulse repetition produces then a periodic radical concentration profile (Fig. 1).

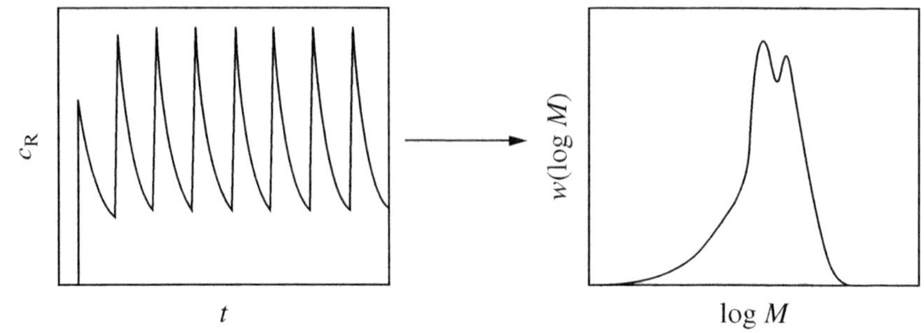

Fig. 1 Illustration of a sample radical concentration (C_R) vs time as influenced by laser pulses during PLP (left) and the resulting molecular weight distribution by SEC (right). *Reprinted with permissions from Beuermann S: Requirements associated with studies into a chain-length dependence of propagation rate coefficients via PLP-SEC experiments,* Macromolecules *35(25):9300–9305, 2002. Copyright 2002 American Chemical Society.*

In between two consecutive pulses during time interval Δt, a population of radicals which have escaped the fate of termination, or any other side reactions such as transfer to monomer, and will have a chain length L which is given by

$$L = K_\mathrm{p}\ [M]\ \Delta t \tag{1}$$

where *[M]* is the monomer concentration. At Δt, when the next laser pulse is fired, the radical concentration suddenly increases again and the probability for termination of existing radicals increases. This results in a significant population of species with chain length L. Molecular weight distribution analysis of the resulting polymer can provide information regarding L and PLP experiments are now commonly combined with size exclusion chromatography (SEC) analysis; the combined procedure is referred to as PLP-SEC. From Eq. (1), we can see that the PLP-SEC analysis results in a direct assessment of K_p if the reaction parameters (*[M]* and Δt) are available. While a significant portion of radicals will terminate with a subsequent laser pulse, a fraction of them can survive multiple pulses. The chain length (L_i) for polymer formed after i laser pulses can then be given as

$$L_i = i\ K_p\ [M]\ \Delta t \tag{2}$$

In fact, the IUPAC recommendation for correct K_p estimation is such that the molecular weight distribution of polymer generated by PLP should have higher order chains (formed by $i > 1$ in Eq. (2)) and should correspond to integer values of i. The description of PLP-SEC here is significantly simplified and more details can be found in (Olaj and Schnöll-Bitai, 1989; Olaj et al., 1987).

The PLP(-SEC) approach has energized the PRE community and over the last three decades, various research groups (e.g., Barner-Kowollik et al., 2010, 2014; Beuermann, 2002; Beuermann and Buback, 2002; Beuermann and García, 2004; Beuermann et al., 1996, 1998, 2001, 2003; Buback, 1996; Gruendling et al., 2010; Hutchinson et al., 1997; Jeličić et al., 2011; Junkers et al., 2010; Kockler et al., 2016; Kuchta et al., 2000; Liang and Hutchinson, 2011; Marien et al., 2017; Nitschke et al., 2020; Rooney et al., 2017; Uhelská et al., 2014; Vir et al., 2018; Willemse et al., 2005; Willemse and Van Herk, 2010) all across the globe have joined in efforts to obtain reliable and consistent propagation rate coefficients for a wide variety of both common and specialized monomers, over expanded temperature and pressure ranges, and along with a multitude of solvents as well as neat (bulk) systems.

The PLP-SEC approach for rate coefficient determination has essentially two separate sequential processes. First, we have the laser induced short burst of propagation followed by termination and secondly, we have the analysis of the oligomer/polymer generated in the first step. While SEC is the most common choice paired for the analysis of PLP, other analytical techniques such as matrix-assisted laser desorption time of flight mass spectrometry (MALDI-TOF MS) (Davis and Zammit, 1996; Willemse and Van Herk, 2010; Zammit et al., 1996), electrospray ionization mass spectrometry (ESI-MS) (Gruendling et al., 2010) and electron spin resonance (ESR) spectroscopy (Willemse et al., 2005) have also been utilized to overcome some of the challenges observed in SEC.

Process conditions for PLP experiments themselves, such as frequency and profile of laser pulses, have also been explored to obtain rate parameters for complex systems. A few laser pulses held for longer time was coined a "railroad" experiment (Buback and Laemmel, 1997) to obtain both propagation and chain transfer reaction rates simultaneously. Reaction rate coefficients obtained from different frequencies for PLP experiments were utilized (Nikitin et al., 2007) to obtain backbiting and mid-chain propagation rate coefficients for the classic n-butyl acrylate system. Frequency variations (Junkers et al., 2010; Kockler et al., 2016) have also been explored and showed the applicability of a high frequency laser system (500 Hz) to obtain propagation rate coefficients for acrylate monomers at elevated temperature. Similar to propagation rate coefficient determination, the

PLP technique has additionally been leveraged to estimate termination rate coefficients. For this, a single pulse (SP) is used along with spectroscopy techniques such as near-infrared (NIR) (Beuermann and Buback, 2002; Beuermann et al., 2008) and electron spin resonance (ESR) (Barth et al., 2009; Buback et al., 2006) spectroscopy to monitor reaction kinetics with time after the laser pulse.

The advancement of PLP-SEC and its variants has been impressive and has positively and enormously impacted PRE. More recently, even branching events have been probed by PLP methods for acrylate monomers (Nikitin et al., 2007; Vir et al., 2018). These topological architectures previously were almost inherently unavailable to the PLP approach as this type of event (chain transfer to polymer) is much less probable than propagation (at least without special approaches where this is deliberately flipped in CFRP) and thus a branch normally would not occur on the time scale of growth between pulses. To address branching events is certainly an encouraging advance, however to the extent of our knowledge direct assessment of the rate coefficients relevant to di- or multi-functional monomers has not been formally attempted. Monomers such as ethylene glycol dimethacrylate and divinyl benzene are so prevalent as crosslinkers for free-radical based systems, it raises the question as to why their relevant parameters are missing. Certainly, one aspect is clearly the challenge in defining proper conditions in a PLP-type method toward isolation for each of the first vinyl incorporation and then the second pendent vinyl. Again, the frequency of utility of the latter during the polymerization is so limited that this fact itself likely precludes it from this analysis. As such, and summarized in the next section, we have approached development of those missing rate coefficients relevant to divinyl monomers through mathematical modeling alongside synthetically produced kinetic data. Here again, we have learned to be especially cognizant of the role of structural features entering the local reaction volume of not only the individual divinyl monomer, but also those of other monomers adjacent to it in the chain. In pursuit of the unknown, modeling again not only serves to reach forward and predict, but also to reach back and teach.

7. An approach to rate parameter estimation for reactive sites on sidechains pendent to the primary chain backbone

The utilization of multi-functional monomers (commonly referred to as crosslinkers) such as ethylene glycol dimethacrylate (EGDMA) and divinyl benzene (DVB) are quite common in free-radical polymerization reactions. These monomers, most often used in small amounts generally less than 1–2% of total monomer, are copolymerized with other common monofunctional monomers such as styrene (STY) and methyl methacrylate (MMA). For EGDMA, for example, symmetry is observed in the monomer structure; as a monomer, the two vinyls are thus equivalent in reactivity and rate parameters. However, once one such vinyl group incorporates into a growing polymer chain, the second vinyl resultingly resides on a sidechain pendent to the macromolecular backbone. Eventual utility of some fraction of these pendent vinyls can of course lead to crosslinking and desired topological network development.

In a copolymerization reaction with a crosslinker monomer, the reaction effectively becomes a terpolymerization case, where along with free monofunctional and difunctional monomers, the pendent group acts as the third monomer (Fig. 2). On careful inspection of such a pendent vinyl group, it can be easily realized that a reaction of another chain-end radical with this pendent group will fall into a polymer-polymer type reaction category; similar to termination reactions. Due to the involvement of larger polymer chains in such reactions, diffusion plays an important role. As described earlier, diffusion of polymeric chains largely depends on polymer chain length and thus each reactive species pair can effectively have a different set of overall effective reaction rate coefficients. In general, for reactions where diffusion plays a key role, the reaction rate coefficients are estimated

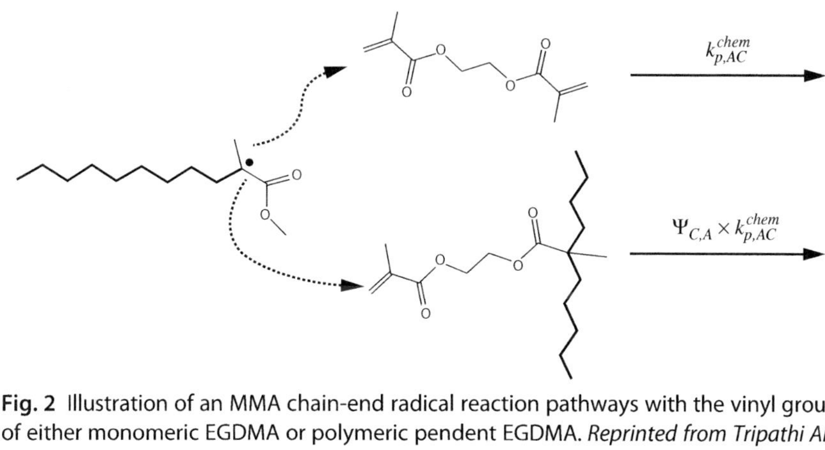

Fig. 2 Illustration of an MMA chain-end radical reaction pathways with the vinyl group of either monomeric EGDMA or polymeric pendent EGDMA. *Reprinted from Tripathi AK, Neenan ML, Sundberg DC, Tsavalas JG: Influence of n-alkyl ester groups on efficiency of crosslinking for methacrylate monomers copolymerized with EGDMA: experiments and Monte Carlo simulations of reaction kinetics and sol–gel structure,* Polymer (Guildf) *96:130–145, 2016. Copyright (2016) with permission from Elsevier.*

from a combination of chemically controlled and diffusion-controlled rate coefficients. Eq. (3) is one approach, which we use here only to serve as an example (D'hooge et al., 2013), taking the form

$$\frac{1}{K_{X,ij}} = \frac{1}{K_X^{chem}} + \frac{1}{K_{X,ij}^{diff} + K_X^{res}} \tag{3}$$

where $K_{X,\ ij}$ is the effective reaction rate coefficient, K_X^{chem} the chemically controlled-rate coefficient, $K_{X,\ ij}^{diff}$ the diffusion controlled-rate coefficient and K_X^{res} the residual rate coefficient to account for radical diffusion due to chain propagation.

The chemically controlled-rate coefficient, K_X^{chem}, is chain-length independent (after oligomeric chain length is achieved (Heuts and Russell, 2006; Smith et al., 2005)) and is applicable specifically to when one or more reacting species can diffuse fast, such as in the propagation step which involves at least one small molecule by definition (K_X^{chem} is deliberately isolated in the PLP method, for example). For polymer-polymer type reactions, this chemically controlled-rate coefficient remains important under dilute conditions (e.g., early conversion during bulk polymerization) where polymer diffusion is relatively unhindered. This leads us to a special case with regard to utility of the pendent vinyl in a divinyl monomer. Here, as that pendent vinyl hinges from a large molecule it can be considered a macromonomer with radical generation at that site dependent on an external source. This could in principle be achieved by initiator radical attack (a small high-diffusion species) or by a chain-end radical on a separate macroradical (linear or otherwise; in either case large low-diffusion species). The translational aspects of the chains, with or without the pendent vinyl units, are for the most part the same. Diffusion is similarly not markedly impacted by backbiting in acrylates as the pendent short chains that develop are almost a negligible outward extension from the backbone if one considers the chain's ability to reptate. This is in principle analogous to the pendent vinyl capped sidechains, where we would envision only minor impact, if any, on diffusion-specific aspects in Eq. (3). However, even in dilute conditions, the chemical-controlled rate coefficient itself is impacted by the penultimate and adjacent neighboring monomer sidechain structures. This brings us to our group's current approach.

We have described the chemical-controlled rate coefficient for a (macro) radical reacting with a pendent vinyl group of a crosslinker ($K_{p,\ AP}^{chem}$) in terms of the chemically-controlled rate coefficient for the same radical reaction with the first vinyl group in the crosslinker as a monomer ($K_{p,AC}^{chem}$) and a reduced reactivity parameter, $\Psi_{C,A}$, as a fraction to be applied to the first (Tripathi et al., 2015, 2016). This is shown in Eq. (4) as well as pictorially in Fig. 2.

$$K_{p,AP}^{chem} = \Psi_{C,A} \cdot K_{p,AC}^{chem} \tag{4}$$

The reactivity of the pendent vinyl group is unsurprisingly expected to be lower than the monomeric version due to new rotational, conformational and steric restrictions when hinged by one end to a polymer chain. Due to this, for symmetrical crosslinkers like EGDMA, the factor $\Psi_{C,A}$ simply takes the form of a fraction between zero and unity.

We started this work by developing such an approach to a reduced reactivity parameter, $\Psi_{C,A}$, (and thus the relevant data to produce a full rate coefficient) for copolymerization of EGDMA with various methacrylate monomers. This was deliberate to first focus on methacrylate-only systems so as to eliminate possibility for competing branching events; topology in this scenario would thus exclusively result from utility of those pendent functional groups. We did this by carefully obtaining experimental kinetic data over the full range of conversion for bulk polymerizations within a differential scanning calorimeter (DSC). Against that data, also including analogous experimental data sets available in the open literature including that of Hamielec and co-workers (Li et al., 1989; Marten and Hamielec, 1979), we applied Monte Carlo simulations (Gillespie, 1976, 1977; Tripathi and Sundberg, 2015) with Eq. (4) substituted into Eq. (3) and varied $\Psi_{C,A}$ for each system until we found minimized error between the experimental and simulated output for kinetics, gel properties such as gel point, and swelling ratio. Fig. (3) shows an example of such a series of perturbations to the $\Psi_{C,A}$ and their associated error in deviation from the experimental data over the whole polymerization. We thus report $\Psi_{C,A}$ from the match with the minimum error of deviation.

Important to note is the fact that with only this parametric estimation by this approach, we generate a value for the rate coefficient, relevant to a single copolymer system (notably not specific to the divinyl monomer alone, but to the pair of mono and divinyl), that is consistent with performance when applied over a range of copolymer ratios; seen in Fig. 4. In each panel of the figure, a single copolymer system is explored (for example, poly(MMA-*co*-EGDMA)) where the characteristic boost in overall reaction rate is observed with increasing EGDMA content with excellent overlap

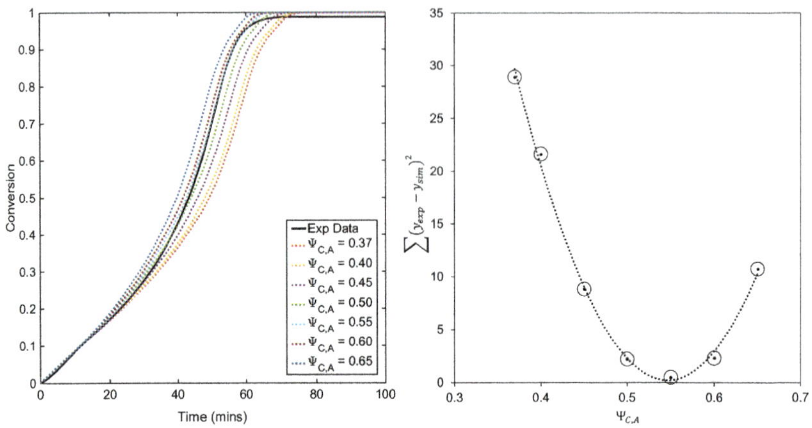

Fig. 3 Perturbations to $\Psi_{C,A}$ and the resulting kinetic profile compared to DSC generated experimental kinetics (left) and their associated error in deviation from the experimental data over the whole polymerization (right).

of the DSC raw experimental data and our simulated form. Fig. 4 (top left) highlights that we are in the process searching not only for an acceptable fit to the early conversion, but to a fit optimized across the full conversion profile capturing various and all ranges of contributions to the polymerization process (for example, the Trommsdorff gel effect for glassy monomers (Trommsdorff et al., 1948)). For each panel in Fig. 4, a single $\Psi_{C,A}$ is relevant to all concentrations of the divinyl (restricted in our explorations to below 10%, yet this is something we are addressing further now, not to be reported on here). Fig. 5 thus shows for a single divinyl EGDMA paired with examples from the homologous series of methacrylates, there is a smooth dependence of the length of the alkyl sidechain on the monovinyl relative to the pendent reactive group in the hinged divinyl.

The $\Psi_{C,A}$ parameter obtained for MMA, EMA, n-BMA and HMA with EGDMA (i.e., up to effectively similar dimensions of pendent monovinyl sidechains with the pendent divinyl hinged sidechain) demonstrates two key points. First, the values of $\Psi_{C,A}$ for those methacrylate-EGDMA systems were all less than 1.0, confirming the hypothesis that the reactivity of the pendent group is lower than that of the free crosslinker monomer. Secondly, the observed trend for the $\Psi_{C,A}$ parameters is attributed to a potential steric effect imposed by neighboring monomers into the available reaction volume (or reaction cavity) surrounding that pendent vinyl. Here in Fig. 5, we see the suggestion of a plateau at longer alkyl lengths,

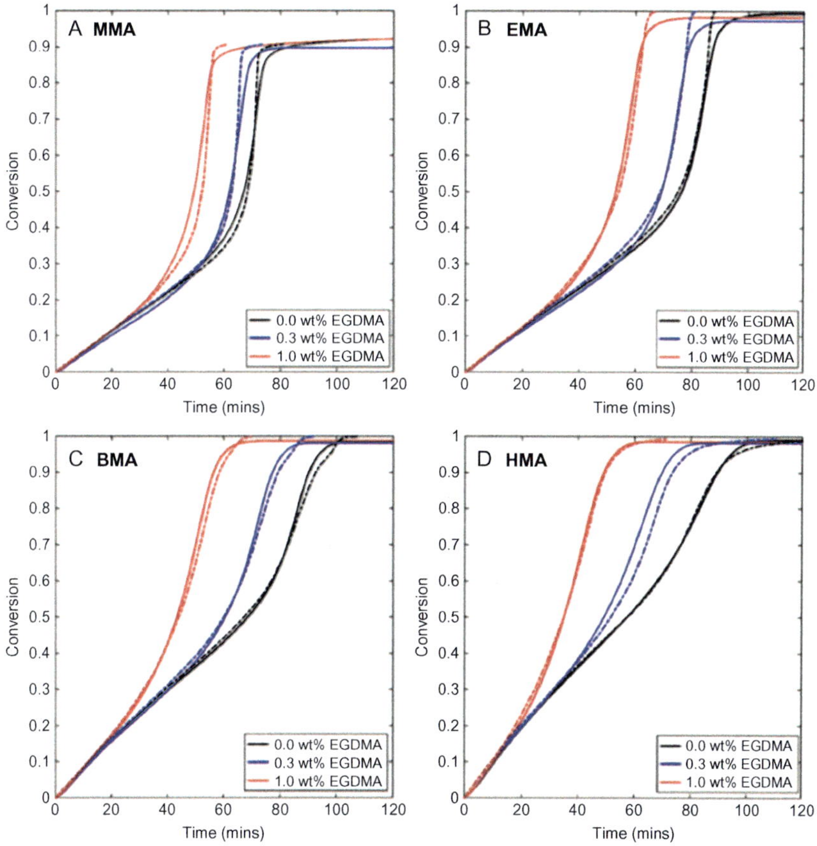

Fig. 4 Plot of experimental (using differential scanning calorimetry, solid lines) and simulated (using Monte Carlo method, dashed lines) conversion vs time for bulk polymerization of (A) methyl methacrylate; (B) ethyl methacrylate; (C) n-butyl methacrylate; (D) n-hexyl methacrylate with 0.3 wt% AIBN initiator at 70 °C at 0.0, 0.3 and 1.0 wt % EGDMA. *Reprinted from Tripathi AK, Neenan ML, Sundberg DC, Tsavalas JG: Influence of n-alkyl ester groups on efficiency of crosslinking for methacrylate monomers copolymerized with EGDMA: experiments and Monte Carlo simulations of reaction kinetics and sol–gel structure,* Polymer (Guildf) *96:130–145, 2016. Copyright (2016) with permission from Elsevier.*

however, while our new contributions are under review at the moment we can foreshadow here that an exciting extension of behavior results once one applies systems with sidechains of even longer length (both for the mono- and divinyl-constituents in the pairs) and thus the ability to take on more coiled rather than extended conformations. Importantly, our $\Psi_{C,A}$ approach remains applicable and captures this extension of

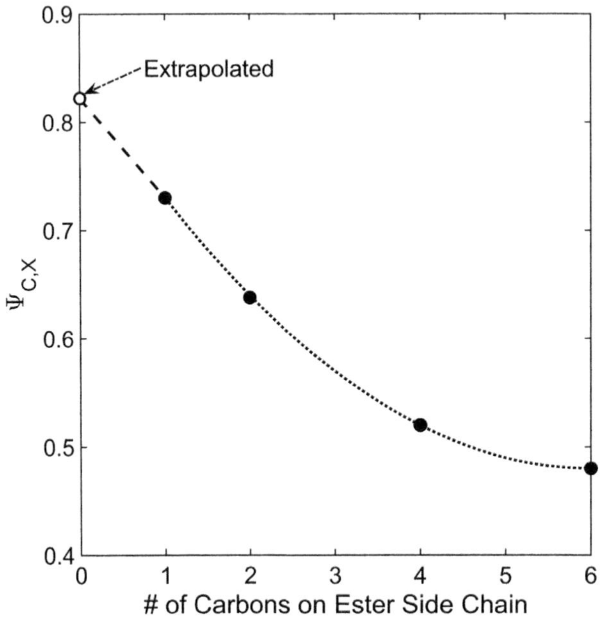

Fig. 5 Estimated reduced reactivity parameter ($\Psi_{C,X}$) for the homologous series of methacrylate monomers with EGDMA vs number of carbons in the respective n-alkyl ester side chains. The data is also extrapolated to the value of zero carbon on the alkyl ester side chain. *Reprinted from Tripathi AK, Neenan ML, Sundberg DC, Tsavalas JG: Influence of n-alkyl ester groups on efficiency of crosslinking for methacrylate monomers copolymerized with EGDMA: experiments and Monte Carlo simulations of reaction kinetics and sol–gel structure,* Polymer (Guildf) *96:130–145, 2016. Copyright (2016) with permission from Elsevier.*

behavior nicely. Moreover, we are similarly in the process of reporting the impact of reactivity ratios of greater disparity between the mono- and di-vinyls in the pairs (we will tease the reader here to say that this has a maybe unsurprising, but very important, effect; again, captured well with this $\Psi_{C,A}$ estimation approach), as well as on the impact of temperature within typical operating ranges (60–95 °C) for common free-radical polymerization. We are anxious to bring those new findings to the community, but for now we can simply foreshadow that the work has indeed been addressed and will be available shortly. The point of focus, however, is that this serves as an example of the PRE mindset applied to "*known unknowns*" so as to generate the ability to predict and optimize systems beyond those we have tried to date.

8. Concluding remarks

In summary, some of the unmet needs of the field (here we only focused attention to divinyl monomer rate coefficients as an example) can be met via simulation (including through new approaches such as graph theory, machine learning, AI). Yet opportunities also remain for creative new analytical methods to capture these "*known unknowns*" experimentally. This need for new and expanded techniques (experimental, theoretical and applied) will only grow as our field blossoms with the harvest of future gardens of polymers embedded with triggers for environmental responsibility (e.g., compostable, biodegradable, and recyclable) as well as advanced function (e.g., stimuli-responsive, self-healing, shape changing). PRE is certain to continue to be needed for our future. Indeed, maturity here is an advantage as we are not new to the game. We have our toolboxes at the ready to address continued development of the world's enormous material markets, but potentially more importantly for those new products soon to come.

References

Arzamendi G, Leiza JR: Molecular weight distribution (soluble and insoluble fraction) in emulsion polymerization of acrylate monomers by Monte Carlo simulations, *Ind Eng Chem Res* 47(16):5934–5947, 2008.

Ballard MJ, Gilbert RG, Napper DH, Pomery PJ, O'Sullivan PW, O'Donnell JH: Propagation rate coefficients from electron spin resonance studies of the emulsion polymerization of methyl methacrylate, *Macromolecules* 19(5):1303–1308, 1986.

Barner-Kowollik C, Bennet F, Schneider-Baumann M, et al.: Detailed investigation of the propagation rate of urethane acrylates, *Polym Chem* 1(4):470–479, 2010.

Barner-Kowollik C, Beuermann S, Buback M, et al.: Critically evaluated rate coefficients in radical polymerization-7. Secondary-radical propagation rate coefficients for methyl acrylate in the bulk, *Polym Chem* 5(1):204–212, 2014.

Barth J, Buback M, Hesse P, Sergeeva T: Chain-length-dependent termination in n-butyl methacrylate and tert-butyl methacrylate bulk homopolymerizations studied via SP-PLP-ESR, *Macromolecules* 42(2):481–488, 2009.

BBC News: *Single-use plastic: China to ban bags and other items (website)*, 2020. https://www.bbc.com/news/world-asia-china-51171491. (Accessed 26 June 2020).

Beuermann S: Requirements associated with studies into a chain-length dependence of propagation rate coefficients via PLP-SEC experiments, *Macromolecules* 35(25): 9300–9305, 2002.

Beuermann S, Buback M: Rate coefficients of free-radical polymerization deduced from pulsed laser experiments, *Prog Polym Sci* 27(2):191–254, 2002.

Beuermann S, García N: A novel approach to the understanding of the solvent effects in radical polymerization propagation kinetics, *Macromolecules* 37(8):3018–3025, 2004.

Beuermann S, Paquet DA, McMinn JH, Hutchinson RA: Determination of free-radical propagation rate coefficients of butyl, 2-ethylhexyl, and dodecyl acrylates by pulsed-laser polymerization, *Macromolecules* 29(12):4206–4215, 1996.

Beuermann S, Buback M, Schmaltz C: Pressure and temperature dependence of butyl acrylate propagation rate coefficients in fluid CO_2, *Macromolecules* 31(23):8069–8074, 1998.

Beuermann S, Buback M, Nelke D: Pressure dependence of the propagation rate coefficient kp for vinyl acetate polymerizations in bulk and in solution of fluid CO_2, *Macromolecules* 34(19):6637–6640, 2001.

Beuermann S, Buback M, Davis TP, et al.: Critically evaluated rate coefficients for free-radical polymerization, 4 propagation rate coefficients for methacrylates with cyclic ester groups, *Macromol Chem Phys* 204(10):1338–1350, 2003.

Beuermann S, Buback M, Hesse P, Kukučková S, Lacík I: Propagation kinetics of free-radical methacrylic acid polymerization in aqueous solution. The effect of concentration and degree of ionization, *Macromol Symp* 248:23–32, 2007.

Beuermann S, Buback M, Hesse P, Hutchinson RA, Kukučková S, Lacík I: Termination kinetics of the free-radical polymerization of nonionized methacrylic acid in aqueous solution, *Macromolecules* 41(10):3513–3520, 2008.

Buback M: Free-radical polymerization kinetics in extended ranges of temperature, pressure and monomer conversion, *Macromol Symp* 111(1):229–242, 1996.

Buback M, Laemmel RA: Simultaneous determination of free-radical propagation and transfer rates from a novel type of PLP-SEC experiment, *Macromol Theory Simul* 6(1): 145–150, 1997.

Buback M, Garcia-Rubio LH, Gilbert RG, et al.: Consistent values of rate parameters in free radical polymerization systems, *J Polym Sci, Polym Lett Ed* 26(7):293–297, 1988.

Buback M, Gilbert RG, Russell GT, et al.: Consistent values of rate parameters in free radical polymerization systems. II. Outstanding dilemmas and recommendations, *J Polym Sci A Polym Chem* 30(5):851–863, 1992.

Buback M, Müller E, Russell GT: SP-PLP-EPR study of chain-length-dependent termination in free-radical polymerization of n-dodecyl methacrylate, cyclohexyl methacrylate, and benzyl methacrylate: evidence of "composite" behavior, *J Phys Chem A* 110(9): 3222–3230, 2006.

Burke AL, Duever TA, Penlidis A: Discriminating between the terminal and penultimate models using designed experiments: an overview, *Ind Eng Chem Res* 36(4): 1016–1035, 1997.

Burnett GM, Melville HW: Propagation and termination coefficients for vinyl acetate photopolymerization, *Nature* 156(3970):661, 1945.

Burnett GM, Melville HW: Determination of the velocity coefficients for polymerization processes. I. The direct photopolymerization of vinyl acetate, *Proc R Soc Lond A Math Phys Sci* 189(1019):456–480, 1947.

Carothers WH: *Synthetic fiber (patent)*, US2130948, 1938.

Chalmers W: The mechanism of macropolymerization reactions, *J Am Chem Soc* 56(4):912–922, 1934.

Davis TP, Zammit MD: The importance of accurate and precise molecular weight measurement in pulsed-laser polymerization, *Macromol Symp* 111(1):243–252, 1996.

D'hooge DR, Reyniers MF, Marin GB: The crucial role of diffusional limitations in controlled radical polymerization, *Macromol React Eng* 7(8):362–379, 2013.

Dostal H, Mark H: The mechanism of polymerisation, *Trans Faraday Soc* 32:54, 1936.

Fawcett EW, Gibson RO, Perrin MW: *Polymerization of olefins (patent)*, US 2,153,553, 1939.

Flory PJ: Molecular size distribution in linear condensation polymers 1, *J Am Chem Soc* 58(10):1877–1885, 1936.

Gillespie DT: A general method for numerically simulating the stochastic time evolution of coupled chemical reactions, *J Comput Phys* 22(4):403–434, 1976.

Gillespie DT: Exact stochastic simulation of coupled chemical reactions, *J Phys Chem* 81(25):2340–2361, 1977.
Goodyear C: *Improvement in the mode of preparing caoutchouc with sulphur for the manufacture of various articles (patent)*, US1090A, 1839.
Gruendling T, Voll D, Guilhaus M, Barner-Kowollik C: A perfect couple: PLP/SEC/ESI-MS for the accurate determination of propagation rate coefficients in free radical polymerization, *Macromol Chem Phys* 211(1):80–90, 2010.
Haehnel AP, Schneider-Baumann M, Hiltebrandt KU, Misske AM, Barner-Kowollik C: Global trends for kp? Expanding the frontier of ester side chain topography in acrylates and methacrylates, *Macromolecules* 46(1):15–28, 2013.
Haehnel AP, Schneider-Baumann M, Arens L, Misske AM, Fleischhaker F, Barner-Kowollik C: Global trends for kp? The influence of ester side chain topography in alkyl (meth)acrylates—completing the data base, *Macromolecules* 47(10): 3483–3496, 2014.
Heuts JPA, Russell GT: The nature of the chain-length dependence of the propagation rate coefficient and its effect on the kinetics of free-radical polymerization. 1. Small-molecule studies, *Eur Polym J* 42(1):3–20, 2006.
Hutchinson RA, Beuermann S, Paquet DA, McMinn JH: Determination of free-radical propagation rate coefficients for alkyl methacrylates by pulsed-laser polymerization, *Macromolecules* 30(12):3490–3493, 1997.
Jeličić A, Yasin M, Beuermann S: Toward the description and prediction of solvent induced variations in methacrylate propagation rate coefficients on the basis of solvatochromic parameters, *Macromol React Eng* 5(5–6):232–242, 2011.
Junkers T, Schneider-Baumann M, Koo SSP, Castignolles P, Barner-Kowollik C: Determination of propagation rate coefficients for methyl and 2-ethylhexyl acrylate via high frequency PLP-SEC under consideration of the impact of chain branching, *Macromolecules* 43(24):10427–10434, 2010.
Kockler KB, Haehnel AP, Junkers T, Barner-Kowollik C: Determining free-radical propagation rate coefficients with high-frequency lasers: current status and future perspectives, *Macromol Rapid Commun* 37(2):123–134, 2016.
Kuchta FD, Van Herk AM, German AL: Propagation kinetics of acrylic and methacrylic acid in water and organic solvents studied by pulsed-laser polymerization, *Macromolecules* 33(10):3641–3649, 2000.
Kuhn W: Über die Kinetik des Abbaues hochmolekularer Ketten, *Ber Dtsch Chem Ges* 63(6):1503–1509, 1930.
Lau W, Westmoreland DG, Novak RW: Direct measurement of propagating radical concentration in a semicontinuous emulsion polymerization, *Macromolecules* 20(2): 457–459, 1987.
Li W-H, Hamielec AE, Crowe CM: Kinetics of the free-radical copolymerization of methyl methacrylate/ethylene glycol dimethacrylate: 1. Experimental investigation, *Polymer (Guildf)* 30(8):1513–1517, 1989.
Liang K, Hutchinson RA: The effect of hydrogen bonding on intramolecular chain transfer in polymerization of acrylates, *Macromol Rapid Commun* 32(14):1090–1095, 2011.
Marien YW, Van Steenberge PHM, Barner-Kowollik C, Reyniers MF, Marin GB, D'Hooge DR: Kinetic Monte Carlo modeling extracts information on chain initiation and termination from complete PLP-SEC traces, *Macromolecules* 50(4):1371–1385, 2017.
Marien YW, Van Steenberge PHM, Pich A, D'Hooge DR: Coupled stochastic simulation of the chain length and particle size distribution in miniemulsion radical copolymerization of styrene and N-vinylcaprolactam, *React Chem Eng* 4(11):1935–1947, 2019.
Marten FL, Hamielec AE: High conversion diffusion-controlled polymerization, In Henderson JN, Bouton TC, editors: *Polymerization reactors and processes*, Washington, DC, 1979, American Chemical Society, pp 43–70.

Nikitin AN, Hutchinson RA, Buback M, Hesse P: Determination of intramolecular chain transfer and midchain radical propagation rate coefficients for butyl acrylate by pulsed laser polymerization, *Macromolecules* 40(24):8631–8641, 2007.

Nitschke A, Riemann L, Kollenbach L, Braun V, Buback M, Vana P: Investigation into the kinetics of n-pentyl methacrylate radical polymerization, *Macromol Chem Phys* 221(1), 2020, 1900345.

O'Driscoll KF, Mahabadi HK: Spatially intermittent polymerization, *J Polym Sci, Polym Chem Ed* 14(4):869–881, 1976.

Olaj OF, Schnöll-Bitai I: Laser-flash-initiated polymerization as a tool for evaluating (individual) kinetic constants of free-radical polymerization-5. Complete analysis by means of a single experiment, *Eur Polym J* 25(7–8):635–641, 1989.

Olaj OF, Bitai I, Gleixner G: The laser flash-initiated polymerization as a tool of evaluating (individual) kinetic constants of free radical polymerization, 1. Outline of method and first results, *Die Makromol Chemie* 186(12):2569–2580, 1985.

Olaj OF, Bitai I, Hinkelmann F: The laser-flash-initiated polymerization as a tool of evaluating (individual) kinetic constants of free-radical polymerization, 2. The direct determination of the rate of constant of chain propagation, *Die Makromol Chemie* 188 (7):1689–1702, 1987.

Rooney TR, Chovancová A, Lacík I, Hutchinson RA: Pulsed laser studies of cationic reactive surfactant radical propagation kinetics, *Polymer* 130:39–49, 2017.

Schulz GV: Über die Beziehung zwischen Reaktionsgeschwindigkeit und Zusammensetzung des Reaktionsproduktes bei Makropolymerisationsvorgängen, *Z Phys Chem* 30B (1):379–398, 1935.

Smith GB, Russell GT, Yin M, Heuts JPA: The effects of chain length dependent propagation and termination on the kinetics of free-radical polymerization at low chain lengths, *Eur Polym J* 41(2):225–230, 2005.

Solomon K: *These 10 countries will inspire you to stop your single-use plastic habit (website)*, 2020 https://www.greatbigstory.com/guides/plastic-bans-around-the-world. Accessed June 26, 2020.

Staudinger H: Über polymerisation, *Ber Dtsch Chem Ges* 53(6):1073–1085, 1920.

Tobita H: Bimodal molecular weight distribution formed in the emulsion polymerization of ethylene, *J Polym Sci A Polym Chem* 40(20):3426–3433, 2002.

Tripathi AK, Sundberg DC: A hybrid algorithm for accurate and efficient Monte Carlo simulations of free-radical polymerization reactions, *Macromol Theory Simul* 24(1): 52–64, 2015.

Tripathi AK, Tsavalas JG, Sundberg DC: Monte Carlo simulations of free radical polymerizations with divinyl cross-linker: pre- and postgel simulations of reaction kinetics and molecular structure, *Macromolecules* 48(1):184–197, 2015.

Tripathi AK, Neenan ML, Sundberg DC, Tsavalas JG: Influence of n-alkyl ester groups on efficiency of crosslinking for methacrylate monomers copolymerized with EGDMA: experiments and Monte Carlo simulations of reaction kinetics and sol–gel structure, *Polymer* 96:130–145, 2016.

Trommsdorff VE, Köhle H, Lagally P: Zur polymerisation des methacrylsäuremethylesters, *Die Makromol Chemie* 1(3):169–198, 1948.

Uhelská L, Chorvát D, Hutchinson RA, Santanakrishnan S, Buback M, Lacík I: Radical propagation kinetics of N-vinylpyrrolidone in organic solvents studied by pulsed-laser polymerization-size-exclusion chromatography (PLP-SEC), *Macromol Chem Phys* 215 (23):2327–2336, 2014.

Vir AB, Marien YW, Van Steenberge PHM, et al.: Access to the β-scission rate coefficient in acrylate radical polymerization by careful scanning of pulse laser frequencies at elevated temperature, *React Chem Eng* 3(5):807–815, 2018.

Willemse RXE, Van Herk AM: Determination of propagation rate coefficients of a family of acrylates with PLP-MALDI-ToF-MS, *Macromol Chem Phys* 211(5):539–545, 2010.

Willemse RXE, Van Herk AM, Panchenko E, Junkers T, Buback M: PLP-ESR monitoring of midchain radicals in n-butyl acrylate polymerization, *Macromolecules* 38(12): 5098–5103, 2005.

Woods M: *Single-use plastic ban enters into effect in France (website)*, 2020. http://www.rfi.fr/en/france/20200101-france-single-use-plastic-ban-enters-effect-environment-pollution. Accessed June 26, 2020.

Zammit MD, Davis TP, Haddleton DM: Determination of the propagation rate coefficient (kp) and termination mode in the free-radical polymerization of methyl methacrylate, employing matrix-assisted laser desorption ionization time-of-flight mass spectrometry for molecular weight distribution analysis, *Macromolecules* 29(1):492–494, 1996.

CHAPTER TWO

On-line monitoring and control of emulsion polymerization reactors

Shaghayegh Hamzehlou and José Maria Asua*
POLYMAT and Kimika Aplikatua Saila, Kimika Fakultatea, University of the Basque Country UPV-EHU, Joxe Mari Korta Zentroa, Donostia-San Sebastian, Spain
*Corresponding author: e-mail address: jm.asua@ehu.eus

Contents

Abstract

Emulsion polymerization is a complex multiphase process that is prone to suffer form run-to-run irreproducibility. It yields a product-by-process material whose performance is determined by a delicate balance between multiple characteristics such as copolymer composition, chemical composition distribution, molar mass distribution, polymer architecture, particle morphology, particle size distribution and surface composition. Therefore, there is a strong incentive to develop on-line monitoring techniques and on-line control strategies that ensure the robust and reproducible production of emulsion polymers. This chapter reviews the practical implementation of on-line monitoring and control of emulsion polymerization reactors, highlighting the areas where improvements are needed.

Advances in Chemical Engineering, Volume 56
ISSN 0065-2377
https://doi.org/10.1016/bs.ache.2020.07.002

1. Introduction

Emulsion polymers have gained industrial interest because on one hand, they substitute polymers produced by less environmentally friendly methods (notably through solvent-based synthesis) and on the other, they present unique application properties (Asua, 1997; Urban and Takamura, 2002). The performance of the emulsion polymers is controlled by multiple characteristics including copolymer composition, chemical composition distribution, molar mass distribution, polymer architecture (linear, branched, crosslinked, …), particle morphology, particle size distribution and surface composition. Furthermore, the emulsion polymers are product-by-process materials whose characteristics are almost completely determined in the reactor.

Emulsion polymers are produced through emulsion polymerization, a multiphase reaction that involves particle nucleation and growth, mass transfer of monomers and radicals between phases and a complex kinetic scheme. This complexity and the practical need of working with technical monomers make the process prone to suffer from run-to-run irreproducibility (Asua, 2004). Therefore, there is a strong incentive to develop on-line control strategies that ensure the robust and reproducible production of emulsion polymers.

Fig. 1 sketches an on-line control scheme. In order to control a given characteristic, its current value in the reactor should be compared with

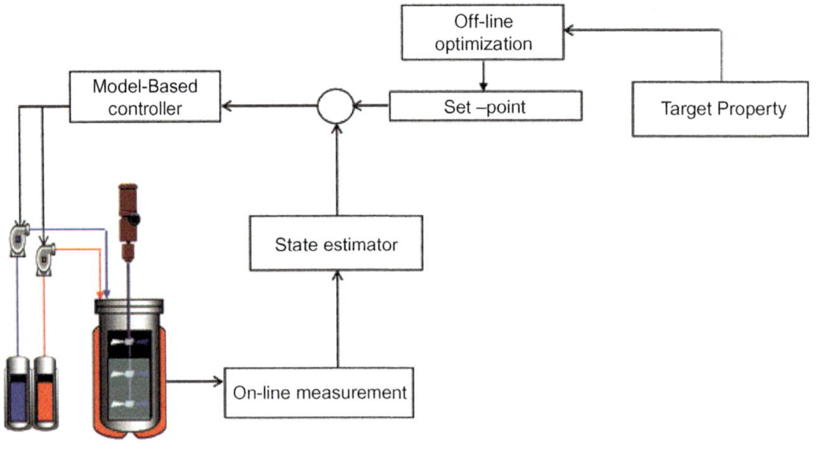

Fig. 1 Close loop control scheme.

the desired value (set point) and for that it has to be either directly measured on-line or estimated from other variables measured on-line. The difficulties in measuring on-line the characteristics of the polymer dispersions have hindered the developments in the on-line control of emulsion polymerization processes. As discussed below, there are few properties that can be measured on-line (Chien and Penlidis, 1990; Frauendorfer et al., 2010; Haven and Junkers, 2017; Kammona et al., 1999) and some important characteristics such as the architecture of branched and crosslinked polymers and the particle morphology are even hard to analyze offline (Makan et al., 2012; Podzimek et al., 2014; Shapiro, 2011).

Nevertheless, some characteristics that cannot be directly measured on-line might be unambiguously estimated from on-line measurements. They are called observable variables. In addition, the non-observable characteristics can be estimated using mathematical models (soft sensors). The uncertainty of the estimation is linked to the precision of the model. When precise models are available, accurate on-line estimations of the non-observable properties are possible. A robust, predictive and fast mathematical model is needed to be able to be used within the time scale of the process.

There is a rich literature of mathematical models for emulsion polymerization that in principle can predict most of the characteristics of the emulsion polymers (Alexopoulos et al., 2013; Arzamendi and Leiza, 2008; Asua, 2003; Asua and De La Cal, 1991; Butte et al., 2002; Ghielmi et al., 1998; Gonzalez-Ortiz and Asua, 1996; Hamzehlou et al., 2014, 2016; Maxwell et al., 1991; Min and Ray, 1974; Pladis et al., 2014; Rawlings and Ray, 1988; Reynhout et al., 2005; Smith and Ewart, 1948; Storti et al., 1992; Stubbs et al., 2008; Tauer and Kuehn, 1995; Thickett and Gilbert, 2006; Tobita, 1995; Vale and Mckenna, 2005; Zubitur et al., 2003). However, due to the complexity of the process, the models are mathematically complex and contain a high number of parameters, which are difficult to estimate and in addition depend on the particular formulation used. Therefore, their use as reliable soft sensors is still a challenge. A problem linked to the complexity of the models is that they frequently require substantial computing times. Artificial neural networks (ANN) can overcome this limitation as they are fast. ANN have been used in emulsion polymerization (Arzamendi et al., 2005; D'Anjou et al., 2003; Fernandes et al., 2004; Tian et al., 2002; Tsen et al., 1996). However, for training the ANN model, large amounts of data, which often are not available, are needed.

In this chapter, first the on-line monitoring and estimation of the polymer latex characteristics in emulsion polymerization are discussed. Then, the practical implementations of the on-line control of different properties of polymer latexes are reviewed. In the final part, the areas where improvements are needed are highlighted and an outlook presented.

2. On-line monitoring and estimation of polymer latex characteristics

In this section, the state-of-the-art for the on-line monitoring and estimation of the polymer latex characteristics (copolymer composition, chemical composition distribution, molar mass distribution, polymer architecture (linear, branched and crosslinked), particle morphology, particle size distribution and surface composition) is reviewed. The review is organized according to the characteristic to be monitored. At the end, the automatic continuous on-line monitoring of polymerization reaction system (ACOMP) that is able to monitor several of these characteristics is discussed.

2.1 Free monomer concentration/copolymer composition/polymerization rate

2.1.1 Direct measurement of the free monomer concentration

Although not directly a polymer characteristic, free monomer concentration is a key variable that allows the unambiguous estimation of copolymer composition, monomer conversion and polymerization rate. Spectroscopic techniques (Raman, Near Infrared (NIRS), NMR and Ion Mobility (IMS)) and gas chromatography are able to provide the direct measurement of the free monomer in the reactor.

Raman and NIRS have benefit from the developments in fiber optics technology which can work under harsh conditions allowing for the in-situ and in-line acquisition of process data, namely no sample treatment is needed. NMR and IMS often require a sample device to transfer the sample to a measuring chamber.

Raman spectroscopy is used widely in monitoring the monomer concentration of different polymerization reactions (Amer, 2009; Reed and Alb, 2014; Van et al., 2001). This technique is also suited for monitoring emulsion polymerization as water has a very weak Raman signal. As individual monomer concentrations can be measured on-line by Raman, monitoring of the copolymer composition is also possible. Elizalde et al. overviewed the use of Raman spectroscopy and the technical problems encountered when using this technique in emulsion polymerization (Elizalde and Leiza, 2009). Raman spectroscopy has been used to monitor the emulsion homopolymerization of styrene using the deconvolution of a single reference band assigned to aromatic ring near 1000 cm^{-1} (Dropsit et al., 2019) instead of the usual approach of employing the ratio of the bands related to monomer (1631 cm^{-1}) and overlapping signals of monomer and polymer (1000 cm^{-1}). However, Raman spectroscopy is more useful for copolymerization. In this case, free monomer monitoring is relatively easy when the peaks of the monomers are sufficiently separated as is the case of the emulsion copolymerization of styrene and butyl acrylate (Asua, 2003). More complex are the copolymerizations in which the Raman spectra of the monomers have overlaps. Reis et al. studied the limitation of the calibration models in the copolymerization of vinyl acetate and butyl acrylate in semi-continuous emulsion polymerization (Reis et al., 2004a). Due to the structural similarity of the monomers, the copolymerization of (meth)acrylates is a particularly difficult case (Martens and Næs, 1989). Successful monitoring of monomer concentration of the emulsion copolymerization of n-butyl acrylate/methyl methacrylate (Elizalde et al., 2004a) at high solids content and starved condition (Asua et al., 2005; Elizalde et al., 2004b) and n-butyl acrylate/vinyl neonanoate (VeoVA9) (van den Brink et al., 2001) have been reported using multivariate calibration techniques (Martens and Næs, 1989). It is worth mentioning here that the performance of the Fourier transform Raman spectroscopy in these cases is dependent on the calibration models and it was shown that maintenance and repairing of the equipment affects the calibration models, making necessary their updating (Elizalde et al., 2005a).

Near-infrared spectroscopy (NIRS) has also been used to monitor the monomer concentration during emulsion polymerization. Santos et al. reviewed the use of NIRS in monitoring the polymerization reactions explaining in detail the challenges of using this technique in heterogeneous systems such as suspension and emulsion polymerization (Santos et al., 2005). Gossen et al. (1993) pioneered the field monitoring the emulsion copolymerization of styrene and methyl methacrylate. Calibration curves generated by partial least squares allowed a reliable monitoring for 30 wt% translucent solids content, but failed for 50 wt% opaque latexes. The challenges and complexities involved in the development of calibration models in heterogeneous systems have been discussed by Reis et al. (2004a, 2007). They showed that fast spectral acquisition can lead to a low signal/noise ratio and that approaches based on principal curves yield reliable predictions. This group monitored the monomer conversion in the copolymerization of styrene and butyl acrylate (Reis et al., 2004b). In addition, as NIRS also provides information on the particle size (see below), the number of polymer particles (Np) and the average number of radicals per particle ($\overline{n}$) were estimated. However, it is worth pointing out that these estimations involve the third power of the particle size, which enhance error propagation leading to uncertain estimates of Np and $\overline{n}$.

Nuclear Magnetic Resonance (NMR) spectroscopy is likely the most often used technique to measure offline the copolymer composition. Landfester showed for the first time the feasibility of using solid-state ^{1}H NMR and ^{13}C NMR for monitoring on-line the conversion of the emulsion polymerization of butyl acrylate carried out within the NMR spectrometer (Landfester et al., 1998). However, its use for on-line monitoring of an emulsion polymerization carried out in a stirred tank reactor is much more challenging because it requires sample extraction. Thus, Vargas et al. pumped the reaction mixture to the NMR probe in a closed loop. In this way they were able to monitor on-line the emulsion polymerization of butyl acrylate (Vargas et al., 2010). In addition to classical proton and carbon NMR spectroscopies, hyperpolarized ^{129}Xe NMR spectroscopy was also employed to monitor styrene polymerizations in miniemulsion, carrying the reaction within the spectrometer (Duewel et al., 2012). The applicability of this method to regular reactors needs a sample extraction, but has not been studied yet. As xenon is usually not present in common chemical substances, no undesired background signal will be present to facilitate the interpretation of the sparse Xe spectra. In conclusion, although from the analytical point of view NMR is an interesting alternative, the challenges associated to its use in real systems seem to be currently insurmountable.

Ion mobility spectrometry (IMS) is an inexpensive technique able to detect very low concentrations of monomers in polymerization reactions. Vautz et al. explored the use of IMS in the emulsion polymerization of vinyl acetate by feeding the gas of the reactor head-space through a continuous flow cell after irradiation with UV light for ionization. The feasibility of the implementation of this technique was tested in a pilot plant (Vautz et al., 2009). In order to get the free monomer in the reaction mixture, it is assumed that the equilibrium gas/aqueous phase/polymer particles is reached for the monomers and the partitioning coefficients of the monomers are needed.

Gas chromatography (GC) was used also to measure on-line the unreacted monomers in emulsion polymerization. Alonso et al. (1986) developed a method based on the GC analysis of the head space of the reactor. This method is robust, but it has the same limitations than IMS. Another alternative is to analyze the liquid phase. Leiza et al. (1993) developed a sampling system that after dilution of the sample, directly injected the liquid phase into the GC. Introducing a standard during the dilution step it was possible to accurately monitor the 55 wt% solids content emulsion terpolymerization of methyl methacrylate, butyl acrylate and vinyl acetate (Leiza et al., 1993). An intrinsic limitation of this system is that the sampling-dilution-analysis process takes some time (18 min for the VAc-MMA-BA system) and therefore the information obtained is delayed. The system required frequent maintenance. The setup was used to monitor other *co*- and ter-polymerization emulsion polymerization systems (Leiza et al., 1992; Urretabizkaia et al., 1994).

In an attempt to avoid the use of a internal standard in the GC analysis, Noël et al. used the combination of GC and on-line densimetry to calculate the partial conversion of each monomers in the batch emulsion polymerization of methyl methacrylate and vinyl acetate (Noël et al., 1995). In this work, only the ratio of the free monomers were obtained from GC and densimetry gave the overall conversion. Using the combination of these two pieces of information it was possible to estimate the partial conversions of both monomers. However, on-line densimetry is not an easy method to implement on-line as the reaction medium should be pumped through the densitometer, which limits the solids content than can be handled (about 10 wt% (Noël et al., 1995)). Furthermore, the solubility of the monomer in water affects the volume additivity assumption needed for densimetry.

2.1.2 Estimation of the free monomer concentration from other measurements

The methods discussed above for the direct measurement of the free monomer are not free from problems as they involved either relatively expensive equipments, or/and frequent maintenance or/and assumptions regarding the partitioning of the monomers among the gas-liquid-solid phases. Therefore, other alternatives such as reaction calorimetry and densimetry have been proposed, reaction calorimetry being most often used as it is easy to implement (mainly at industrial scale) and robust.

Reaction calorimetry can be used to monitor the heat generated in the exothermic polymerization reactions which is proportional to the rate of the polymerization for homopolymerizations (Moritz, 1989). However, for copolymerizations, the overall heat generated in the reaction is the result of the heats generated by both monomers, and the individual polymerization rate of each monomer cannot be unambiguously determined from the overall polymerization heat, i.e., they are not observable. Urretabizkaia et al. (1993) demonstrated that the simple Mayo-Lewis equation (Mayo and Lewis, 1944) can be used as a reliable soft sensor that allows an accurate estimation of the individual polymerization rates, and therefore of the copolymer composition. The information needed is the enthalpy of the polymerization of each monomer and the reactivity ratios. Since then, reaction calorimetry has been frequently used to monitor free radical emulsion co- and ter-polymerizations (de Buruaga et al., 1996, 2000; Gugliotta et al., 1995a,b; Guinot et al., 2000; Saenz de Buruaga et al., 1997; Vega et al., 2002). However, it is worth mentioning that the technique shows limitations under starved conditions, which are commonly used in emulsion polymerization. While the conversion of the main monomers and therefore the copolymer composition are accurately estimated, the error in the estimation of the amounts of both unreacted and minor monomers (e.g., the acrylic acid used in many formulations) may be high, which is a drawback for some applications (Asua, 2004).

Densimetry using a U-tube densimeter has been employed to monitor on-line the conversion in emulsion polymerization. The measured density is correlated with the exact density of all components in the reactor. In the case of copolymerization with composition drift, for estimation of polymer density and monomer mixture, the use of an additional devise to obtain the evolution of copolymer composition is needed. An alternative is using models which correlate the copolymer composition with the conversion which needs some parameters to be known such as reactivity ratios and partition coefficients. Canegallo et al. have used densimetry for on-line monitoring of different copolymerization systems (styrene/methyl methacrylate, acrylonitrile/methyl methacrylate and vinyl acetate/methyl methacrylate) in batch emulsion polymerization using model to estimate the copolymer composition (Canegallo et al., 1993). The same method has been used for on-line monitoring of the conversion by densimetry in a semibatch emulsion copolymerization of styrene and butyl acrylate (Galimberti et al., 1998). Although some successful examples of the use of densimetry in emulsion polymerization have been reported, its practical use is limited due to several challenges arises during emulsion polymerization such as the presence of monomer droplets, flocculation of the latex, and viscosity increase which make it difficult to obtain meaningful measurement of density (Reed and Alb, 2014).

2.1.3 Comparison between spectroscopy and calorimetry

Elizalde et al. compared Raman spectroscopy and calorimetry for monitoring monomer concentration in the high solids content semibatch emulsion copolymerization of different comonomer systems (VAc/BA and BA/MMA). They concluded that both methods perform well, although Raman spectroscopy provided more precise results under starved conditions (Elizalde et al., 2005b) (see Fig. 2). Reis et al. also evaluated NIR and Raman spectroscopy for different homo- and *co*-polymerization systems in semi-continuous emulsion polymerization and showed that by selecting the appropriate calibration model, both methods give good and comparable estimation of the monomer concentration during the reaction (Reis et al., 2004c).

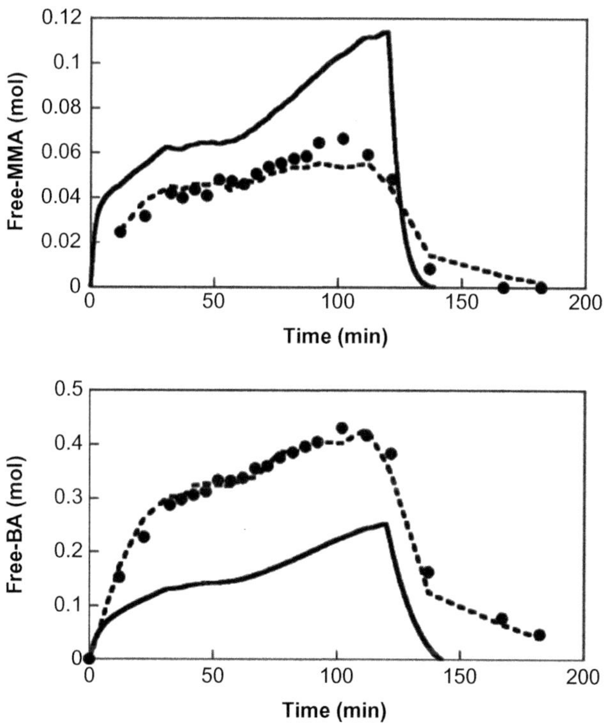

Fig. 2 Comparison of the free monomer measured by GC (dots), Raman spectroscopy (dashed line) and calorimetry (solid line) in the starved semi-continuous emulsion polymerization of MMA/BA. *Reprinted with permission from reference Elizalde O, Azpeitia M, Reis MM, Asua JM, Leiza JR: Monitoring emulsion polymerization reactors: calorimetry versus Raman spectroscopy,* Ind Eng Chem Res *44:7200–7207, 2005b. doi:10.1021/ie050451y.*

Duewel et al. compared the evolution of conversion in miniemulsion polymerization of styrene by calorimetry and ^{129}Xe NMR spectroscopy and showed that both methods gives comparable results (Duewel et al., 2012).

2.2 Particle size average and distribution

Particle size (dp) and particle size distribution (PSD) strongly affect the emulsion polymerization process and the final properties of the latex. Average particle size in emulsion polymerization can be monitored by NIR and Raman spectroscopy using correlation of spectra with the average particle size of the latex. Reis et al. used in-situ NIR spectroscopy to monitor the average particle size in the semibatch emulsion copolymerization of styrene and butyl acrylate (Reis et al., 2003a, 2004b). In spite of some successful cases, in general NIRS provided poor dp prediction, which was attributed to the fact that the wavelength of the NIR radiation is usually much larger than typical emulsion particles (Gossen et al., 1993; Santos et al., 2005).

Raman spectroscopy has also been used to measure dp in semibatch emulsion copolymerization of styrene and acrylic acid (Reis et al., 2003b). It was shown that despite varying monomer concentration and solids content (up to 48 wt% of polymer content) during the polymerization, a very good agreement between reference measurements and predictions of dp was obtained using a multivariate linear calibration model. The same authors studied the chemometric challenges of the measurement of dp by Raman showing that non-linear calibration models provided a relatively good estimation of the particle size (Fig. 3) (Reis et al., 2007).

ANN has also used to monitor the dp. Tsen et al. developed a hybrid ANN model based on scarce experimental data and first principle models to obtain high data set for training the ANN model for monitoring and control of the batch emulsion polymerization of vinyl acetate (Tsen et al., 1996). Conversion and average particle size were monitored using stacked ANN as soft sensor in the emulsion polymerization of vinyl acetate and Veova10 emulsion copolymerization at high solids content (55 wt%) in a loop reactor (Araújo et al., 2001). It is worth pointing out that the predictions of this type of soft sensors are limited to the region where the ANN was trained, namely they cannot be safely used outside this region.

On-line monitoring of particle size distribution is more challenging. Methods often used to measure PSD, such as capillary hydrodynamic

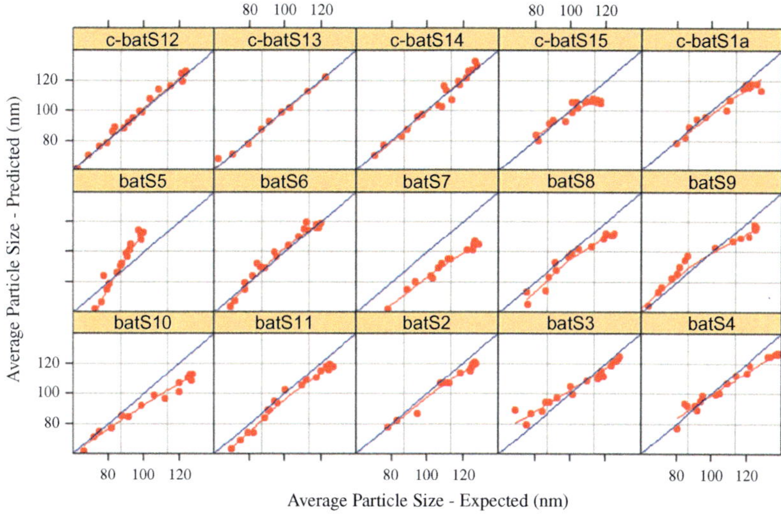

Fig. 3 Average particle size predicted by Raman spectroscopy using non-linear models in the batch emulsion polymerization of styrene. *Reprinted with permission from reference Reis MM, Araújo PHH, Sayer C, Giudici R: Spectroscopic on-line monitoring of reactions in dispersed medium: chemometric challenges,* Anal Chim Acta *595:257–265, 2007. doi:10.1016/j.aca.2007.04.048.*

fractionation (CHDF) (Elizalde et al., 2000; Miller et al., 1995; Schneider and Mckenna, 2002) usually require sampling from the reactor and long analysis times. Doyle et al. used CHDF for monitoring the PSD in the semibatch emulsion copolymerization of VAc/BA (Dokucu et al., 2008a,b). In a different approach, Belhabri et al. developed a method to estimate the PSD from the spectral analysis of the autocorrelation function of the back-scattered light intensity. The authors applied this technique to analyze the PSD of dispersions of alumina in water (Belhabri et al., 2004) finding a good agreement between the PSD obtained by this on-line method and the ones measured by microscopy and Coulter LS230. However, this method was applied to the dispersion of alumina in water at low solids content (5%) and 40 vol% oil-in-water emulsion. However, its application to emulsion polymerization process, where monomer droplets and polymer particles can coexist or bimodal PSDs are present has not been studied yet.

In the absence of proper on-line sensor, attempts to design of an observer for PSD combining a process model of the reaction and on-line measurements of the conversion have been reported (Edouard et al., 2005).

The main drawback is that mathematical modeling of PSD involves many unknown parameters affecting the accuracy of such observers.

Photon Density Wave spectroscopy (PDW) is a dilution-free, calibration-free and fast measurement that can provide the particle size in turbid dispersions in mono and bidisperse systems (Bressel et al., 2013, 2015), and therefore is very promising for on-line monitoring of particle size in emulsion polymerization. However, its practical performance in an emulsion polymerization system in the presence of nucleation and growth is yet to be studied.

2.3 Molar mass

Average molar mass and molar mass distribution of the polymer is not observable on-line as no on-line sensor is available. The offline methods for characterizing molar mass averages and distributions such as Gel Permeation Chromatography (GPC), Size Exclusion Chromatography with Multi-Angle Light Scattering (SEC/MALS) and Asymmetrical Flow Field-Flow Fractionation (AF4) need sample preparation and rather long characterization times. In addition, the analysis of branched and crosslinked polymers is still an unsolved issue. However, in some specific cases, it is possible to use a combination of available on-line measurements and a soft sensor to estimate the molar mass distribution. This is the case of linear polymers when the main chain termination event is by chain transfer to chain transfer agent (CTA). Under these circumstances, the kinetic chain length of the polymer formed at each time in the reactor is given by.

$$x_n = \frac{k_p[M]}{k_{tr,CTA}[CTA]} \tag{1}$$

where kp and $k_{tr,\ CTA}$ are the propagation and transfer to CTA rate coefficients, respectively, [M] is the monomer concentration and [CTA] is the chain transfer agent concentration. Knowing the kinetic rate coefficients, [CTA] can be estimated from on-line measurement of the unreacted monomer. Using this concept, Vicente et al. were able to monitor the molar mass average and distribution of linear polymers using reaction calorimeter (Vicente et al., 2001a, 2003). A similar idea was used later to monitor on-line the molar mass of the reversible addition–fragmentation chain transfer (RAFT) emulsion polymerization of styrene using calorimetry (Altarawneh et al., 2009).

2.4 Radical concentration

Radical concentration is not a characteristic of the latex, but it is critical in the formation of the polymer, and therefore in some cases such as when studying polymerization kinetics, it can be interesting to monitor this variable. This can be done by Electron Spin Resonance spectroscopy (ESR) that allows the detection of different radical species and has been used for the direct observation of propagating radicals in the emulsion polymerization of MMA (Ballard et al., 1984). It should be noted that in the case of high mobile such as BA chain-end radicals, the steady state concentration of occluded radicals remains below the detection limit of the spectrometer during the whole polymerization process (Cutting and Tabner, 1997). ESR was used to estimate the termination rate constant during the unseeded batch emulsion polymerization of MMA and to study the effect of different reaction parameters such as initiator and surfactant concentration and temperature on the kinetics of the reaction (Cutting, 1993). Using a continuous sampling system, Chang et al.monitored in real-time the batch emulsion polymerization of methyl methacrylate (MMA) (Parker et al., 1996).

2.5 ACOMP (automatic continuous on-line monitoring of polymerization reaction)

The ACOMP technology stands out because it allows simultaneous monitoring of several characteristics during the polymerization process (Reed, 2014). It is based on a continuous withdraw of a small stream from reactor and its dilution and transfer to different detectors. In this way, continuous measurement of conversion, copolymer composition, molar mass distribution, average particle size and intrinsic viscosity is possible. ACOMP is versatile and avoids contact of sensitive detectors with the harsh environment of the reactor. However, the price to pay is the continuous waste stream and delays in getting the information. ACOPM is able to detect undesired situations such as reactor runaway and microcoagulation. Reed et al. described in detail the principles and applications of ACOMP and its potential for industrial application in different polymerization processes including emulsion polymerization (Alb and Reed, 2009, 2010; Reed, 2019). The ACOMP technology in emulsion polymerization, is based on analyzing separately the polymer and the colloidal characteristics. The polymer is diluted in organic solvent such as tetrahydrofuran (THF) and is used to measure the organic phase characteristics such as molar mass of the polymer, conversion and polymer intrinsic viscosity. Multi-angle light scattering (MALS) detector, a UV/visible diode array spectrophotometer, differential refractometer (RI), single capillary viscometer are the detectors used. The colloidal properties are measured by diluting the latex in water and using Mie scattering, dynamic light scattering, turbidity or heterogeneous time dependent static light scattering (HTDSLS) as detectors. This methodology was used by Reed et al. in batch emulsion homo-(Alb and Reed, 2008a) and *co*-polymerizations of BA and MMA (Alb and Reed, 2008b). Fig. 4 shows the schematic of the "Front-End instrumentation used in dual stream

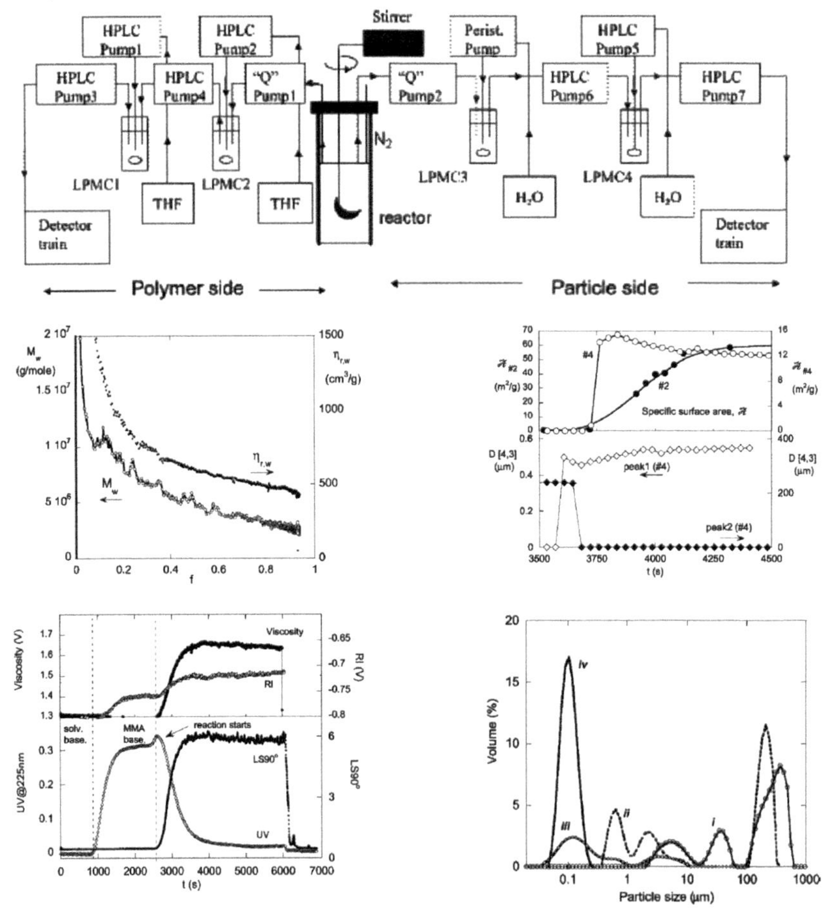

Fig. 4 Schematic of the "Front-End instrumentation used in dual stream extraction in ACOMP during emulsion polymerization reaction". *Reprinted with permission from reference Alb AM, Reed WF: Simultaneous monitoring of polymer and particle characteristics during emulsion polymerization,* Macromolecules *41:2406–2414, 2008a. doi:10.1021/ma7027825.*

extraction in ACOMP during emulsion polymerization reaction". Conversion, average molar mass and intrinsic viscosity are measured from the polymer side and monomer droplet and polymer particle size is measured from colloid side.

2.6 Remaining challenges

Although there has been a tremendous progress in the development of on-line monitoring, both through hard and soft sensors, still some important characteristics cannot be monitored on-line. One of these characteristics is the molecular weight distribution and the polymer architecture of non-linear branched and crosslinked polymers. These materials encompass a substantial fraction of the waterborne coatings and adhesives, which are the main applications of emulsion polymers (Grand View Research, 2019). Although it is recognized that MWD and polymer architecture strongly affect application properties (Ballard et al., 2013; Daniloska et al., 2014; Qie and Dube, 2012, 2011), and significant advances have been done in the development of analytical tools and methods (Agirre et al., 2013; Mehravar et al., 2018; Saalwächter et al., 2013) even offline, the accurate characterization is still an unsolved issue.

Another challenge is the particle morphology. Hybrid composite waterborne particles offer the possibility to enhance the performance and to enlarge the range of properties of single phase polymer dispersions. Thus, polymer-polymer particles have been used to overcome the dilemma of the waterborne coatings: simultaneously achieving strong mechanical properties and low minimum film forming temperature (Hasanzadeh et al., 2014; Juhu and Lan, 1995; Khan et al., 2009; Limousin et al., 2019). Precise off-line characterization of the particle morphology is not completely solved and the most advanced methods (Rajabalinia et al., 2019a) are lengthy and not suitable for on-line monitoring. In the next future, the most pragmatic alternative is the use of mathematical models as soft sensors. However, although there are efficient mathematical models (Hamzehlou et al., 2016) that describe quite well the evolution of the particle morphology during the process (Hamzehlou et al., 2017; Rajabalinia et al., 2019b) and have even been used for optimization (Faust et al., 2019), still there is plenty of work to do before their use as soft sensors to on-line monitor particle morphology.

A characteristic that is often neglected, but it is critical in many applications is surface composition, and currently there is no method able to on-line monitor this characteristic.

3. On-line control of polymer characteristics in emulsion polymerization

Once the polymer characteristics are monitored, the on-line control reduces to calculate the control actions and to implement them (Congalidis and Richards, 1998; Dimitratos et al., 1994; Richards and Congalidis, 2006). In this chapter, no attempt to review the different control models used will be made, limiting ourselves to review the practical implementations. The section is organized following the same order as for on-line monitoring.

3.1 Copolymer composition

Copolymer composition is likely the easiest characteristic to control online because the monomer conversions and the polymerization rates can be accurately determined on-line. The evolution of this area has been parallel to that of the monitoring techniques. The first on-line control was based on gas chromatography measurements that were successful for emulsion copolymerization (Leiza et al., 1992) and terpolymerization (Urretabizkaia et al., 1994), although the delay in the on-line monitoring should be considered in the control scheme. The development of reaction calorimetry allowed an easier and faster on-line control of the polymer composition for both copolymers (de Buruaga et al., 1996; Gugliotta et al., 1995b; Saenz de Buruaga et al., 1997) and terpolymers (De Buruaga et al., 2000). Finally, on-line control of polymer composition using spectroscopy was developed (van den Brink et al., 2001). Nowadays, control of the composition of emulsion polymers is a solved issue if we consider relatively water insoluble monomers that are used in large fractions. This is not the case of many functional monomers that are used in small fractions and often have a significant solubility in the aqueous phase.

3.2 Molar mass distribution (MMD)

Because of the lack of methods for on-line and even offline monitoring of branched and crosslinked polymers, on-line control of MWD has been limited to linear polymers (Echevarría et al., 1998a,b; Vicente et al., 2001a). This was achieved by controlling the ratio [M]/[CTA] that determines the length of the polymer chains obtained at each moment. The control strategy was to produce first the high molecular weights and later the smaller molecular weights that required a higher concentration of CTA. Following this procedure, a good control of MWDs was achieved as shown in Fig. 5.

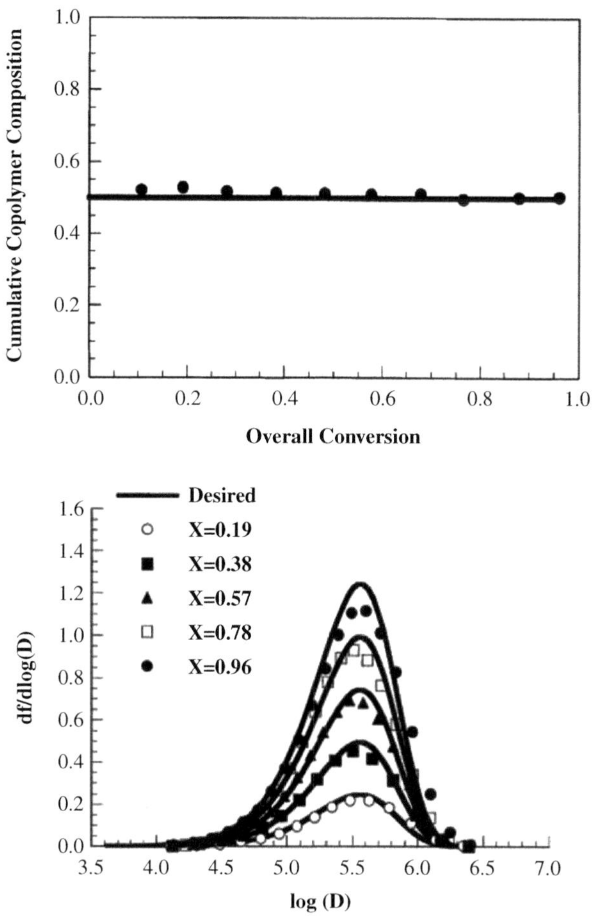

Fig. 5 Simultaneous control of copolymer composition and MWD. *Reprinted by permission from reference Vicente M, Leiza JR, Asua JM: Maximizing production and polymer quality (MWD and composition) in emulsion polymerization reactors with limited capacity of heat removal,* Chem Eng J *58:215–222, 2003.*

Combination of this approach with that discussed in Section 3.1, has led to the simultaneous control of copolymer composition and molecular weight distribution (Vicente et al., 2001b, 2002a, 2003).

3.3 Particle size distribution

The interest in controlling the PSD is because it affects the emulsion polymerization process as well as the application properties such as adhesion (do Amaral et al., 2005) and film formation (Fortini et al., 2016;

Makepeace et al., 2017; Routh and Keddie, 2010). A goal often found in industrial practice is to maximize the solids content of the dispersion maintaining low the viscosity. This is achieved by using bimodal PSDs (Boutti et al., 2005; Do Amaral et al., 2004; Mariz et al., 2011). Dokucu et al. implemented the on-line control of the particle size distribution in the semibatch emulsion copolymerization of vinyl acetate and butyl acrylate using frequent measurements of the density to determine monomer conversion and infrequent measurements of the PSD by capillary hydrodynamic fractionation (Dokucu et al., 2008a,b). The method was able to tackle with disturbances in the surfactant and monomer amounts.

3.4 Remaining challenges

The on-line control of molecular weight distribution of branched and crosslinked polymers (and the associated polymer architecture), particle morphology, and surface composition are basically unexplored areas. This is related to the lack of both on-line sensors and mathematical models that could eventually act as soft sensors. Therefore, in practice, these characteristics are controlled following predefined polymerization paths given in terms of variables such as monomer conversion and that are designed based on heuristic knowledge. In academic literature, attempts to open-control branched and crosslinked polymer have been reported (Vicente et al., 2002a,b). An attempt to control particle morphology using open-loop control has been recently reported (Gerlinger et al., 2019). Examples of manipulating surface composition can also be found (Bilgin et al., 2017; Chimenti et al., 2019). However, in spite that these are successful examples, it is worth to point out that open-loop control is not appropriate for emulsion polymerization because this process is prone to suffer from poor run-to-run reproducibility (Asua, 2004).

4. Conclusions and outlook

Emulsion polymers are product-by-process materials whose characteristics are almost completely determined in the reactor. The performance of these materials depends on multiple characteristics including copolymer composition, chemical composition distribution, molar mass distribution, polymer architecture (linear, branched and crosslinked), particle morphology, particle size distribution and surface composition. On-line control of copolymer composition, MWD of linear polymers and up to some point PSD is a solved problem, mainly because either sensors for accurate on-line monitoring of these variables or robust soft sensors are available. However, the lack of both on-line sensors and mathematical models that could eventually act as soft sensors precludes the on-line control of molecular weight distribution of branched and crosslinked polymers (and the associated polymer architectures), particle morphology, and surface composition.

Nobody can predict how the technology will evolve in the coming years, but it looks that in short term, the main developments for on-line hard sensors will be made in the area of PSD, the developments for polymer architecture, particle morphology and surface composition being more unlikely. Therefore, the advances in these areas will be associated to the development of accurate, robust and fast mathematical models for soft sensors.

In addition, there is an aspect that is basically ignored in scientific literature dealing with on-line control of emulsion polymerization, but that has tremendous consequences. An emulsion polymer has multiple characteristics and should meet multiple application properties that depend on all the characteristics. In most cases, when a given characteristic is modified, many application properties are also modified and not always in the right direction. A clear example can be found in the pressure sensitive adhesives. An increase in molecular weight improves shear resistance but is deleterious for tack. This means the control strategy should considered all the characteristics at the same time, which is in contrast with the works found in literature that focus on the control of a very few characteristics, without paying attention to the effect of the control actions on the rest of properties. The simultaneous on-line control of all of the characteristics that are significant for a given emulsion polymer is a pending task in this field.

References

Agirre A, Santos JI, Leiza JR: Toward understanding the architecture (branching and MWD) of crosslinked acrylic latexes, *Macromol Chem Phys* 214:589–598, 2013. https://doi.org/10.1002/macp.201200482.

Alb AM, Reed WF: Simultaneous monitoring of polymer and particle characteristics during emulsion polymerization, *Macromolecules* 41:2406–2414, 2008a. https://doi.org/10.1021/ma7027825.

Alb AM, Reed WF: Recent advances in automatic continuous online monitoring of polymerization reactions (ACOMP), *Macromol Symp* 271:15–25, 2008b. https://doi.org/10.1002/masy.200851103.

Alb AM, Reed WF: Online monitoring of molecular weight and other characteristics during semibatch emulsion polymerization under monomer starved and flooded conditions, *Macromolecules* 42:8093–8101, 2009. https://doi.org/10.1021/ma901434w.

Alb AM, Reed WF: Fundamental measurements in online polymerization reaction monitoring and control with a focus on ACOMP, *Macromol React Eng* 4:470–485, 2010. https://doi.org/10.1002/mren.200900079.

Alexopoulos AH, Pladis P, Kiparissides C: Nonhomogeneous mixing population balance model for the prediction of particle size distribution in large scale emulsion polymerization reactors, *Ind Eng Chem Res* 52:12285–12296, 2013.

Alonso M, Recasens F, Puigjaner L: Estimation of copolymer composition from online headspace analysis in batch emulsion polymerization, *Chem Eng Sci* 41:1039–1044, 1986.

Altarawneh IS, Gomes VG, Srour MH: Online polymer molecular weight and conversion monitoring via calorimetric measurements in RAFT emulsion polymerization, *Polym Int* 58:1427–1434, 2009. https://doi.org/10.1002/pi.2678.

Amer MS, editor: *Raman spectroscopy for soft matter applications*, New York, 2009, Wiley & Sons.

Araújo PHH, Sayer C, De La Cal JC, Asua JM, Lima EL, Pinto JC: Utilization of neural networks as soft sensors to monitor emulsion polymerization reactions (average particle diameter and conversion), *Lat Am Appl Res* 31:525–531, 2001.

Arzamendi G, Leiza JR: Molecular weight distribution (soluble and insoluble fraction) in emulsion polymerization of acrylate monomers by Monte Carlo simulations, *Ind Eng Chem Res* 47:5934–5947, 2008. https://doi.org/10.1021/ie701752f.

Arzamendi G, D'Anjou A, Graña M, Leiza JR, Asua JM: Model reduction in emulsion polymerization using hybrid first principles/artificial neural networks models, 2 long chain branching kinetics, *Macromol Theory Simul* 14:125–132, 2005. https://doi.org/10.1002/mats.200400064.

Asua JM, editor: *Polymeric dispersions: principles and applications*, 1 ed, Netherland, 1997, Springer.

Asua JM: A new model for radical desorption in emulsion polymerization, *Macromolecules* 36:6245–6251, 2003. https://doi.org/10.1021/ma0345192.

Asua JM: Emulsion polymerization: from fundamental mechanisms to process developments, *J Polym Sci A Polym Chem* 42:1025–1041, 2004.

Asua JM, De La Cal JC: Entry and exit rate coefficients in emulsion polymerization of styrene, *J Appl Polym Sci* 42:1869–1877, 1991.

Asua M, Leiza JR, Elizalde O: Monitoring of high solids content starved-semi-batch emulsion copolymerization reactions by Fourier transform Raman spectroscopy, *Appl Spectrosc* 59:1270–1279, 2005.

Ballard MJ, Gilbert RG, Napper DH, Pomery PJ, O'Donnell JH: Application of electron spin resonance spectroscopy to emulsion polymerization, *Macromolecules* 17:504–506, 1984.

Ballard N, Carretero P, Asua JM: Effect of reaction temperature on adhesive properties of waterborne polyurethane/acrylic hybrids synthesized by semicontinuous miniemulsion polymerization, *Macromol React Eng* 7:504–514, 2013. https://doi.org/10.1002/mren.201300122.

Belhabri M, Legrand J, Le Sauze N, Arhaliass A: On-line determination of particle size distributions in flowing dispersions, *Part Part Syst Charact* 21:365–371, 2004. https://doi.org/10.1002/ppsc.200400950.

Bilgin S, Tomovska R, Asua JM: Effect of ionic monomer concentration on latex and film properties for surfactant-free high solids content polymer dispersions, *Eur Polym J* 93:480–494, 2017. https://doi.org/10.1016/j.eurpolymj.2017.06.029.

Boutti S, Graillat C, Mckenna TF: High solids content emulsion polymerisation without intermediate seeds. Part III. Reproducibility and influence of process conditions, *Polymer (Guildf)* 46:1223–1234, 2005. https://doi.org/10.1016/j.polymer.2004.11.045.

Bressel L, Hass R, Reich O: Particle sizing in highly turbid dispersions by Photon Density Wave spectroscopy, *J Quant Spectrosc Radiat Transf* 126:122–129, 2013. https://doi.org/10.1016/j.jqsrt.2012.11.031.

Bressel L, Wolter J, Reich O: Particle sizing in highly turbid dispersions by Photon Density Wave spectroscopy: bidisperse systems, *J Quant Spectrosc Radiat Transf* 162:213–220, 2015. https://doi.org/10.1016/j.jqsrt.2015.01.025.

Butte A, Storti G, Morbidelli M: Evaluation of the chain length distribution in free-radical polymerization, 2 emulsion polymerization, *Macromol Theory Simul* 11:37–52, 2002.

Canegallo S, Storti G, Morbidelli M, Carra S: Densimetry for on-line conversion monitoring in emulsion homo- and co-polymerization, *Appl Polym Sci* 47:961–979, 1993.

Chien DCH, Penlidis A: On-line sensors for polymerization reactors, *J Macromol Sci C Polym Rev* 30:1–42, 1990.

Chimenti S, Vega JM, García-lecina E, Grande H-J, Paulis M, Leiza JR: In-situ phosphatization and enhanced corrosion properties of films made of phosphate functionalized nanoparticles, *React Funct Polym* 143; 2019, 104334.

Congalidis JP, Richards JR: Process control of polymerization reactors: an industrial perspective, *Polym React Eng* 6:71–111, 1998. https://doi.org/10.1080/10543414.1998.10744484.

Cutting GR: Radical termination and radical concentrations during the batch emulsion polymerization, *Macromolecules* 26:951–955, 1993.

Cutting GR, Tabner BJ: Radical concentrations and reaction temperature profiles during the (batch) core-shell emulsion polymerization of methyl methacrylate and butyl acrylate, studied by Electron Spin Resonance Spectroscopy, *Eur J Org Chem* 33:213–217, 1997.

D'Anjou A, Torrealdea FJ, Leiza R, Asua M, Arzamendi G: Model reduction in emulsion polymerization using hybrid first-principles/artificial neural network models, *Macromol Theory Simul* 12:42–56, 2003.

Daniloska V, Carretero P, Tomovska R, Asua JM: High performance pressure sensitive adhesives by miniemulsion photopolymerization in a continuous tubular reactor, *Polymer (Guildf)* 55:5050–5056, 2014. https://doi.org/10.1016/j.polymer.2014.08.038.

de Buruaga IS, Arotcarena M, Armitage PD, Gugliotta LM, Leiza JR, Asua JM: On-line calorimetric control of emulsion polymerization reactors, *Chem Eng Sci* 51:2781–2786, 1996.

De Buruaga IS, Leiza JR, Asua JM: Model-based control of emulsion terpolymerization based on calorimetric measurements, *Polym React Eng* 8:39–75, 2000.

Dimitratos J, Eliçabe G, Georgakis C: Control of emulsion polymerization reactors, *AIChE J* 40:1993–2021, 1994. https://doi.org/10.1002/aic.690401210.

Do Amaral M, Van Es S, Asua JM: Effect of the particle size distribution on the low shear viscosity of high-solid-content latexes, *J Polym Sci A Polym Chem* 42:3936–3946, 2004. https://doi.org/10.1002/pola.20269.

do Amaral M, Roos A, Asua JM, Creton C: Assessing the effect of latex particle size and distribution on the rheological and adhesive properties of model waterborne acrylic pressure-sensitive adhesives films, *J Colloid Interface Sci* 281:325–338, 2005. https://doi.org/10.1016/j.jcis.2004.08.142.

Dokucu MT, Park MJ, Doyle FJ: Multi-rate model predictive control of particle size distribution in a semibatch emulsion copolymerization reactor, *J Process Control* 18:105–120, 2008a. https://doi.org/10.1016/j.jprocont.2007.04.002.

Dokucu MT, Park MJ, Doyle FJ: Reduced-order methodologies for feedback control of particle size distribution in semi-batch emulsion copolymerization, *Chem Eng Sci* 63:1230–1245, 2008b. https://doi.org/10.1016/j.ces.2007.07.021.

Dropsit E, Hoppe S, Chapron D, Durand A, Bourson P: In situ conversion monitoring of styrene emulsion polymerization by deconvolution of a single reference band near 1000 cm^{-1}, *J Raman Spectrosc* 50:1938–1948, 2019. https://doi.org/10.1002/jrs.5738.

Duewel M, Vogel N, Weiss CK, Landfester K, Spiess HW, Münnemann K: Online monitoring of styrene polymerization in miniemulsion by hyperpolarized ^{129}Xenon NMR spectroscopy, *Macromolecules* 45:1839–1846, 2012. https://doi.org/10.1021/ma202605n.

Echevarría A, Leiza JR, De La Cal JC, Asua JM: Molecular-weight distribution control in emulsion polymerization, *AIChE J* 44:1667–1679, 1998a. https://doi.org/10.1002/aic.690440718.

Echevarría A, Leiza JR, De la Cal JC, Asua JM: On-line control of molecular weight distribution in semibatch emulsion polymerization using CTA, *IFAC Dyn Control Process Syst* 31:213–218, 1998b. https://doi.org/10.1016/s1474-6670(17)44930-5.

Edouard D, Sheibat-Othman N, Hammouri H: Observer design for particle size distribution in emulsion polymerization, *AIChE J* 51:3167–3185, 2005. https://doi.org/10.1002/aic.10559.

Elizalde O, Leiza JR: Raman application in emulsion polymerization systems. In Amer MS, editor: *Raman Spectroscopy for soft matter Applications*, 2009, Wiley & Sons, pp 95–144.

Elizalde O, Leal GP, Leiza JR: Particle size distribution measurements of polymeric dispersions: a comparative study, *Part Part Syst Charact* 17:236–243, 2000.

Elizalde O, Arzamendi G, Leiza JR, Asua JM: Seeded semibatch emulsion copolymerization of n-butyl acrylate and methyl methacrylate, *Ind Eng Chem Res* 43:7401–7409, 2004a. https://doi.org/10.1021/ie0400649.

Elizalde O, Leiza JR, Asua JM: On-line monitoring of all-acrylic emulsion polymerization reactors by Raman spectroscopy, *Macromol Symp* 206:135–148, 2004b. https://doi.org/10.1002/masy.200450211.

Elizalde O, Asua JM, Leiza JR: Monitoring of emulsion polymerization reactors by Raman spectroscopy: calibration model maintenance, *Appl Spectrosc* 59:1280–1285, 2005a. https://doi.org/10.1366/000370205774430954.

Elizalde O, Azpeitia M, Reis MM, Asua JM, Leiza JR: Monitoring emulsion polymerization reactors: calorimetry versus Raman spectroscopy, *Ind Eng Chem Res* 44:7200–7207, 2005b. https://doi.org/10.1021/ie050451y.

Faust JMM, Hamzehlou S, Leiza JR, Asua JM, Mhamdi A, Mitsos A: Dynamic optimization of a two-stage emulsion polymerization to obtain desired particle morphologies, *Chem Eng J* 359:1035–1045, 2019. https://doi.org/10.1016/J.CEJ.2018.11.081.

Fernandes FAN, Lona LMF, Penlidis A: Inverse modeling applications in emulsion polymerization of vinyl acetate, *Chem Eng Sci* 59:3159–3167, 2004. https://doi.org/10.1016/j.ces.2004.04.003.

Fortini A, Martín-fabiani I, Lesage J, et al: Dynamic stratification in drying films of colloidal mixtures, *Phys Rev Lett* 116:, 2016, 118301https://doi.org/10.1103/PhysRevLett.116.118301.

Frauendorfer E, Wolf A, Hergeth WD: Polymerization online monitoring, *Chem Eng Technol* 33:1767–1778, 2010. https://doi.org/10.1002/ceat.201000265.

Galimberti F, Siani A, Morbidelli M, Storti G: Applications of composition control to the styrene-butyl acrylate emulsion copolymerization, *Chem Eng Commun* 163:69–95, 1998. https://doi.org/10.1080/00986449808912345.

Gerlinger W, Asua JM, Chaloupka T, et al: Dynamic optimization and non-linear model predictive control to achieve targeted particle morphologies, *Chem Ing Tech* 91:323–335, 2019. https://doi.org/10.1002/cite.201800118.

Ghielmi A, Storti G, Morbidelli M, Ray WH: Molecular weight distribution in emulsion polymerization: role of active chain compartmentalization, *Macromolecules* 31:7172–7186, 1998. https://doi.org/10.1021/ma980657t.

Gonzalez-Ortiz LJ, Asua JM: Development of particle morphology in emulsion polymerization. 3. Cluster nucleation and dynamics in polymerizing systems, *Macromolecules* 29:4520–4527, 1996. https://doi.org/10.1021/ma00113a016.

Gossen PD, MacGregor JF, Pelton RH: Composition and particle diameter for styrene/methyl methacrylate copolymer latex using UV and NIR spectroscopy, *Appl Spectrosc* 47:1852–1870, 1993. https://doi.org/10.1366/0003702934066154.

Grand View Research: *Emulsion polymers market size, share & trends analysis report by product, by application and segment forecasts, 2019–2025,* 2019.

Gugliotta LM, Arotçarena M, Leiza JR, Asua JM: Estimation of conversion and copolymer composition in semicontinuous emulsion polymerization using calorimetric data, *Polymer (Guildf)* 36:2019–2023, 1995a. https://doi.org/10.1016/0032-3861(95)91447-F.

Gugliotta LM, Leiza JR, Arotcarena M, Armitage PD, Asua JM: Copolymer composition control in unseeded emulsion polymerization using calorimetric data, *Ind Eng Chem Res* 34:3899–3906, 1995b. https://doi.org/10.1021/ie00038a029.

Guinot P, Othman N, Févotte G, Mckenna TF: On-line monitoring of emulsion copolymerisations using hardware sensors and calorimetry, *Polym React Eng* 8:115–134, 2000. https://doi.org/10.1080/10543414.2000.10744544.

Hamzehlou S, Ballard N, Carretero P, et al: Mechanistic investigation of the simultaneous addition and free-radical polymerization in batch miniemulsion droplets: Monte Carlo simulation versus experimental data in polyurethane/acrylic systems, *Polymer (Guildf)* 55:4801–4811, 2014. https://doi.org/10.1016/j.polymer.2014.07.024.

Hamzehlou S, Leiza JR, Asua JM: A new approach for mathematical modeling of the dynamic development of particle morphology, *Chem Eng J* 304:655–666, 2016. https://doi.org/10.1016/j.cej.2016.06.127.

Hamzehlou S, Aguirre M, Leiza JR, Asua JM: Dynamics of the particle morphology during the synthesis of waterborne polymer-inorganic hybrids, *Macromolecules* 50:7190–7201, 2017. https://doi.org/10.1021/acs.macromol.7b01488.

Hasanzadeh I, Mahdavian AR, Salehi-mobarakeh H: Particle size and shell composition as effective parameters on MFFT for acrylic core—shell particles prepared via seeded emulsion polymerization, *Prog Org Coat* 77:1874–1882, 2014. https://doi.org/10.1016/j.porgcoat.2014.06.018.

Haven JJ, Junkers T: Online monitoring of polymerizations: current status, *Eur J Org Chem* 2017:6474–6482, 2017. https://doi.org/10.1002/ejoc.201700851.

Juhu D, Lan J: Film formation from dispersion of core-shell latex particles, *Macromolecules* 28:1306–1308, 1995.

Kammona O, Chatzi EG, Kiparissides C: Recent developments in hardware sensors for the on-line monitoring of polymerization reactions, *J Macromol Sci, Rev Macromol Chem Phys* 39:57–134, 1999. https://doi.org/10.1081/mc-100101417.

Khan AK, Ray BC, Maiti J, Dolui SK: Preparation of core-shell latex from co-polymer of styrene-butyl acrylate-methyl methacrylate and their paint properties, *Pigm Resin Technol* 38:159–164, 2009. https://doi.org/10.1108/03699420910957015.

Landfester K, Spiegel S, Born R, Spiess HW: On-line detection of emulsion polymerization by solid-state NMR spectroscopy, *Colloid Polym Sci* 276:356–361, 1998. https://doi.org/10.1007/s003960050251.

Leiza JR, De La Cal JC, Asua JM: On-line copolymer composition control in the semi-continuous emulsion copolymerization of ethyl acrylate and methyl methacrylate, *Polym React Eng* 1:461–498, 1992.

Leiza JR, la Cal JCD, Montes M, Asua JM: On-line monitoring of conversion and polymer composition in emulsion polymerization systems, *Process Control Qual* 4:197–210, 1993.

Limousin E, Ballard N, Asua JM: Soft core-hard shell latex particles for mechanically strong VOC-free polymer films, *J Appl Polym Sci* 136:, 2019, 47608https://doi.org/10.1002/app.47608.

Makan AC, Otte T, Pasch H: Analysis of high molar mass branched polybutadienes by SEC-MALLS and AF4-MALLS, *Macromolecules* 45:5247–5259, 2012.

Makepeace DK, Fortini A, Markov A, et al: Stratification in binary colloidal polymer films: experiment and simulations, *Soft Matter* 13:6969–6980, 2017. https://doi.org/10.1039/C7SM01267E.

Mariz IdFA, Leiza JR, De la Cal JC: Competitive particle growth: a tool to control the particle size distribution for the synthesis of high solids content low viscosity latexes, *Chem Eng J* 168:938–946, 2011. https://doi.org/10.1016/j.cej.2011.02.023.

Martens H, Næs T: *Multivariate calibration,* New York, 1989, Wiley.

Maxwell IA, Morrison BR, Napper DH, Gillbert RG: Entry of free radicals into latex particles in emulsion polymerization, *Macromolecules* 24:1629–1640, 1991.

Mayo FR, Lewis FM: Copolymerization I. A basis for comparing the behavior of monomers in copolymerization. The copolymerization of styrene and methyl methacrylate, *J Am Chem Soc* 66:1594–1601, 1944.

Mehravar E, Agirre A, Ballard N, et al: Insights into the network structure of cross-linked polymers synthesized via miniemulsion nitroxide-mediated radical polymerization, *Macromolecules* 51:9740–9748, 2018. https://doi.org/10.1021/acs.macromol.8b01648.

Miller CM, Sudol ED, Silebi CA, El-Aasser MS: Capillary hydrodynamic fractionation (CHDF) as a tool for monitoring the evolution of the particle size distribution during miniemulsion polymerization, *J Colloid Interface Sci* 172:249–256, 1995.

Min KW, Ray WH: On the mathematical modeling of emulsion polymerization reactors, *J Macromol Sci C Polym Rev* C11:177–255, 1974.

Moritz H: Polymerization calorimetry-a powerful tool for reaction control. In Reichert K, Geiseler W, editors: *Polymer reaction Engineering*, Weinheim, 1989, VCH Verlag, pp 248–266.

Noël LFJ, Brouwer ECP, Van Herk AM, German AL: On-line gas chromatography and densimetry to obtain partial conversions of both monomers in emulsion copolymerization, *J Appl Polym Sci* 57:245–254, 1995. https://doi.org/10.1002/app.1995.070570301.

Parker HY, Westmoreland DG, Chang HR: ESR study of MMA batch emulsion polymerization in real time: effects of particle size, *Macromolecules* 29:5119–5127, 1996. https://doi.org/10.1021/ma951652o.

Pladis P, Alexopoulos AH, Kiparissides C: Mathematical modeling and simulation of vinylidene fluoride emulsion polymerization, *Ind Eng Chem Res* 53:7352–7364, 2014. https://doi.org/10.1021/ie403548m.

Podzimek S, Machotova J, Snuparek J, Vecera M, Prokupek L: Characterization of molecular structure of acrylic copolymers prepared via emulsion polymerization using A4F-MALS technique, *J Appl Polym Sci* 131:1–7, 2014. https://doi.org/10.1002/app.40995.

Qie L, Dubé M: Manipulating latex polymer microstructure using chain transfer agent and cross-linker to modify PSA performance and viscoelasticity, *Macromol React Eng* 5:117–128, 2011. https://doi.org/10.1002/mren.201000046.

Qie L, Dube MA: Influence of polymer microstructure on the performance of post-treated latex-based pressure sensitive adhesives, *J Appl Polym Sci* 124:349–364, 2012. https://doi.org/10.1002/app.34921.

Rajabalinia N, Hamzehlou S, Modin E, Chuvilin A, Leiza JR, Asua JM: Coupling HAADF-STEM tomography and image reconstruction for the precise characterization of particle morphology of composite polymer latexes, *Macromolecules* 52:5298–5306, 2019a.

Rajabalinia N, Hamzehlou S, Leiza JR, Asua JM: Experimental validation of a mathematical model for the evolution of the particle morphology of waterborne polymer-polymer hybrids: paving the way to the design and implementation of optimal polymerization strategies, *Chem Eng J* 363:259–269, 2019b. https://doi.org/10.1016/j.cej.2019.01.140.

Rawlings JB, Ray WH: The modeling of batch and continuous emulsion polymerization reactors. Part II: comparison with experimental data from continuous stirred tank reactors, *Polym Eng Sci* 28:257–274, 1988.

Reed WF: Background and principles of automatic continuous online monitoring of polymerization reactions (ACOMP). In Reed WF, Alb AM, editors: *Monitoring polymerization reactions: from fundamentals to applications*, 2014, Wiley, pp 231–246.

Reed WF: Automatic continuous online monitoring and control of polymerization reactions and related methods, *Encycl Anal Chem* 1–55, 2019. https://doi.org/10.1002/9780470027318.a9288.pub2.

Reed WF, Alb AM, editors: *Monitoring polymerization reactions. From fundamental to applications*, 2014, Wiley.

Reis MM, Araújo PHH, Sayer C, Giudici R: Correlation between polymer particle size and in-situ NIR spectra, *Macromol Rapid Commun* 24:620–624, 2003a. https://doi.org/10.1002/marc.200350004.

Reis MM, Araújo PHH, Sayer C, Giudici R: Evidences of correlation between polymer particle size and Raman scattering, *Polymer (Guildf)* 44:6123–6128, 2003b. https://doi.org/10.1016/S0032-3861(03)00669-4.

Reis MM, Araújo PHH, Sayer C, Giudici R: Development of calibration models for estimation of monomer concentration by Raman spectroscopy during emulsion polymerization: facing the medium heterogeneity, *J Appl Polym Sci* 93:1136–1150, 2004a. https://doi.org/10.1002/app.20474.

Reis MM, Araújo PHH, Sayer C, Giudici R: In situ near-infrared spectroscopy for simultaneous monitoring of multiple process variables in emulsion copolymerization, *Ind Eng Chem Res* 43:7243–7250, 2004b. https://doi.org/10.1021/ie034277u.

Reis MM, Araújo PHH, Sayer C, Giudici R: Comparing near infrared and Raman spectroscopy for on-line monitoring of emulsion copolymerization reactions, *Macromol Symp* 206:165–178, 2004c. https://doi.org/10.1002/masy.200450213.

Reis MM, Araújo PHH, Sayer C, Giudici R: Spectroscopic on-line monitoring of reactions in dispersed medium: chemometric challenges, *Anal Chim Acta* 595:257–265, 2007. https://doi.org/10.1016/j.aca.2007.04.048.

Reynhout XEE, Meuldijk J, Drinkenburg BAH, Iedema PD, Wulkow M: A novel method to model emulsion polymerization kinetics: the explicit radical-particle size distribution approach, *Polym-Plast Technol Eng* 44:707–740, 2005. https://doi.org/10.1081/PTE-200060384.

Richards JR, Congalidis JP: Measurement and control of polymerization reactors, *Comput Chem Eng* 30:1447–1463, 2006. https://doi.org/10.1016/j.compchemeng.2006.05.021.

Routh AF, Keddie J: *Fundamentals of latex film formation*, 2010, Springer, pp. XVIII, 308. https://doi.org/10.1007/978-90-481-2845-7.

Saalwächter K, Chassé W, Sommer JU: Structure and swelling of polymer networks: insights from NMR, *Soft Matter* 9:6587–6593, 2013. https://doi.org/10.1039/c3sm50194a.

Saenz de Buruaga I, Echevaria A, Armitage PD, de la Cal J, Leiza JR, Asua JM: On-line control of a semibatch emulsion polymerization reactor based on calorimetry, *AIChE J* 43:1069–1081, 1997. https://doi.org/10.1002/aic.690430420.

Santos AF, Silva FM, Lenzi MK, Pinto JC: Monitoring and control of polymerization reactors using NIR spectroscopy, *Polym-Plast Technol Eng* 44:1–61, 2005. https://doi.org/10.1080/PTE-200046030.

Schneider M, Mckenna TF: Comparative study of methods for the measurement of, *Part Part Syst Charact* 19:28–37, 2002.

Shapiro YE: Structure and dynamics of hydrogels and organogels: an NMR spectroscopy approach, *Prog Polym Sci* 36:1184–1253, 2011. https://doi.org/10.1016/j.progpolymsci.2011.04.002.

Smith WV, Ewart RH: Kinetics of emulsion polymerization, *J Chem Phys* 16:592–599, 1948. https://doi.org/10.1063/1.1746951.

Storti G, Polotti G, Cociani M, Mirbidelli M: Molecular weight distribution in emulsion polymerization. I. The homopolymerization case, *J Polym Sci A Polym Chem* 30:731–750, 1992.

Stubbs J, Carrier R, Sundberg DC: Monte Carlo simulation of emulsion polymerization kinetics and the evolution of latex particle morphology and polymer chain architecture, *Macromol Theory Simul* 17:147–162, 2008. https://doi.org/10.1002/mats.200700061.

Tauer K, Kuehn I: Modeling particle formation in emulsion polymerization: an approach by means of the classical nucleation theory, *Macromolecules* 28:2236–2239, 1995.

Thickett SC, Gilbert RG: Mechanism of radical entry in electrosterically stabilized emulsion polymerization systems, *Macromolecules* 39:6495–6504, 2006. https://doi.org/10.1021/ma061326g.

Tian Y, Zhang J, Morris J: Optimal control of a batch emulsion copolymerisation reactor based on recurrent neural network models, *Chem Eng Process* 41:531–538, 2002.

Tobita H: Molecular weight distribution in random crosslinking of polymer chains, *J Polym Sci B* 33:1191–1202, 1995.

Tsen AY-D, Jang SS, Shan D, Wong H, Joseph B: Predictive control of quality in batch polymerization using hybrid ANN models, *AIChE J* 42:455–465, 1996.

Urban D, Takamura K: *Polymer dispersions and their industrial applications,* Weinheim, FRG, 2002, Wiley-VCH Verlag GmbH & Co. KGaAhttps://doi.org/10.1002/3527600582.

Urretabizkaia A, Sudol ED, El-Aasser MS, Asua JM: Calorimetric monitoring of emulsion copolymerization reactions, *J Polym Sci A Polym Chem* 31:2907–2913, 1993. https://doi.org/10.1002/pola.1993.080311204.

Urretabizkaia A, Leiza JR, Asua JM: On-line terpolymer composition control in semicontinuous emulsion polymerization, *AIChE J* 40:1850–1864, 1994.

Vale HM, Mckenna TF: Modeling particle size distribution in emulsion polymerization reactors, *Prog Polym Sci* 30:1019–1048, 2005. https://doi.org/10.1016/j.progpolymsci.2005.06.006.

van den Brink M, Pepers M, van Herk AM, German AL: On-line monitoring and composition control of the emulsion copolymerization of VeoVA 9 and butyl acrylate by Raman spectroscopy, *Polym React Eng* 9:101–133, 2001. https://doi.org/10.1081/PRE-100103273.

Van M, Brink D, Hansen J, De Peinder P, Van Herk AM, German AL: Measurement of partial conversions during the solution copolymerization of styrene and butyl acrylate using on-line Raman spectroscopy, *J Appl Polym Sci* 79:426–436, 2001. https://doi.org/10.1002/1097-4628(20010118)79:3<426::AID-APP50>3.0.CO;2-V.

Vargas MA, Cudaj M, Hailu K, Sachsenheimer K, Guthausen G: Online low-field 1H NMR spectroscopy: monitoring of emulsion polymerization of butyl acrylate, *Macromolecules* 43:5561–5568, 2010. https://doi.org/10.1021/ma1006599.

Vautz W, Mauntz W, Engell S, Baumbach JI: Monitoring of emulsion polymerisation processes using ion mobility spectrometry—a pilot study, *Macromol React Eng* 3:85–90, 2009. https://doi.org/10.1002/mren.200800043.

Vega JR, Gugliotta LM, Meira GR: Emulsion copolymerization of acrylonitrile and butadiene. Semibatch strategies for controlling molecular structure on the basis of calorimetric measurements, *Polym React Eng* 10:59–82, 2002. https://doi.org/10.1081/PRE-120003922.

Vicente M, Benamor S, Gugliotta LM, Leiza JR, Asua JM: Control of molecular weight distribution in emulsion polymerization using on-line reaction calorimetry, *Ind Eng Chem Res* 40:218–227, 2001a. https://doi.org/10.1021/ie000387e.

Vicente M, Leiza JR, Asua JM: Simultaneous control of copolymer composition and MWD in emulsion copolymerization, *AIChE J* 47:1594–1606, 2001b. https://doi.org/10.1002/aic.690470712.

Vicente M, Leiza R, Asua JM: Control of microstructural properties in emulsion polmerization systems, *Macromol Symp* 182:291–303, 2002a. https://doi.org/10.1017/CBO9781107415324.004.

Vicente M, Leiza JR, Asua JM: Challenges in on-line control of composition and molecular weight distribution of nonlinear emulsion copolymers. In Daniels ES, Sudol ED, El-Aasse MS, editors: *Polymer colloids: science and technology latex systems ACS symposium series*, vol. 801, 2002b, American Chemical Society, pp 113–125.

Vicente M, Leiza JR, Asua JM: Maximizing production and polymer quality (MWD and composition) in emulsion polymerization reactors with limited capacity of heat removal, *Chem Eng J* 58:215–222, 2003.

Zubitur M, Armitage PD, Amor SB, Leiza JR, Asua JM: Mathematical modeling of multimonomer (vinylic, divinylic, acidic) emulsion copolymerization systems, *Polym React Eng* 11:627–662, 2003.

CHAPTER THREE

Exploiting the pulsed laser polymerization-size exclusion chromatography technique to retrieve kinetic parameters in radical polymerization: State-of-the-art and future challenges

Yoshi W. Marien[a,*], Mariya Edeleva[a], Paul H.M. Van Steenberge[a], and Dagmar R. D'hooge[a,b,*]

[a]Laboratory for Chemical Technology (LCT), Department of Materials, Textiles and Chemical Engineering, Ghent University, Ghent, Belgium
[b]Centre for Textile Science and Engineering (CTSE), Department of Materials, Textiles and Chemical Engineering, Ghent University, Ghent, Belgium
*Corresponding authors: e-mail address: yoshi.marien@ugent.be; dagmar.dhooge@ugent.be

Contents

Advances in Chemical Engineering, Volume 56
ISSN 0065-2377
https://doi.org/10.1016/bs.ache.2020.08.001

Abstract

Pulsed laser polymerization (PLP) has emerged as a leading technique to determine radical polymerization characteristics. Here we present a concise overview of the potential of this polymerization technique mainly in combination with size exclusion chromatography (SEC) analysis. Focus is not only on the traditional determination of intrinsic propagation reactivities for radical polymerizations with one dominant macroradical type based on inflection point data but also on (i) the determination of such intrinsic reactivities if multiple dominant radical types are present (*e.g.*, for radical polymerization with branch formation and copolymerization); and (ii) the exploitation of the complete SEC trace to obtain information on the quantum yield, chain initiation and (apparent) termination reactivity. General guidelines are provided how to increase the accuracy of PLP-SEC measurements and to combine such measurements with advanced kinetic modeling and *ab initio* calculations to further increase our understanding of radical polymerization chemistry and to construct chemical structure-reactivity relationships.

1. Introduction

A myriad of polymeric materials used in everyday life (*e.g.*, coatings, household goods, and construction items) is produced by free radical polymerization (FRP), which occurs *via* a chain-growth mechanism with conventional radical initiator dissociation, chain initiation, propagation, and termination as the basic reactions (Asua, 2007; Matyjaszewski and Davies, 2002; Nesvadba, 2012). Additional side reactions (*e.g.*, chain transfer to polymer and macroaddition) affect the polymer molecular structure, which in turn can influence the macroscopic properties of the polymer end-product (Junkers and Barner-Kowollik, 2008; Peck and Hutchinson, 2004; Purmova et al., 2008; Van Steenberge et al., 2011). The design and optimization of industrial FRP processes is very challenging as a complex interplay between chemical reactions and diffusional limitations takes place (Barner-Kowollik and Russell, 2009; Constales et al., 2017; Derboven et al., 2013; Peklak et al., 2006). A model-based design strategy combined with experimental validation is thus recommended (D'hooge et al., 2015; Mastan et al., 2015; Saldívar-Guerra, 2019; Zhou and Luo, 2016).

A necessary condition for a successful model-based design is the consideration of all kinetically relevant elementary reactions and knowledge of the corresponding intrinsic kinetic parameters, *i.e.*, the chemical rate coefficients (k_{chem} values) or Arrhenius parameters (Beuermann and Buback, 2002; D'hooge et al., 2016). Many research efforts have been directed toward the determination of these parameters. A number of kinetic studies have considered multi-response regression analysis or tuning to polymerization data such as monomer conversion, average chain lengths, branching density and macromonomer content (De Roo et al., 2005; D'hooge et al., 2010; Hamzehlou et al., 2016; Haven et al., 2017; Kiparissides et al., 1980; Laki et al., 2020; Lena et al., 2018; Nikitin et al., 2010; Plessis et al., 2001; Zetterlund et al., 2001). Focus has also been on the use of *ab initio* calculations, originally in the gas phase and more recently also correcting for solvent effects (Desmet et al., 2018; Mavroudakis et al., 2015; Noble and Coote, 2013; Ruiperez, 2019; Soroush and Rappe, 2019).

Alternatively, specific experimental techniques such as pulsed laser polymerization (PLP) have been applied in which photo-initiator radical fragments are generated by multiple pulses and a specific growth pattern is associated by the presence of such pulses. The strength of PLP lies in the control over the average radical life time so that individual reactions can be studied in a step-wise approach, as only reactions with a time scale smaller than the average radical life time are kinetically relevant (Barner-Kowollik et al., 2008; Marien et al., 2016; Nikitin et al., 2007; Vir et al., 2019; Wenn and Junkers, 2016). As illustrated in this chapter, the PLP technique combined with size exclusion chromatography (SEC) analysis of the molar mass distribution (MMD) (or chain length distribution (CLD)) has revolutionized the accurate determination of in the rule intrinsic rate coefficients. This PLP-SEC method is complementary with PLP-matrix assisted laser desorption and ionization—time of flight—mass spectroscopy (MALDI-TOF-MS) (Willemse and van Herk, 2010) and single pulse—PLP (SP-PLP) with the subsequent tracking of the monomer conversion variation *via* near infrared (NIR) spectroscopy (Buback et al., 1986) or of the radical concentrations *via* electron paramagnetic resonance (EPR) spectroscopy (Barth et al., 2010; Kattner and Buback, 2015). Taking into account its ease of use and consequently wide application during the last decades focus is in this chapter mainly limited to the PLP-SEC technique.

As will be highlighted in Section 2, the PLP-SEC technique has originally been introduced for the determination of the (intrinsic) propagation rate coefficient in radical polymerizations with only one dominant macroradical, such as the macroradical formed by head-to-tail addition in FRP of styrene (Buback et al., 1995), methacrylates (Beuermann et al., 1997, 2000) and acrylates (Asua et al., 2004; Barner-Kowollik et al., 2014), for the last at sufficiently low polymerization temperature. Upon guaranteeing the accuracy of the PLP-SEC technique, as discussed in Section 2.1, this has allowed to detect trends for the reactivity of the dominant (bulk) head-to-tail propagation mode within monomer families based on PLP inflection point data as illustrated in detail in Section 2.2. In Section 3, an extension to radical polymerizations with more than one dominant propagation type is made, further referred to as the determination of multi-mode propagation reactivities. Focus is on industrially relevant case studies, namely acrylate FRP (Section 3.1), vinyl acetate FRP (Section 3.2), and free radical copolymerization (Section 3.3). Here it is also highlighted that PLP inflection point data can be used to determine k_{chem} values for other reactions than propagation such as backbiting and β-scission, making this chapter complementary to earlier reviews with a main focus on acrylates (Ballard and Asua, 2018) or propagation reactivities (Beuermann and Buback, 2002; Drawe and Buback, 2016; Kockler et al., 2016b; Van Herk, 2000). In Section 4, attention is specifically paid to a potential difference for PLP as studied in bulk or in solution. As shown in Section 5 by using the complete PLP-MMD instead of only inflection point data kinetic information on photodissociation, chain initiation and termination can be obtained as well, summarizing recent work with a leading role of our group (Marien et al., 2017a,b). Finally, in Section 6 general conclusions and perspectives are provided.

2. One mode propagation reactivities from PLP inflection point data

The PLP-SEC technique has originally been introduced by Aleksandrov et al. (1977) and Olaj et al. (1987) for the direct determination of the intrinsic propagation rate coefficient $k_{p,chem}$ in radical polymerizations with one dominant propagation mode. In PLP laser pulses with a laser pulse frequency ν, corresponding to a dark time $\Delta t = \nu^{-1}$, are applied to a mixture of monomer, photo-initiator and possibly solvent. These laser pulses lead to the quasi-instantaneous generation of photo-initiator derived radicals. A part of these radicals propagate until they terminate at a next laser pulse on a time scale $j\Delta t$ ($j = 1, 2, \ldots$). As confirmed by both deterministic and stochastic simulations (Drawe and Buback, 2016; Marien et al., 2017a; Nikitin et al., 2019; Sarnecki and Schweer, 1995) these time scales correspond to inflection points at chain lengths L_j ($j = 1, 2, \ldots$) in a multimodal SEC trace (Fig. 1), allowing the direct determination of $k_{p,chem}$ *via*:

$$k_{p,chem} = L_j [M]_0^{-1} (j\Delta t)^{-1} \tag{1}$$

In Eq. (1), $[M]$ can be taken equal to the initial monomer concentration $[M]_0$ due to the low monomer conversions targeted in PLP experiments (typically

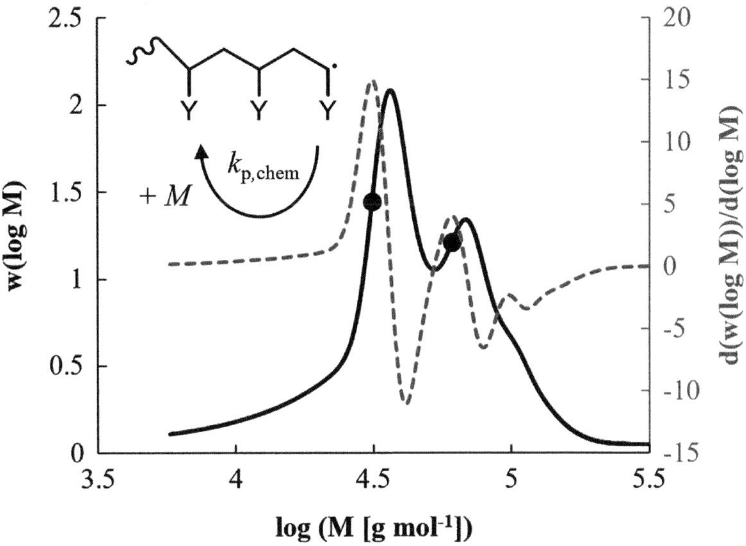

Fig. 1 Principle of the PLP-SEC technique for the determination of $k_{p,chem}$ in radical polymerization with one dominant propagation mode (example of head-to-tail propagation with Y the head of the monomer); under well-defined PLP conditions the depicted multimodal SEC trace results so that $k_{p,chem}$ is reliably determined *via* Eq. (1) from the chain lengths L_j ($j = 1, 2, \ldots$) corresponding to the inflection points (black dots) of the SEC trace (full black line) at the end of the PLP experiment. The inflection points correspond to the peak maxima of the derivative of the SEC trace (dashed purple line).

<5 mol%). In what follows, it is highlighted that the accuracy of Eq. (1) is affected by both characteristics of the PLP experiment itself such as the so-called termination rate limit and the selected type of photo-initiator, and by characteristics of the *a posteriori* SEC characterization, in particular SEC calibration and column broadening. Focus in this review is on low pressure PLP-SEC experiments. For high pressure PLP measurements the reader is referred to Beuermann and Buback (2002).

2.1 Factors affecting the accuracy of the PLP-SEC technique

2.1.1 Termination rate limit: Use of inflection point or peak maximum data?

A key characteristic of a PLP experiment determining the overall chain growth pattern and, hence, the validity of Eq. (1) is the fraction of radicals terminating between two laser pulses, β. The evaluation of β is non-trivial bearing in mind that the concentration of radicals generated by the laser pulses ($\Delta[R_0]$) decreases along a PLP experiment and that at the molecular level termination is chain length dependent, either intrinsic or by diffusional limitations (Achilias, 2007; Barner-Kowollik and Russell, 2009; Derboven et al., 2015; Johnston-Hall and Monteiro, 2008). It has been verified in various simulation studies (Buback et al., 1996; Drawe and Buback, 2016; Nikitin et al., 2019; Sarnecki and Schweer, 1995) that under conditions leading to intermediate β values (*e.g.*, *ca.* 0.2–0.8 (Drawe and Buback, 2016)) the location of the inflection points provides the most accurate measure for $k_{p,chem}$ determination *via* Eq. (1). In contrast, under conditions in the so-called high termination rate limit (*e.g.*, $\beta > 0.8$) the location of the peak maxima provides a more accurate measure (Desmet et al., 2017; Sarnecki and Schweer, 1995). This aspect is highlighted at the end of Section 3.2 for PLP of vinyl acetate (VAc).

For completeness it is mentioned here that Nikitin et al. (2019) recently showed that both in the low termination rate limit (*e.g.*, $\beta < 0.2$) or high termination rate limit (*e.g.*, $\beta > 0.8$) transforming the typically used log-MMD representation $w(\log M)$ to $g(\log M)$ (Eq. 2) accentuates the multimodal nature of the SEC trace.

$$g(\log M) = \frac{w(\log M)}{M^{\gamma}} \tag{2}$$

In Eq. (2), the exponent γ can take non-integer values.

2.1.2 Stability of the inflection point toward other chemical reactions

The positioning of the inflection points of a PLP-SEC trace can also be influenced by other reactions than (head-to-tail) propagation. As explained in Section 3, which focuses on PLP with multiple propagation reactivities as linked to separate macroradical types, this can be exploited as an advantage to determine the associated k_{chem} values as their time scales and, hence, the relative importance of the extra reactions can be altered, *e.g.*, by partly replacing monomer by its saturated analogue, increasing the PLP temperature or lowering ν.

The rule of thumb is to first identify the region of ν values in which one still can study the main propagation reaction (Asua et al., 2004; Barner-Kowollik et al., 2014). In a next step, one can switch to other PLP conditions. Here it is important to note that upon studying a potential solvent effect on $k_{p,chem}$ (Section 4)—even for a PLP with one dominant macroradical type—one first needs to identify if there is an impact of chain transfer to solvent. In other words, an apparent solvent effect can be observed by a shift of the inflection point beyond $[M]_0$ (Eq. 1) if chain transfer to solvent is also contributing.

2.1.3 Isothermicity or not?

To apply Eq. (1) it is important to validate if a given set temperature is actually reached and maintained during the PLP experiment. As such the reaction volume used in PLP is small and a direct temperature control system is considered. However for too high photo-initiator amounts, laser pulse energies, pulse numbers and laser pulse frequencies, and too exothermic monomers, non-isothermicity can even be obtained during PLP experiments, as highlighted in a recent theoretical study of Nikitin et al. (2016).

2.1.4 Selection of photo-initiator

Recent work of Marien et al. (2017a) has showed that to obtain well-defined consecutive PLP peaks, as relevant for the use of Eq. (1) for different *j* values, it is beneficial to use a photo-initiator with only one chain initiating radical fragment. This is the case for the commonly used photo-initiator 2,2-dimethoxy-2-phenylacetophenone (DMPA). As demonstrated *via* EPR, ESI-MS and MALDI-TOF-MS one of its initiator fragments participates in termination reactions but only to a very limited degree to chain initiation reactions (Barner-Kowollik et al., 2002; Fischer et al., 1990; Szablan et al., 2007b,c).

2.1.5 SEC calibration and column broadening

A fifth factor affecting the accuracy of $k_{p,chem}$ determination using the PLP-SEC technique is SEC calibration. In case narrow monodisperse standards exist for the polymer under consideration (*e.g.*, polystyrene or polymethacrylate), a calibration line can be directly made based on the known (average) molar masses of the narrow standards. If no such standards exist, which is likely, three major approaches have been proposed (Grubisic et al., 1967; Houwink, 1940; Mark, 1938; Szablan et al., 2007a; Zammit et al., 1998). A first approach has as input Mark-Houwink parameters, as available for many polymers (Mark et al., 2003), and requires a (mass) concentration-sensitive detector, typically a differential refractive index (DRI) detector to identify the relative contribution of aliquots containing polymer of a given molar mass based on the column separation. In a second and third approach an additional molar mass-sensitive detector is employed, namely a differential viscometer (DV) or low angle laser light scattering (LALLS) detector. In the DV approach the intrinsic viscosity is measured per aliquot coming out of the SEC column, which is used to determine the molar mass for that aliquot *via* a universal calibration curve. In the LALLS approach, the molar mass is determined directly, so without using a universal calibration curve.

It should be noted that the DV and LALLS approach require accurate knowledge of the amount of injected polymer and a good quality of the integrated detector response signals, the latter which could be affected by, *e.g.*, a noisy base line (Coote et al., 1997) and issues regarding calibration for the short chain lengths (Payne et al., 2013). Moreover, the calibration of the DV and LALLS detector is rather difficult and small calibration errors resulting, *e.g.*, from poorly tuned inter detector delays can lead to large systematic errors on the determined molar masses (Zammit et al., 1998). These errors can however be detected and eliminated as, for example, outlined by Zammit et al. (1998) by (i) simultaneously using the (independently calibrated) SEC-DV and SEC-LALLS method and comparing the obtained molar masses; and (ii) benchmarking SEC-DV and SEC-LALLS calibration by using polymers for which narrow standards with known (average) molar masses are available.

Besides calibration the quality of $k_{\mathrm{p,chem}}$ determination *via* PLP-SEC can also be affected by SEC-broadening. Based on principles introduced by Tung and Runyon (1969) and Billiani et al. (1988), Buback et al. (1996) suggested that the experimental SEC-broadened log-MMD, $w_{\mathrm{SEC}}(\log M)$, can be predicted by Gaussian smoothing of the unbroadened simulated log-MMD, $w(\log M)$. This mathematically entails a convolution of $w(\log M)$ with a Gaussian distribution with broadening parameter $\sigma_{\mathrm{v}}b$.

$$w_{SEC}(\log M) = \frac{1}{(2\pi)^{0.5}\sigma_v b}\int_0^{+\infty} \exp\left(-\frac{\left(\log(M) - \log(\tilde{M})\right)^2}{2(\sigma_v b)^2}\right) w(\log \tilde{M})\, dlog(\tilde{M}) \quad (3)$$

2.2 Trends and family type behavior

The PLP-SEC technique can also be applied to detect whether certain trends within monomer families or family type behavior can be identified for $k_{\mathrm{p,chem}}$. Initial focus has been on methacrylates and acrylates with linear ester groups and PLP conditions leading to one dominant macroradical type. For example, based on earlier literature data (Buback et al., 1997; Hutchinson et al., 1997a; Zammit et al., 1998) it has been reported in a well-received IUPAC contribution (Beuermann et al., 2000) that for linear alkyl methacrylates bulk $k_{\mathrm{p,chem}}$ values increase with increasing length of the ester side chain. This trend has been confirmed in further PLP-SEC (Haehnel et al., 2013) and PLP-MALDI-TOF-MS (Willemse and van Herk, 2010) studies.

The same increasing trend has been observed for linear alkyl acrylates at lower temperatures in combination with sufficiently high frequencies (Beuermann and Buback, 2002; Haehnel et al., 2014a). Haehnel et al. suggested that these increases in $k_{\text{p,chem}}$ can be explained by a decreasing concentration of the ester moieties, resulting in a reduced stabilization of the macroradical in the transition state.

It has been further observed that the bulk $k_{\text{p,chem}}$ values of several branched thus sterically hindered alkyl methacrylates are within PLP experimental error and, hence, can be described by a joint Arrhenius fit (Haehnel et al., 2014a). Haehnel et al. suggested that the observed family type behavior can be explained by considering entropic/steric effects. More in detail, Haehnel et al. indicated that the α-methyl group leads to strong steric interactions with the branched alkyl group, independent of the specific ester side chain, causing large rotational barriers and leading to an increased entropy loss compared to linear ester side chains during propagation (Haehnel et al., 2014a). In contrast, for branched alkyl acrylates this family type behavior nor a strictly increasing trend of $k_{\text{p,chem}}$ with increasing number of carbon atoms in the ester side chain could be observed. Branching appears to play an important role although no clear trend could be detected (Haehnel et al., 2014a).

A switch from branched to cyclic side chains can although again lead to a family behavior, as witnessed in a IUPAC contribution regarding the determination of bulk $k_{\text{p,chem}}$ of methacrylates with cyclic ester side chains (Beuermann et al., 2003), based on literature data (Buback and Kurz, 1998; Hutchinson et al., 1998; Zammit et al., 1998). As a final example of family type behavior, Kockler and coworkers identified several nitrogen-containing methacrylates showing this behavior (Kockler et al., 2015, 2016a).

3. Multi-mode propagation reactivities from PLP inflection point data

PLP-SEC experiments with more than one dominant propagation mode involve at least two macroradical species types. In such a case, Eq. (1) needs to be rewritten with an apparent propagation rate coefficient $k_{\text{p,app}}$ reflecting a mixed propagation reactivity:

$$k_{p,app} = L_j[M]_0^{-1}(j\Delta t)^{-1} \quad (4)$$

Note that in combination with a detailed kinetic model including regression analysis Eq. (4) is a post-processing step and, hence, no direct physical/ fundamental meaning is required. It suffices to obtain an accessible parameter which is dependent on the PLP reaction conditions and thus the occurrence of certain reactions. As such one thus moves away from a mere heuristic so-called model-free approach.

In this section, three industrially relevant cases of PLP-SEC with multiple dominant propagation modes are highlighted, namely PLP applied to acrylates, vinyl acetate (VAc) and copolymerization mixtures. In each of these cases, we put forward a specific approach according to the state-of-the-art of the PLP field. In the first case, the strength of a combined experimental and kinetic modeling PLP approach is highlighted for one of most studied monomer types, typically with *n*-butyl acrylate (*n*BuA) as reference monomer. In the second case, we demonstrate that PLP kinetic simulations can also be combined with *ab initio* calculations regarding the intrinsic absolute or relative reactivity of the considered reactions, specifically if the overall reaction scheme contains many competitive reactions. In the final and third case, we demonstrate that with an inherently large number of kinetic parameters also other (experimental) techniques beyond PLP-SEC are recommended. Hence, depending on the situation less or more tools are required with the most complex case highlighted in Fig. 2. Here one can focus on deterministic kinetic or kinetic Monte Carlo (*k*MC) modeling techniques.

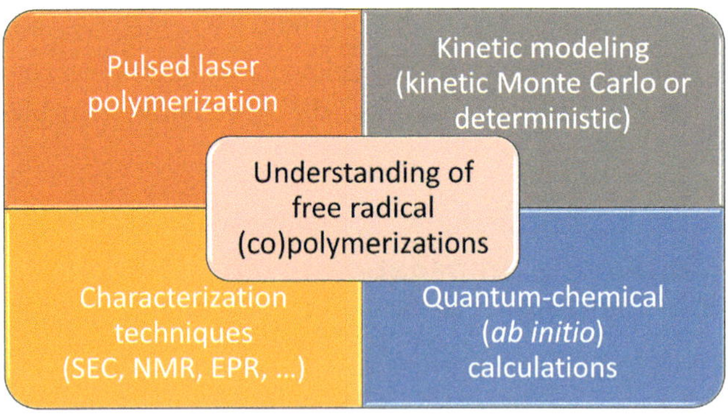

Fig. 2 The combination of PLP experiments, various experimental characterization techniques, kinetic modeling techniques (kinetic Monte Carlo or deterministic), and quantum chemical (*e.g.*, *ab initio*) calculations allows to obtain a detailed understanding of free radical (co)polymerization kinetics and specifically the determination of its kinetic parameters. Sections 3.1–3.3 highlight the relevance of these combinations with gradually more combinations included.

3.1 Acrylate PLP

Depending on the acrylate PLP conditions, in particular the polymerization temperature and ν, fewer or more elementary reactions affect the observed $k_{p,app}$ (Eq. 4). At "low" polymerization temperatures (*e.g.*, <323 K) in general the reactions in the left part of Fig. 3A affect the overall propagation behavior. At "high" polymerization temperatures in general also the reactions in the right part of Fig. 3A need to be considered. In what follows, the reaction scheme and estimation of the associated intrinsic rate coefficients under respectively "low" and "high" temperature PLP are addressed.

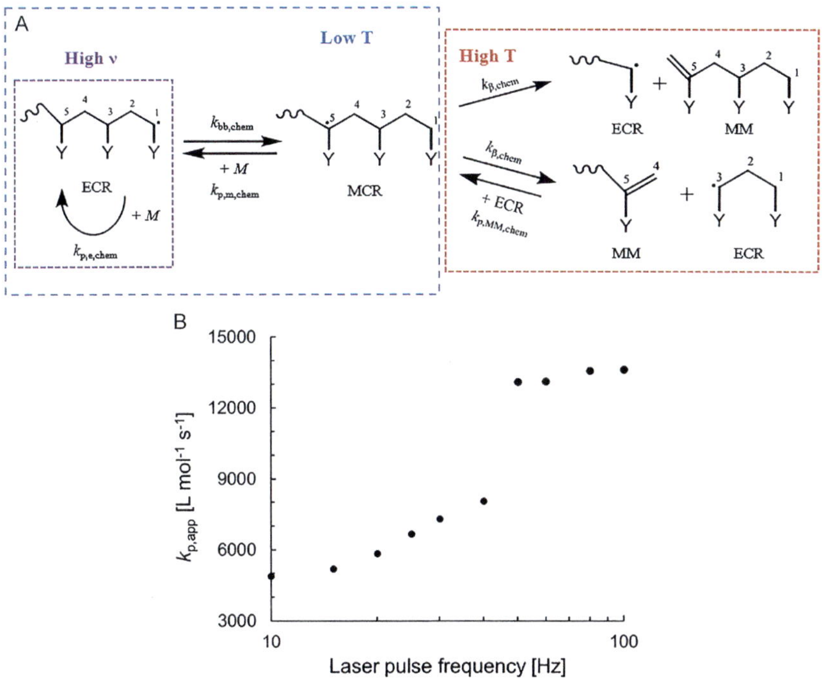

Fig. 3 (A) Overview of the key reactions in PLP of acrylates with p: propagation; bb: backbiting; β: β-scission; e: end-chain radical; m: mid-chain radical. The left box is relevant already at "low" temperature (for a high frequency (ν) only the purple box). The additional second right box only at higher temperature so that a step-wise PLP study can be conducted, as explained in Sections 3.1.1 and 3.1.2; macropropagation (intrinsic rate coefficient $k_{p,MM,chem}$) is only shown with a (secondary) end-chain radical (ECR) but can be ignored under typical PLP conditions (Vir et al., 2018); (B) Typical sigmoidal frequency variation in an acrylate PLP experiment (evaluation at inflection point position thus use of Eq. (4); $j=1$) at "low" temperature with at higher ν a tendency to only secondary propagation and at lower frequencies a mixed propagation behavior.

3.1.1 "Low" temperature PLP: Estimation of intrinsic backbiting and tertiary propagation intrinsic reactivities

As illustrated in the left part of Fig. 3A in "low" temperature PLP a conventional secondary end-chain radical (ECR; intrinsic propagation rate coefficient $k_{p,e,chem}$) can be converted to a tertiary mid-chain radical (MCR; intrinsic propagation rate coefficient $k_{p,m,chem}$) *via* intramolecular chain transfer to polymer or backbiting (intrinsic rate coefficient $k_{bb,chem}$), which dominantly occurs *via* a six-membered transition state (Cuccato et al., 2013; Liu et al., 2014; Yu and Broadbelt, 2012). Upon propagation of an MCR again an ECR is formed but with now a short chain branch in its overall molecular structure. Since $k_{p,m,chem}$ is much lower than $k_{p,e,chem}$ (*e.g.*, *ca.* a factor 1000 lower for *n*BuA at 333 K (Barth et al., 2010; Nikitin et al., 2007; Vir et al., 2019)) the formation of MCRs *via* backbiting leads to a rate retardation.

This rate retardation can be controlled *via* PLP-SEC experiments with varying ν (Marien et al., 2016; Nikitin et al., 2007; Wenn and Junkers, 2016). At high ν (*e.g.*, $\geq$500 Hz) high $k_{p,app}$ values approaching $k_{p,e,chem}$ are obtained as the time scale for backbiting is sufficiently larger than the average radical lifetime ($k_{bb,chem}^{-1} > \nu^{-1}$). Hence, $k_{p,e,chem}$ can be directly determined from high ν experiments using Eq. (1) (purple box in Fig. 3A; left part). At low ν (*e.g.*, $\nu < 200$ Hz) much lower $k_{p,app}$ values are obtained (see conceptually in Fig. 3B) at which $k_{p,app}$ is affected by $k_{p,e,chem}$, $k_{p,m,chem}$, $k_{bb,chem}$, and $[M]_0$. By gradually lowering ν a more mixed propagation behavior can be mimicked so that $k_{bb,chem}$ and $k_{p,m,chem}$ can be determined at a given polymerization temperature provided that $k_{p,e,chem}$ has already been determined at high ν. Note that fitting based on PLP data is preferred over multi-response regression analysis based on (semi-)batch FRP experiments in which a uniform and constant temperature profile is more difficult to obtain, due to the larger reactor scale and higher monomer conversions. In contrast, in PLP experiments, temperature non-uniformities can be strongly minimized by working with low sample volumes and a low number of pulses, as explained above (Nikitin et al., 2016).

Several PLP-based methods have been developed for the estimation of $k_{bb,chem}$ and $k_{p,m,chem}$. For completeness we include here also PLP-based methods that are not pure PLP-SEC. For example, Plessis et al. used ^{13}C nuclear magnetic resonance (NMR) analysis for the determination of the branching level of the polymer formed in bulk and solution PLP experiments. By assuming that the rate of formation of MCRs *via* backbiting is equal to the rate of MCR disappearance *via* tertiary propagation (and neglecting intermolecular chain transfer to polymer in view of the low monomer conversions) these authors could relate the (short chain) branching level to $k_{\mathrm{bb,chem}}$ and $k_{\mathrm{p,e,chem}}$ as (Plessis et al., 2003):

$$f_{br} = \frac{k_{bb,chem}}{k_{p,e,chem}[M]_0 + k_{bb,chem}} \tag{5}$$

Regression of Eq. (5) to experimental branching level data allowed the estimation of $k_{\mathrm{bb,chem}}$ plugging in an appropriate $k_{\mathrm{p,e,chem}}$. Alternatively, $k_{\mathrm{bb,chem}}$ and $k_{\mathrm{p,m,chem}}$ have been determined by fitting the ECR and MCR concentration profiles as measured by EPR after applying SP-PLP (Barth et al., 2010).

In addition to these PLP-NMR and SP-PLP-EPR methods, PLP-SEC experiments at varying ν have been exploited (*cf.* Fig. 3B). A first approach has been developed by Nikitin et al. in which $k_{\mathrm{bb,chem}}$ is determined from the specific laser pulse frequency ν_0 which corresponds to the onset of the decrease in $k_{\mathrm{p,app}}$ with decreasing ν. A semi-empirical relationship (Nikitin et al., 2007) is used to capture this onset of the sigmoidal variation of $k_{\mathrm{p,app}}$ with ν:

$$k_{bb,chem} = \alpha\nu_0 \tag{6}$$

with α a proportionality coefficient that needs to be determined *in silico*. In a next step Nikitin et al. (2007) assessed $k_{\mathrm{p,m,chem}}$ at low ν by approximating the average propagation rate coefficient $< k_{p,chem} >$ from FRP kinetics (also with a balancing of the backbiting and tertiary propagation rate) by $k_{p,app}$ (Nikitin et al., 2003):

$$k_{p,app} \approx < k_{p,chem} > = k_{p,e,chem} - \frac{k_{p,e,chem} - k_{p,m,chem}}{1 + \frac{k_{p,m,chem}[M]_0}{k_{bb,chem}}} \tag{7}$$

Simulations and experiments have however revealed that $k_{\mathrm{p,app}}$ keeps decreasing at decreasing ν (Marien et al., 2016; Nikitin et al., 2007; Wenn and Junkers, 2016). Hence, if $< k_{p,chem} >$ is evaluated from a sigmoidal fit to experimental $k_{\mathrm{p,app}}$ data $< k_{p,chem} >$ is underestimated. Nikitin et al. therefore formally evaluated Eq. (7) with a higher value than the one obtained *via* the sigmoidal fit to experimental $k_{\mathrm{p,app}}$ data. It should be further realized that the determination of α in Eq. (6) requires knowledge of $\Delta[R_0]$ and $\sigma_{\mathrm{v}} b$ as well as of the still-to-be-determined values of $k_{\mathrm{bb,chem}}$ and $k_{\mathrm{p,m,chem}}$. Hence, an iterative solution strategy is needed.

A second approach for the determination of $k_{bb,chem}$ and $k_{p,m,chem}$ from PLP-SEC experiments at varying ν has been developed by Wenn and Junkers (2016). These authors considered the sideward shift of the sigmoidal variation in Fig. 3B for varying values of $k_{p,m,chem}$ to match a fixed arbitrary experimental value of $k_{p,app}$. In a next step, Wenn and Junkers assessed $k_{bb,chem}$ values from Eq. (7). Alternatively, $k_{bb,chem}$ could first be determined from the sideward shift and $k_{p,m,chem}$ from Eq. (7). As mentioned by Wenn and Junkers $< k_{p,chem} >$ is underestimated, in agreement with the discussion above.

A third approach for $k_{bb,chem}$ and $k_{p,m,chem}$ estimation from variable ν PLP-SEC experiments has been developed by Marien et al. (2016). These authors performed regression to low ν $k_{p,app}$ data only. In addition to the consideration of multiple low frequencies, similar to Plessis et al. (2003), a variation of the solvent volume fraction was used to probe the competition between unimolecular and bimolecular reactions, leading to increased parameter sensitivity and decreased parameter correlation. These authors first illustrated their method for the estimation of $k_{bb,chem}$ in case $k_{p,m,chem}$ is known both *in silico* and experimentally. Moreover, the simultaneous estimation of $k_{bb,chem}$ and $k_{p,m,chem}$ was demonstrated *in silico* by performing regression to simulated $k_{p,app}$ data perturbed with a random Gaussian noise determined *via* random sampling from a Gaussian distribution with a standard deviation to mimic an experimental PLP-SEC error of 20%. The input values of both $k_{bb,chem}$ and $k_{p,m,chem}$ are retrieved with high accuracy, demonstrating the robustness of the method toward experimental outliers.

3.1.2 "High" temperature PLP: Estimation of intrinsic β-scission reactivities

In high temperature acrylate radical polymerization (*e.g.*, >350 K) the reactions in the right part of Fig. 3 can come into play, in addition to the reactions in the left part of Fig. 3. Under these conditions, β-scission of MCRs (intrinsic rate coefficient $k_{\beta,chem}$) may occur, forming ECRs and macromonomers. Also thermal self-initiation and macromonomer propagation may become important. However, Vir et al. (2018) demonstrated that these reactions can be ignored under PLP conditions (at least for *n*BuA as monomer). *Via* regression to low ν bulk $k_{p,app}$ data (Eq. 4 with $j=1$) these authors also demonstrated for *n*BuA at 413 K that a higher value of $k_{\beta,chem}$ results than determined/proposed previously in literature (Cuccato et al., 2013; Hamzehlou et al., 2016; Junkers and Barner-Kowollik, 2009; Nikitin et al., 2010).

This high intrinsic reactivity of β-scission has been further supported in a follow-up study introducing a step-wise method for the estimation of backbiting, MCR propagation and β-scission acrylate Arrhenius parameters (Vir et al., 2019). In a first step, regression to low temperature (*e.g.*, <323 K) $k_{p,app}$ data (Eq. 4 with $j=1$) at varying low ν and solvent volume fractions was performed for the estimation of the Arrhenius parameters for backbiting and MCR propagation under conditions at which β-scission is negligible. These Arrhenius parameters were used as input for the second step in which the Arrhenius parameters for β-scission were estimated at elevated temperature, again *via* regression to variable low ν $k_{p,app}$ data (Eq. 4 with $j=1$) at various solvent volume fractions. To avoid a possible solvent effect, *n*-butyl propionate was considered as a chemically similar solvent to *n*BuA. Although the method of Vir et al. works at fixed solvent volume fraction, the consideration of various solvent fractions reduced the correlation between k_{bb} and $k_{p,m}$ estimates. Hence, this approach allows an inherent evaluation of its level of success according to rules from the statistics field.

3.2 Vinyl acetate PLP

Also vinyl acetate (VAc) FRP features multiple macroradical species, leading to multi-mode propagation reactivities, as shown in Fig. 4A. In contrast to common radical polymerizations it is not afforded to only focus on head-to-tail propagation. An explicit differentiation between head and tail radicals is required so that four addition types need to be accounted for in a general PLP of VAc: head-to-tail (ht), head-to-head (hh), tail-to-head (th), and tail-to-tail (tt) propagation. In addition, as demonstrated by *ab initio* calculations (Desmet et al., 2017), the tail radical can undergo backbiting (bb) with the resulting MCR upon propagation again dominantly forming a head radical (p,m).

A ν dependence of $k_{p,app}$ (Eq. 4) has been experimentally observed for PLP of VAc considering inflection point data, similar to the simulated trend displayed by the green symbols in Fig. 4B, although this ν dependence is less pronounced as in Fig. 3B (acrylate case). An increase in $k_{p,app}$ of *ca.* 15% has been observed at 323 K if ν is increased from low (25–100 Hz) to high (300–500 Hz) values (Barner-Kowollik et al., 2017). Several hypotheses have been put forward to explain this ν dependence (Barner-Kowollik et al., 2017), namely: (i) a shift away from the high termination rate limit so that one has the issue of the use of inflection point or peak maxima (Hutchinson et al., 1994); (ii) an intrinsic chain length dependency for

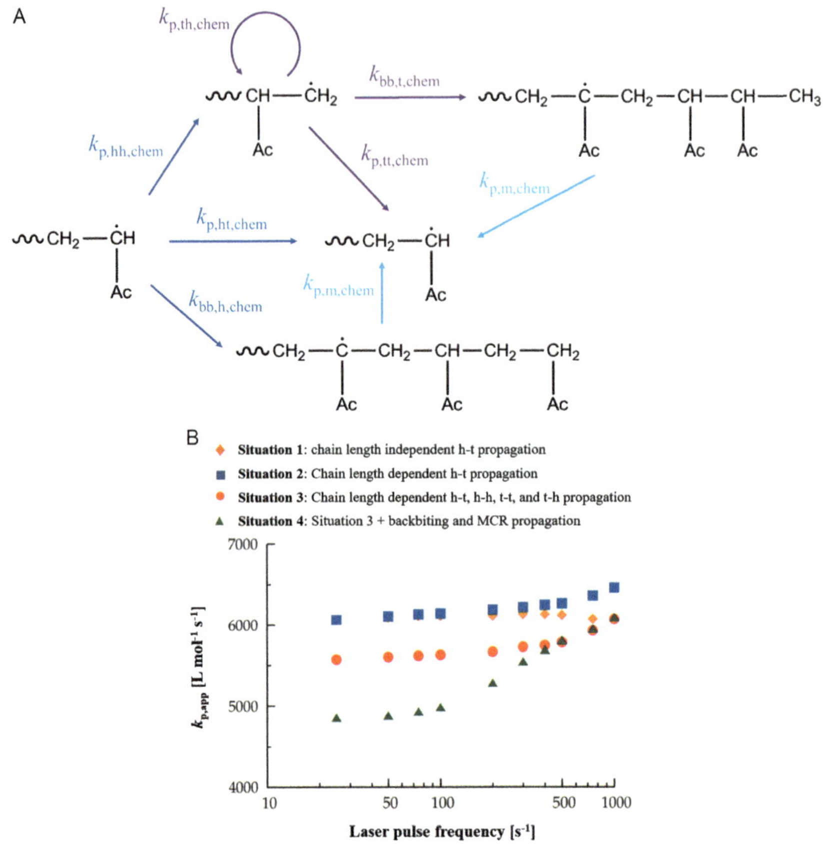

Fig. 4 (A) Multiple macroradical species types and related reactions in radical polymerization of VAc; (B) Simulated frequency (ν) dependencies under PLP conditions at 323 K (evaluation at inflection point position using Eq. (4); $j = 1$) testing various assumptions (situation 1–4); parameters from Desmet et al. (2017). Only situation 4 is consistent with experimental trends for PLP of VAc (Barner-Kowollik et al., 2017); h: head; t: tail; p: propagation (if m with mid-chain radical); bb: backbiting.

propagation as shorter chains results for higher ν (Heuts and Russell, 2006; Nikitin and Evseev, 1999; Olaj et al., 2000); (iii) the occurrence of head-to-head propagation in addition to conventional head-to-tail propagation at lower ν (Flory and Leutner, 1948; Monyatsi et al., 2014); (iv) the formation of MCRs *via* backbiting only at lower ν (Junkers et al., 2009).

Direct discrimination between the various hypotheses is impossible based on experimental PLP-SEC data only as there too many competitive reaction steps and also too many hypotheses to evaluate. Hence, Desmet et al. (2017) developed a detailed *ab initio* based *k*MC model accounting for all the reactions in Fig. 4A with a head-to-tail propagation rate coefficient consistent with high ν data and the *ab initio* calculations further highlighting that tail (macro)radicals are at 323 K five times less reactive than head (macro)radicals and can also backbite. The intrinsic chain length dependence for propagation was formally accounted for *via*:

$$k_{p,chem}(i) = k_{p,chem}\left[1 + C_1 \exp\left(-\frac{\ln(2)}{i_{1/2}} i\right)\right] \tag{8}$$

with $k_{p,chem}$ the long chain limit propagation rate coefficient, i the chain length, and C_1 and $i_{1/2}$ system specific parameters (Heuts and Russell, 2006).

Notably, this detailed kinetic model predicts experimentally observed amounts of head-to-head defects and quantitatively the experimentally observed trend for ν behavior (green symbols in Fig. 4B; situation 4). A combination of hypothesis (iii), the changing relevance of head-to-head propagation, and hypothesis (iv), the changing relevance of backbiting, is demonstrated as a very plausible joint explanation for the overall experimental behavior. This is also illustrated in Fig. 4B by the extra simulated situations (situation 1–3). Here situation 1 (orange symbols) only covers chain length independent head-to-tail propagation and no significant ν dependence is observed, as the β values do not alter much upon shifting from low to high ν. If 500 pulses are applied, β varies from 0.98 in the beginning of the PLP experiment to 0.97 at the end for $\nu = 100$ Hz, while for $\nu = 500$ Hz β varies from 0.95 to 0.92. Hence, only a limited shift away from the high termination limit is observed upon increasing ν, indicating that hypothesis (*i*) does not provide an adequate explanation for the experimentally observed ν dependence of $k_{p,app}$. These overall high β values put although the additional question if an evaluation with peak maxima instead of inflection point data is recommended, as opposed to what is currently recommended by IUPAC. Before addressing this we first discuss the simulation results associated with situation 2 and 3 in Fig. 4B.

It follows that with situation 2 (blue symbols in Fig. 4B), still only considering head-to-tail propagation but now considering chain length dependent propagation (use of Eq. 8), only the $k_{p,app}$ values at the higher frequencies are slightly changed. Hence, hypothesis (ii) cannot be used to explain the main decrease at lower ν. Such decrease can at least be partially explained by including other propagation types, specifically head-to-head propagation, as becomes clear upon focusing on situation 3 (red symbols in Fig. 4B) in agreement with modeling work from Monyatsi et al. (2014).

In that work, the full ν could be accounted for but there the authors neglected backbiting so that the parameter tuning directed toward an overestimation of the difference in intrinsic propagation reactivity of tail and head radicals (factor 40 *vs* 5). The *ab initio* calculations of Desmet et al. (2017) inherently covered for more consistency for propagation pre-exponential factors and additionally highlighted that backbiting of tail radicals (as opposed to the acrylate field with head radicals) is also needed in the kinetic description. Upon comparing the simulation results from situation 3 and 4 it becomes clear that indeed this reaction is prone to a ν change as well, explaining the last extra drop at the lowest ν values in Fig. 4B.

It should be noted that VAc MCRs, as formed by backbiting, could not be detected in the EPR study of Kattner and Buback (2014). On the other hand, these MCRs have been detected *via* EPR for acrylates (Barth et al., 2009; Willemse et al., 2005). However, acrylate head macroradicals can directly undergo backbiting, while the work of Desmet et al. (2017) indicates that VAc tail macroradicals first need to be formed *via* head-to-head propagation. This likely leads to overlap of spectral features, thereby clouding the comparison between MCR radical presence in acrylate and VAc radical polymerization. Moreover, lower (simulated) MCR concentrations are obtained in low ν PLP experiments of VAc compared to acrylates, which also complicates the detection of MCR species in polymerizations of VAc. This illustrates the use of kinetic simulations in the overall appreciation of experimental analysis.

Finally, as highlighted above, it is worthwhile to verify if peak maxima should not be used instead of inflection point data. Also this has been covered in the work of Desmet et al. (2017) noting that a very low frequencies one expects FRP like behavior. For this purpose an expression for $< k_{p,chem} >$ has been derived based on the quasi-steady state approximation (QSSA) for the involved radical types. Assuming the initial monomer concentration ($[M] = [M]_0$), as relevant in well-conducted PLP experiments, it follows that:

$$
\begin{aligned}
< k_{p,chem} >= & \Big[k_{p,ht,chem} k_{p,tt,chem} + 2 k_{p,hh,chem} k_{p,tt,chem} + k_{p,hh,chem} k_{p,th,chem} \\
& + \frac{k_{bb,t,chem}}{[M]_0} \left(k_{p,ht,chem} + 2 k_{p,hh,chem} \right) \Big] \\
& \left[\left(k_{p,hh,chem} + k_{p,tt,chem} \right) + \frac{k_{bb,t,chem}}{[M]_0} \left(1 + \frac{k_{p,hhchem}}{k_{p,m,chem}} \right) \right]^{-1}
\end{aligned} \tag{9}
$$

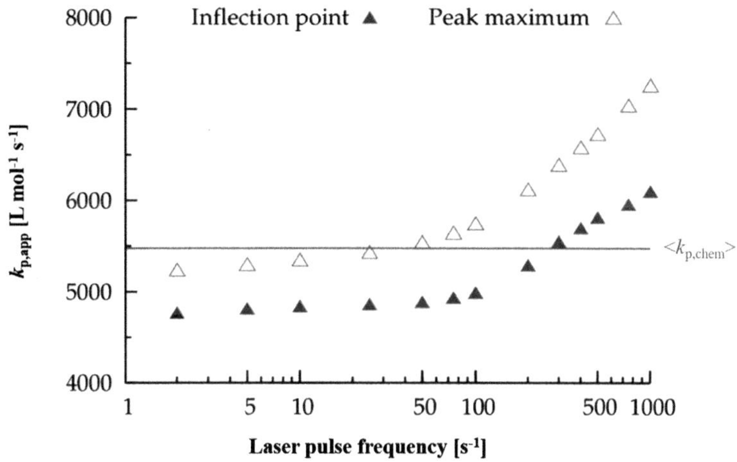

Fig. 5 Comparison of $<k_{p,chem}>$ as calculated using Eq. (9) (full green line) and $k_{p,app}$ as calculated *via* Eq. (4) with L_1 based on the location of the inflection points (green filled symbols) or peak maxima (unfilled symbols) of PLP-SEC traces obtained *via* *k*MC simulation (*cf.* situation 4 in Fig. 4 as consistent with experimental trends (Barner-Kowollik et al., 2017)); focus is on a polymerization temperature of 323 K. *Figure reproduced from Desmet GB, Marien YW, Van Steenberge PHM, et al.: Ab initio based kinetic Monte Carlo analysis to unravel the propagation kinetics in vinyl acetate pulsed laser polymerization,* Polym Chem *8:7143–7150, 2017 with permission from the Royal Society of Chemistry.*

Upon ignoring backbiting this equation reduces to the one mentioned in Monyatsi et al. (2014), highlighting the overall consistency of the derivation.

A comparison of $<k_{p,chem}>$ and $k_{p,app}$ obtained *via* simulation using the detailed *k*MC model of Desmet et al. (2017) using Eq. (4) with L_1 based either on the location of the inflection points (filled green triangles) and peak maxima (unfilled green triangles) in PLP of VAc at 323 K is shown in Fig. 5. As expected, at high ν, a mismatch between $k_{p,app}$ and $<k_{p,chem}>$ is obtained independent of the calculation method for $k_{p,app}$, consistent with the invalidity of the QSSA at high ν, as also observed for acrylates (Marien et al., 2017a). More importantly, if the location of the peak maxima is used to calculate $k_{p,app}$ a good agreement between $k_{p,app}$ (unfilled triangles) and $<k_{p,chem}>$ (full green line) is obtained at low ν. If the location of the inflection points is used instead to calculate $k_{p,app}$ (filled symbols) $<k_{p,chem}>$ is *ca.* 10% lower than $<k_{p,chem}>$. Hence, the detailed simulations give preference to the use of peak maxima in future PLP studies of VAc, consistent with earlier work of Sarnecki and Schweer (1995).

3.3 PLP with comonomers

A third case of radical polymerization with multiple dominant propagation modes is free radical copolymerization with for instance two comonomers M_A and M_B. The common approach to describe the corresponding propagation reactions is to either assume that the (intrinsic) propagation reactivities of the macroradicals only depend on the terminal unit so that a terminal model can be considered or on both the terminal and penultimate unit so that a penultimate monomer unit (PMU) model can be considered (Coote and Davis, 1999; Fierens et al., 2018). A further distinction can be made between the so-called explicit and implicit PMU model, with the latter model relying on the assumption that the radical selectivities are unaffected by the penultimate unit.

As illustrated in Fig. 6, focusing for simplicity only on ECRs, this leads to 4 distinguishable propagation rate coefficients in the terminal model and respectively 6 and 8 in the implicit and explicit PMU model. In the terminal model one practically focuses on two monomer reactivity ratios $r_A = k_{p,AA,chem}/k_{p,AB,chem}$ and $r_B = k_{p,BB,chem}/k_{p,BA,chem}$. Similarly, in the explicit PMU model one introduces four r values (r_A, r_B, r_A', r_B') and two radical reactivity ratios (s_A and s_B): $r_i = \frac{k_{p,iii,chem}}{k_{p,iij,chem}}$, $r_i' = \frac{k_{p,jii,chem}}{k_{p,jij,chem}}$, $s_i = \frac{k_{p,jii,chem}}{k_{p,iii,chem}}$ with $i \neq j$ and $i, j = \mathrm{A, B}$. Note that in the implicit PMU model it holds that $r_i = r_i'$.

Defining the fraction of monomer i in the reaction mixture as f_i, in a general form $< k_{p,chem} >$ can be expressed as:

$$< k_{p,chem} > = \frac{\bar{r}_A f_A^2 + 2 f_A f_B + \bar{r}_B f_B^2}{\dfrac{\bar{r}_A f_A}{\bar{k}_{p,AA}} + \dfrac{\bar{r}_B f_B}{\bar{k}_{p,BB}}} \tag{10}$$

Terminal model	Penultimate unit model	
$R_{i,A} + M_A \xrightarrow{k_{p,AA,chem}} R_{i+1,A}$	$R_{i,AA} + M_A \xrightarrow{k_{p,AAA,chem}} R_{i+1,AA}$	$R_{i,BA} + M_A \xrightarrow{k_{p,BAA,chem}} R_{i+1,AA}$
$R_{i,A} + M_B \xrightarrow{k_{p,AB,chem}} R_{i+1,B}$	$R_{i,AA} + M_B \xrightarrow{k_{p,AAB,chem}} R_{i+1,AB}$	$R_{i,BA} + M_B \xrightarrow{k_{p,BAB,chem}} R_{i+1,AB}$
$R_{i,B} + M_A \xrightarrow{k_{p,BA,chem}} R_{i+1,A}$	$R_{i,AB} + M_A \xrightarrow{k_{p,ABA,chem}} R_{i+1,BA}$	$R_{i,BB} + M_A \xrightarrow{k_{p,BBA,chem}} R_{i+1,BA}$
$R_{i,B} + M_B \xrightarrow{k_{p,BB,chem}} R_{i+1,B}$	$R_{i,AB} + M_B \xrightarrow{k_{p,ABB,chem}} R_{i+1,BB}$	$R_{i,BB} + M_B \xrightarrow{k_{p,BBB,chem}} R_{i+1,BB}$

Fig. 6 Propagation reactions and their corresponding intrinsic rate coefficients in a terminal model (purple box) and penultimate unit model (blue box) for a conventional radical copolymerization with only end-chain radicals; *R*: radical; *i*: chain length; comonomers M_A and M_B.

with $\bar{r}_i$ and $\bar{k}_{p,ii}$ given by

$$\bar{r}_i = r_i'\left(\frac{f_i r_i + f_j}{f_i r_i' + f_j}\right) \text{ with } i \neq j, i = \text{A, B} \tag{11}$$

$$\bar{k}_{p,ii} = k_{p,iii,chem}\left(\frac{f_i r_i + f_j}{f_i r_i + \frac{f_j}{s_i}}\right) \text{ with } i \neq j, i = \text{A, B} \tag{12}$$

For the implicit PMU model these equations can be simplified since $\bar{r}_i = r_i$ (i = A, B). In the terminal model, these equations are further simplified since $\bar{k}_{p,ii} = k_{p,ii,chem}$ (i = A, B) with $k_{p,ii,chem}$ the homopropagation rate coefficient. Interestingly, *via* well-defined PLP-SEC experiments at low frequencies $<k_{\text{p,chem}}>$ can be assessed using Eq. (4) (Coote et al., 1997; Davis et al., 1989; Hutchinson et al., 1997b; Liang et al., 2009).

However, for certain comonomer pairs, *e.g.*, for copolymerizations of styrene with methacrylates, the terminal model parameters determined *via* regression to $<k_{\text{p,chem}}>$ data deviate from the ones determined *via* regression to NMR-based average copolymer composition data. In those situations, a satisfactory description of both $<k_{\text{p,chem}}>$ and the average copolymer composition using a single set of terminal model parameters is impossible (Fukuda et al., 1985; Schweer, 1993). This mismatch could already be formally overcome by using the implicit PMU model (Davis et al., 1990; Li et al., 2006; Liang et al., 2009; Wang and Hutchinson, 2008).

On the long run it is although recommended to use explicit PMU models as these are the most fundamental ones, specifically if one also wants to understand the different faith of ECRs and MCRs in radical copolymerization. It should however be admitted that currently only limited experimental evidence is available to demonstrate the need to consider the explicit PMU model. This is understandable as the discrimination between the implicit and explicit PMU model is difficult, due to the large number of kinetic parameters leading to difficult reliable parameter estimations (Coote and Davis, 1999; Fierens et al., 2018). On the other hand kinetic modeling studies combined with *ab initio* data have already put forward that strong deviations in polymerization characteristics are obtained, even for radical copolymerization with similar r and r' values (Fierens et al., 2017). Hence, it is clear that we further need to combine dedicated experimental techniques such as PLP, detailed *k*MC or deterministic kinetic models and *ab initio* calculations to obtain a complete understanding of radical copolymerization mechanisms, as illustrated in Fig. 2.

4. Bulk *vs* solution PLP inflection point data

An interesting research angle of the PLP-SEC technique is the investigation of potential solvent effects on $k_{p,chem}$ (Buback, 2015; Haehnel et al., 2014b, 2015). Most reference $k_{p,chem}$ data have been recorded in bulk but radical polymerizations are not automatically conducted in bulk. Sometimes solvents are used to lower the viscosity at higher monomer conversions and in certain cases one aims at or needs benign solvents such as water. For common monomers and in case polarity changes are limited between the monomer and the organic solvent (*e.g.*, styrene/toluene) a limited to negligible solvent effect (*e.g.*, max 15%), has been recorded (Beuermann, 2009; Buback, 2015; Haehnel et al., 2014b, 2015) taking into account that typical PLP experimental errors are also around 10%. However, special attention needs to be paid to the determination of $k_{p,chem}$ data determined in aqueous media for which hydrogen bonding and pre-complex formation can take place (Ganachaud et al., 2000; Stach et al., 2008; Vachaudez et al., 2013). Also for radical polymerizations in ionic liquids the situation is non-trivial as the ionic strength is too high not to play a role then for transition states.

For example, a strong increase of $k_{p,chem}$ with decreasing initial monomer concentration has been reported *via* PLP-SEC experiments for N-vinylpyrrolidone (Stach et al., 2008), N-vinylformamide (Stach et al., 2010), N-isopropylacrylamide (Ganachaud et al., 2000), and (non-ionized) methacrylic acid (Beuermann et al., 2006, 2007; Kuchta et al., 2000) in water. Deviations from this increasing trend have been reported for fully ionized methacrylic acid, for which a reverse trend has been observed (Lacík et al., 2009). For nonionized acrylic acid the increasing trend of $k_{p,chem}$ with decreasing initial monomer concentration has again been observed, with the exception of a maximum with *ca.* 3 m% of acrylic acid initially present (Lacík et al., 2001, 2003). Significantly higher (apparent) activation energies were also found at initial concentrations below or equal to 3 m%. In PLP of acrylic acid secondary ECRs may undergo backbiting (Buback et al., 2007) leading to tertiary MCRs and, hence, a more complicated propagation behavior (*cf.* multiple radical types and propagation modes covered in Section 3). For methacrylic acid (Beuermann et al., 2006), N-vinylpyrrolidone (Stach et al., 2008) and N-vinylformamide (Stach et al., 2010) such backbiting has not been reported and $E_{A,p}$ remains more or less constant taking into account experimental error. Hence, for these monomers the variation of $k_{p,chem}$ with initial aqueous monomer concentration has been attributed to a variation of the pre-exponential factor ($A_{p,chem}$), although Deglmann et al. (2018) hinted that the constancy of the activation energy in several cases is only an experimental artifact. Notably recent kinetics studies (Idowu and Hutchinson, 2019; Rooney and Hutchinson, 2018; Schier et al., 2017) use more advanced kinetic descriptions to acknowledge solvent effects such as hydrogen bridging focusing on monomers such as 2-hydroxyethyl acrylate (HEA) and 2-methoxyethyl acrylate (MEA).

Also for polymerizations in ionic liquids significant effects of the initial monomer concentration on the determined $k_{p,chem}$ *via* PLP-SEC have been observed. In particular for MMA an increase of $k_{p,chem}$ with decreasing initial monomer concentration has been observed for various ionic liquids (Harrisson et al., 2002, 2003; Jeličić et al., 2009a,b; Woecht et al., 2008). Notably, in the work of Jeličić et al. the obtained $E_{A,p,MMA}$ for various ionic liquids and initial monomer concentrations are rather similar and close to the IUPAC benchmark value (Beuermann et al., 1997).

5. Determination of quantum yield and other chemical reactivities from the complete PLP-SEC trace

Whereas Sections 2–4 detail the various treatments in literature to measure and predict intrinsic propagation rate coefficients, Section 5 treats emerging developments on how to use more than merely the inflections points of a PLP-SEC trace. In all cases, SEC broadening is accounted for so that reported effects are additionally to that and are genuine phenomena.

5.1 Understanding of photodissociation and determination of quantum yield

Based on previous principles (Castignolles et al., 2006; Odian, 2004), Marien et al. (2017b) introduced Eq. (13) to calculate the volume-averaged value of the concentration of photo-initiator radicals generated at each laser pulse, $\Delta[R_0]$, in small PLP reaction vials:

$$\Delta[R_0] = 2\Phi \frac{E_{\text{pulse}}\lambda}{hcN_A V}[1 - \exp(-2.303\varepsilon[I_2]L)] \tag{13}$$

This equation highlights that $\Delta[R_0]$ depends on the sample volume V and optical path length L (*cf.* the Bouguer-Beer-Lambert law (Ricci et al., 1994)) and that $\Delta[R_0]$ can be most easily varied experimentally *via* the pulse energy E_{pulse} and the photo-initiator concentration $[I_2]$. The latter decreases leading to a substantial decrease of $\Delta[R_0]$ if the number of applied pulses becomes too high (Castignolles et al., 2006; Marien et al., 2017a).

Eq. (13) also highlights that $\Delta[R_0]$ is proportional to the photodissociation quantum yield Φ, which expresses the probability that the absorption of a photon by a photo-initiator molecule leads to a dissociation event (Allonas et al., 2003; Noble et al., 2016). Notably, kinetic modeling studies showed that in particular the peak heights of a PLP-SEC trace are sensitive to $\Delta[R_0]$ (Buback et al., 1996; Marien et al., 2017a). Very recently Marien et al. (2017b) exploited this sensitivity by performing regression to the ratio of the height of the first (h_1) and second (h_2) peak in the PLP-SEC traces obtained with varying pulse energies, as demonstrated in Fig. 7.

5.2 Determination of chain initiation reactivities

In addition to information on Φ, Marien et al. (2017a) demonstrated that also information on the chain initiation reactivity can be extracted from complete PLP-SEC traces. Focusing on DMPA, which dissociates into a benzoyl radical with chain initiation reactivity $k_{p,I,chem}$ and a dimethoxybenzyl radical with

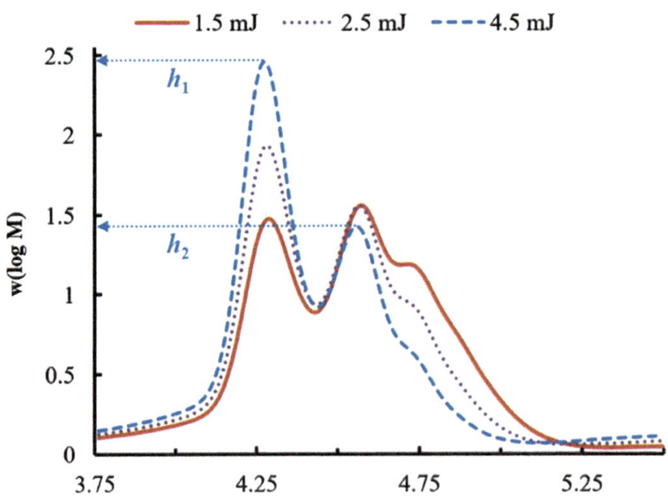

Fig. 7 Simulated sensitivity of the height of the first peak (h_1) and second (h_2) peak in PLP-SEC traces obtained at three pulse energy (E_{pulse}) values; full red line: 1.5 mJ, dotted purple line: 2.5 mJ, dashed blue line: 4.5 mJ; kinetic Monte Carlo (*k*MC) model parameters from Marien et al. (2017b). From the height ratio, the quantum yield can be determined using Eq. (13), as explained in the work of Marien et al.

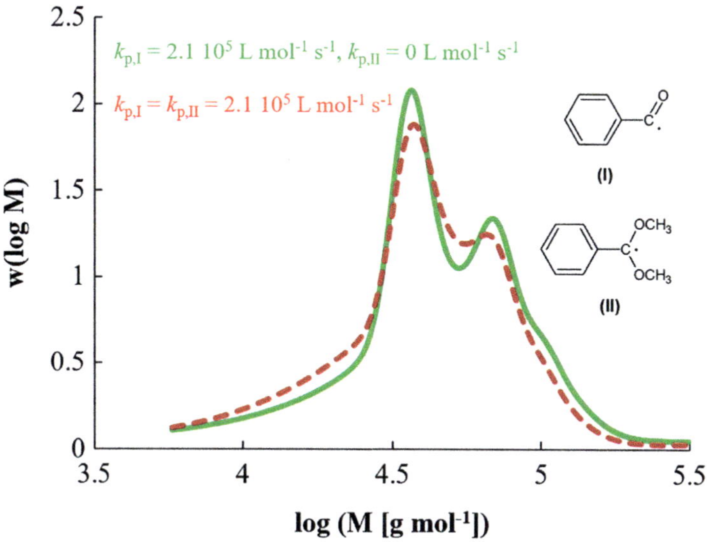

Fig. 8 Kinetic Monte Carlo (*k*MC) modeling of complete PLP-SEC traces allows to identify the disparate chain initiation reactivities of the 2,2-dimethoxy-2-phenylacetophenone (DMPA) initiator fragments benzoyl radical ($k_{p,I,chem}$) and dimethoxybenzyl radical ($k_{p,II,chem}$) at 306 K; $\nu = 500$ Hz full green line, as consistent with experimental data (Marien et al., 2017a), from *k*MC simulation with $k_{p,I,chem} = 2.1 \times 10^5$ L mol^{-1} s^{-1} and $k_{p,II,chem} = 0$ L mol^{-1} s^{-1}; dashed red line from *k*MC simulation with $k_{p,I,chem} = k_{p,II,chem} = 2.1 \times 10^5$ L mol^{-1} s^{-1}; other model parameters as reported in Marien et al. (2017a) with SEC broadening accounted for.

chain initiation reactivity $k_{p,II,chem}$ upon irradiation at low to intermediate temperature, and based on experimental evidence from EPR spectroscopy (Fischer et al., 1990) and kinetic modeling results for SP-PLP experiments (Buback et al., 2000), these authors accounted for the very low chain initiation probability of the dimethoxybenzyl radicals by assuming a vanishing value for its chain initiation rate coefficient.

As illustrated in Fig. 8, deviations are obtained for a PLP simulation with $k_{p,I,chem} \neq 0$ L mol^{-1} s^{-1} and $k_{p,II,chem} = 0$ L mol^{-1} s^{-1} (full green line) compared to a simulation in which both initiator fragments are equally reactive ($k_{p,I,chem} = k_{p,II,chem} \neq 0$ L mol^{-1} s^{-1}; dashed red line). In the latter simulation less sharp peaks are obtained, which is disadvantageous for the identification of higher order inflection points (Eq. 1 or 4), which are needed for a consistency check in case the determination of $k_{p,chem}$ is envisaged. Hence, the many successes for PLP experiments with DPMA can be attributed to the disparate chain initiation reactivities of its photo-initiator fragments.

5.3 Determination of (apparent) chain length dependent termination reactivities

Various kinetic studies have demonstrated that also information on the termination reactivity can be extracted from PLP-SEC experiments. As indicated in the work of Barner-Kowollik and Russell (2009) one can expect here to always measure apparent termination rate coefficients with only in the limit of small chain lengths the typical intrinsic values for termination between radicals. A subscript "app" needs thus to be used here referring to chain length dependent diffusional limitations. This subscript "app" has thus a different meaning as in Eq. (4) where it relates to a mixed behavior on the intrinsic level.

Initially the average value of the apparent termination rate coefficient $<k_{t,app}>$ in PLP experiments thus at low monomer conversions has been a key point of interest (Buback and Lämmel, 1998; Olaj and Schnöll-Bitai, 1989). It is however more interesting to obtain information on the functional form of the chain length dependence of $k_{t,app}$. For example, Shipp et al. (2003) compared the performance of a Smoluchowski based termination model (Smoluchowski, 1917) and a geometric mean single power law model to describe complete PLP-SEC traces. More recently, Marien et al. (2017a) compared the situations of no chain length dependence and the situations with a detailed chain length dependence for homo-termination *via* a single power law model or *via* low conversion composite k_t-model equations (Johnston-Hall and Monteiro, 2008) and a geometric mean for cross-termination rate coefficients:

$$k_t^{app}(i,i) = k_t^{app}(1,1)i^{-\alpha_S}\ i \le i_c \tag{14}$$

$$k_t^{app}(i,i) = k_t^{app}(1,1)i_c^{-\alpha_S+\alpha_L}i^{-\alpha_L}\ i > i_c \tag{15}$$

$$k_t^{app}(i,j) = \left[k_t^{app}(i,i)k_t^{app}(j,j)\right]^{0.5} \tag{16}$$

As shown in Marien et al. (2017a) only such detailed apparent termination model is consistent with experimental PLP data. As can be observed in Fig. 9, the (apparent) termination rate is overestimated in case no chain length dependence is taken into account (*i.e.*, $k_{t,app}(i,j) = k_{t,app}(1,1)\ \forall\ i,j$), consistent with the lesser importance of the formation of longer chains, comparing the associated SEC trace (dashed red line) with the reference thus experimentally benchmarked case (full green line). If on the other hand the decrease of $k_{t,app}$ with increasing chain length is accounted for, the formation of longer chains is simulated. However, if $k_{t,app}(i,i)$ is calculated using a single power law evaluated with the chain length dependence of long chains only

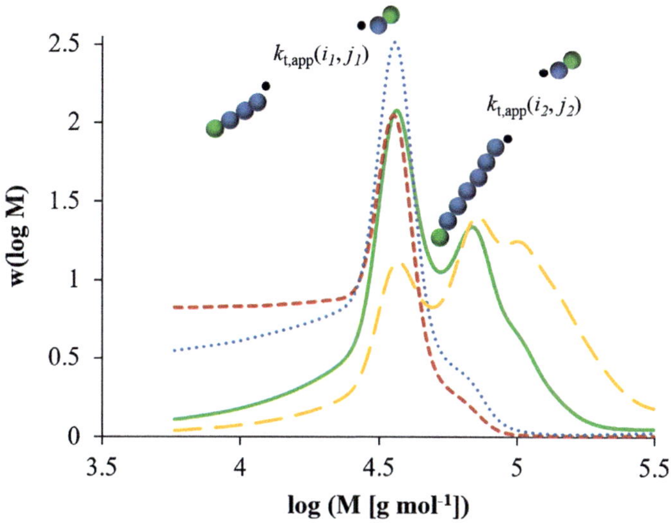

Fig. 9 Simulated SEC traces if (i) no chain length dependence of $k_{t,app}$ is accounted for (dashed red line; $k_{t,app}(i,j) = k_{t,app}(1,1)\ \forall\ i, j$), (ii) $k_{t,app}(i,i)$ is expressed *via* a single power law with only Eq. (14) (long yellow dashes); (iii) the update of (ii) with $\alpha_S = \alpha_L$, (dotted blue line, only Eq. 14); and (iv) $k_{t,app}(i,j)$ is calculated fully based on Eqs. (14–16) (full green line; consistent with experimental data as shown in Marien et al. (2017a)): other kinetic Monte Carlo parameters are also taken from the same reference. Only the green line covers the balance between short-long termination and the multimodality of the radical population at low monomer conversion.

(*i.e.*, using Eq. 14 $\forall\ i$ but with α_L instead of α_S; dotted blue line), only a limited increase of the fraction of long chains is simulated compared to the situation with chain length independence. In contrast, if the power law model is evaluated with the chain length dependence of short chains (Eq. 14 thus with $\alpha_S\ \forall\ i$; yellow long dashes) the fraction of long chains is overestimated. These last two PLP simulation results thus underpin the importance of distinguishing between the chain length dependence of shorter radicals, which is more controlled by center-of-mass diffusion, and longer radicals, which is more controlled by segmental diffusion, at least at the low monomer conversions considered. In other words the multimodality of the PLP-SEC trace can be used to assess the associated α values (Eqs. 14 and 15).

Complementary to these PLP-SEC studies, SP-PLP-EPR studies have been performed for the determination of the model parameters of the composite k_t-model. These parameters have mostly been determined at low monomer conversions, for example, for methyl methacrylate (Barth and Buback, 2009), *n*-pentyl methacrylate (Nitschke et al., 2020), *n*BuA

(Barth et al., 2010), and VAc and vinyl pivalate (Kattner and Buback, 2014). However, recently, Kattner and Buback also applied the SP-PLP-EPR technique up to high monomer conversions, considering bulk styrene polymerization (Kattner and Buback, 2017). Alternatively, also SP-PLP-RAFT experiments have been considered to study chain length dependent termination (Buback et al., 2005). Furthermore, also information on the mode of termination, *i.e.*, recombination or disproportionation, has been obtained from PLP-SEC experiments (Ballard et al., 2016; Nikitin and Hutchinson, 2007; Sarnecki and Schweer, 1995). In particular, Nikitin and coworkers developed a methodology to obtain the fraction of recombination and disproportionation *via* analysis of SEC traces obtained in the high termination limit (Nikitin et al., 2013; Nikitin and Hutchinson, 2007).

The above formulated insights thus indicate that PLP-SEC analysis can provide valuable complementary information to strengthen other experimental techniques for the determination of apparent termination rate coefficients focusing more at higher monomer conversion dependencies such as the reversible addition-fragmentation chain transfer—chain length dependent—termination (RAFT-CLD-T) technique (Derboven et al., 2015; Johnston-Hall et al., 2007, 2008). Specifically it allows to facilitate the extrapolation toward the regions of high chain lengths at low monomer conversions as difficult to access for these techniques but automatically encountered during PLP experiments.

6. Conclusions and perspectives

The toolbox to interpret PLP-SEC experiments has evolved from the early 1970s analytic and heuristic models for basic reaction schemes to advanced stochastic kinetic models involving many side reactions nowadays, even fully accounting for chain length dependent propagation and (apparent) termination kinetics. The advent of PLP setups with a broad ν range has additionally allowed clever design of experimental PLP conditions to suppress side reactions and to focus on the step-wise determination and regression of kinetic parameters in different temperature regimes. Ever increasing computational power has further allowed a stochastic prediction of discrete reaction events and to test hypotheses to understand and accurately treat experimental PLP data. All these advances combined have enabled to expand the focus on inflection point data to a full-fledged exploitation of the complete PLP-SEC trace, leading to a wealth of kinetic and PLP information. This chapter specifically highlighted the potential of the complete PLP-SEC trace for the determination of the quantum yield, chain initiation reactivities of the photo-initiator radical fragments, and apparent chain length dependent termination rate coefficients at low monomer conversions as encountered in PLP.

Many challenges although remain such as (i) the detailed understanding of the effect of the reaction environment on the intrinsic reactivities specifically in situations with complex interactions (*e.g.*, hydrogen bonding) and (ii) the quantification of kinetic parameters in radical copolymerization of less common comonomer pairs and in any case the kinetic parameters of the related copolymerization side reactions. Interestingly, as illustrated in the present chapter, is here the synergy with *ab initio* calculations. For more complex reaction schemes we can plug in certain rate coefficients as obtained by such calculations with the ultimate aim of a complete first principles based kinetic description. Such detailed insights will also pave the way for the construction of very fundamental chemical structure–reactivity relationships.

References

Achilias DS: A review of modeling of diffusion controlled polymerization reactions, *Macromol Theory Simul* 16:319–347, 2007.

Aleksandrov A, Genkin VN, Kitaĭ M, et al.: Kinetics of laser-initiated polymerization and molecular-weight distribution of the resultant polymer, *Sov J Quantum Electron* 7:547–550, 1977.

Allonas X, Lalevee J, Fouassier JP: Investigation of cleavage processes in photoinitiators: from experiments to molecular modeling, *J Photochem Photobiol A* 159:127–133, 2003.

Asua JM, editor: *Polymer reaction engineering*, Oxford, 2007, Blackwell Publishing Ltd.

Asua JM, Beuermann S, Buback M, et al.: Critically evaluated rate coefficients for free-radical polymerization, 5—propagation rate coefficient for butyl acrylate, *Macromol Chem Phys* 205:2151–2160, 2004.

Ballard N, Asua JM: Radical polymerization of acrylic monomers: an overview, *Prog Polym Sci* 79:40–60, 2018.

Ballard N, Hamzehlou S, Ruipérez F, et al.: On the termination mechanism in the radical polymerization of acrylates, *Macromol Rapid Commun* 37:1364–1368, 2016.

Barner-Kowollik C, Russell GT: Chain-length-dependent termination in radical polymerization: subtle revolution in tackling a long-standing challenge, *Prog Polym Sci* 34:1211–1259, 2009.

Barner-Kowollik C, Vana P, Davis TP: Laser-induced decomposition of 2,2-dimethoxy-2-phenylacetophenone and benzoin in methyl methacrylate homopolymerization studied via matrix-assisted laser desorption/ionization time-of-flight mass spectrometry, *J Polym Sci A Polym Chem* 40:675–681, 2002.

Barner-Kowollik C, Gunzler F, Junkers T: Pushing the limit: pulsed laser polymerization of n-butyl acrylate at 500 hz, *Macromolecules* 41:8971–8973, 2008.

Barner-Kowollik C, Beuermann S, Buback M, et al.: Critically evaluated rate coefficients in radical polymerization—7. Secondary-radical propagation rate coefficients for methyl acrylate in the bulk, *Polym Chem* 5:204–212, 2014.

Barner-Kowollik C, Beuermann S, Buback M, et al.: Critically evaluated rate coefficients in radical polymerization—8. Propagation rate coefficients for vinyl acetate in bulk, *Macromol Chem Phys* 218:1600357, 2017.

Barth J, Buback M: SP-PLP-EPR investigations into the chain-length-dependent termination of methyl methacrylate bulk polymerization, *Macromol Rapid Commun* 30: 1805–1811, 2009.

Barth J, Buback M, Hesse P, et al.: EPR analysis of n-butyl acrylate radical polymerization, *Macromol Rapid Commun* 30:1969–1974, 2009.

Barth J, Buback M, Hesse P, et al.: Termination and transfer kinetics of butyl acrylate radical polymerization studied via SP-PLP-EPR, *Macromolecules* 43:4023–4031, 2010.

Beuermann S: Solvent influence on propagation kinetics in radical polymerizations studied by pulsed laser initiated polymerizations, *Macromol Rapid Commun* 30:1066–1088, 2009.

Beuermann S, Buback M: Rate coefficients of free-radical polymerization deduced from pulsed laser experiments, *Prog Polym Sci* 27:191–254, 2002.

Beuermann S, Buback M, Davis TP, et al.: Critically evaluated rate coefficients for free-radical polymerization, 2. Propagation rate coefficients for methyl methacrylate, *Macromol Chem Phys* 198:1545–1560, 1997.

Beuermann S, Buback M, Davis TP, et al.: Critically evaluated rate coefficients for free-radical polymerization, 3—propagation rate coefficients for alkyl methacrylates, *Macromol Chem Phys* 201:1355–1364, 2000.

Beuermann S, Buback M, Davis TP, et al.: Critically evaluated rate coefficients for free-radical polymerization, 4—propagation rate coefficients for methacrylates with cyclic ester groups, *Macromol Chem Phys* 204:1338–1350, 2003.

Beuermann S, Buback M, Hesse P, et al.: Free-radical propagation rate coefficient of non-ionized methacrylic acid in aqueous solution from low monomer concentrations to bulk polymerization, *Macromolecules* 39:184–193, 2006.

Beuermann S, Buback M, Hesse P, et al.: Critically evaluated rate coefficients for free-radical polymerization. Part 6: propagation rate coefficient of methacrylic acid in aqueous solution, *Pure Appl Chem* 79:1463–1469, 2007.

Billiani J, Rois G, Lederer K: A new procedure for simultaneous calibration of separation and axial-dispersion in sec, *Chromatographia* 26:372–376, 1988.

Buback M: Solvent effects on acrylate k(p) in organic media: a response, *Macromol Rapid Commun* 36:1979–1983, 2015.

Buback M, Kurz CH: Free-radical propagation rate coefficients for cyclohexyl methacrylate, glycidyl methacrylate and 2-hydroxyethyl methacrylate homopolymerizations, *Macromol Chem Phys* 199:2301–2310, 1998.

Buback M, Lämmel RA: Termination rate coefficients derived from PLP-SEC data, *Macromol Theory Simul* 7:197–202, 1998.

Buback M, Hippler H, Schweer J, et al.: Time-resolved study of laser-induced high-pressure ethylene polymerization, *Makromol Chem Rapid* 7:261–265, 1986.

Buback M, Gilbert RG, Hutchinson RA, et al.: Critically evaluated rate coefficients for free-radical polymerization, 1. Propagation rate coefficient for styrene, *Macromol Chem Phys* 196:3267–3280, 1995.

Buback M, Busch M, Lammel RA: Modeling of molecular weight distribution in pulsed laser free-radical homopolymerizations, *Macromol Theory Simul* 5:845–861, 1996.

Buback M, Geers U, Kurz CH: Propagation rate coefficients in free-radical homopolymerizations of butyl methacrylate and dodecyl methacrylate, *Macromol Chem Phys* 198:3451–3464, 1997.

Buback M, Busch M, Kowollik C: Chain-length dependence of free-radical termination rate deduced from laser single-pulse experiments, *Macromol Theory Simul* 9:442–452, 2000.

Buback M, Junkers T, Vana P: Laser single pulse initiated raft polymerization for assessing chain-length dependent radical termination kinetics, *Macromol Rapid Commun* 26: 796–802, 2005.

Buback M, Hesse P, Lacik I: Propagation rate coefficient and fraction of mid-chain radicals for acrylic acid polymerization in aqueous solution, *Macromol Rapid Commun* 28: 2049–2054, 2007.

Castignolles P, Nikitin AN, Couvreur L, et al.: Pulsed laser polymerization of alkyl acrylates: potential effects of the oxygen presence and high laser power, *Macromol Chem Phys* 207:81–89, 2006.

Constales D, Yablonsky GS, D'hooge DR, et al.: Chapter 10—polymers: design and production, In Constales D, Yablonsky GS, D'hooge DR, et al., editors: *Advanced data analysis & modelling in chemical engineering*, Amsterdam, 2017, Elsevier, pp 307–349.

Coote ML, Davis TP: The mechanism of the propagation step in free-radical copolymerisation, *Prog Polym Sci* 24:1217–1251, 1999.

Coote ML, Zammit MD, Davis TP, et al.: Copolymerization propagation kinetics of styrene and methyl methacrylate—revisited. 1. Pulsed laser polymerization study, *Macromolecules* 30:8182–8190, 1997.

Cuccato D, Mavroudakis E, Dossi M, et al.: A density functional theory study of secondary reactions in n-butyl acrylate free radical polymerization, *Macromol Theory Simul* 22: 127–135, 2013.

Davis TP, O'Driscoll KF, Piton MC, et al.: Determination of propagation rate constants for the copolymerization of methylmethacrylate and styrene using a pulsed laser technique, *J Polym Sci, Part C: Polym Lett* 27:181–185, 1989.

Davis TP, Odriscoll KF, Piton MC, et al.: Copolymerization propagation kinetics of styrene with alkyl methacrylates, *Macromolecules* 23:2113–2119, 1990.

De Roo T, Wieme J, Heynderickx GJ, et al.: Estimation of intrinsic rate coefficients in vinyl chloride suspension polymerization, *Polymer* 46:8340–8354, 2005.

Deglmann P, Hungenberg K-D, Vale HM: Dependence of propagation rate coefficients in radical polymerization on solution properties: a quantitative thermodynamic interpretation, *Macromol React Eng* 12:1800010, 2018.

Derboven P, D'hooge DR, Stamenovic MM, et al.: Kinetic modeling of radical thiol-ene chemistry for macromolecular design: importance of side reactions and diffusional limitations, *Macromolecules* 46:1732–1742, 2013.

Derboven P, D'hooge DR, Reyniers M-F, et al.: The long and the short of radical polymerization, *Macromolecules* 48:492–501, 2015.

Desmet GB, Marien YW, Van Steenberge PHM, et al.: Ab initio based kinetic Monte Carlo analysis to unravel the propagation kinetics in vinyl acetate pulsed laser polymerization, *Polym Chem* 8:7143–7150, 2017.

Desmet GB, De Rybel N, Van Steenberge PHM, et al.: Ab-initio-based kinetic modeling to understand raft exchange: the case of 2-cyano-2-propyl dodecyl trithiocarbonate and styrene, *Macromol Rapid Commun* 39:1700403, 2018.

D'hooge DR, Reyniers MF, Stadler FJ, et al.: Atom transfer radical polymerization of isobornyl acrylate: a kinetic modeling study, *Macromolecules* 43:8766–8781, 2010.

D'hooge DR, Van Steenberge PHM, Derboven P, et al.: Model-based design of the polymer microstructure: bridging the gap between polymer chemistry and engineering, *Polym Chem* 6:7081–7096, 2015.

D'hooge DR, Van Steenberge PHM, Reyniers M-F, et al.: The strength of multi-scale modeling to unveil the complexity of radical polymerization, *Prog Polym Sci* 58:59–89, 2016.

Drawe P, Buback M: The PLP-SEC method: perspectives and limitations, *Macromol Theory Simul* 25:74–84, 2016.

Fierens SK, Van Steenberge PHM, Reyniers M-F, et al.: How penultimate monomer unit effects and initiator influence ICAR ATRP of n-butyl acrylate and methyl methacrylate, *AICHE J* 63:4971–4986, 2017.

Fierens SK, Van Steenberge PHM, Reyniers MF, et al.: Analytical and advanced kinetic models for characterization of chain-growth copolymerization: the state-of-the-art, *React Chem Eng* 3:128–145, 2018.

Fischer H, Baer R, Hany R, et al.: 2,2-dimethoxy-2-phenylacetophenone—photochemistry and free-radical photofragmentation, *J Chem Soc Perkin Trans* 2:787–798, 1990.

Flory PJ, Leutner FS: Occurrence of head-to-head arrangements of structural units in polyvinyl alcohol, *J Polym Sci* 3:880–890, 1948.

Fukuda T, Ma YD, Inagaki H: Free-radical copolymerization. 3. Determination of rate constants of propagation and termination for the styrene methyl-methacrylate system—a critical test of terminal-model kinetics, *Macromolecules* 18:17–26, 1985.

Ganachaud F, Balic R, Monteiro MJ, et al.: Propagation rate coefficient of poly (n-isopropylacrylamide) in water below its lower critical solution temperature, *Macromolecules* 33:8589–8596, 2000.

Grubisic Z, Rempp P, Benoit H: A universal calibration for gel permeation chromatography, *J Polym Sci Part B: Polym Lett* 5:753–759, 1967.

Haehnel AP, Schneider-Baumann M, Hiltebrandt KU, et al.: Global trends for k(p)? Expanding the frontier of ester side chain topography in acrylates and methacrylates, *Macromolecules* 46:15–28, 2013.

Haehnel AP, Schneider-Baumann M, Arens L, et al.: Global trends for k(p)? The influence of ester side chain topography in alkyl (meth)acrylates—completing the data base, *Macromolecules* 47:3483–3496, 2014a.

Haehnel AP, Wenn B, Kockler K, et al.: Solvent effects on acrylate k(p) in organic media?—a systematic PLP-SEC study, *Macromol Rapid Commun* 35:2029–2037, 2014b.

Haehnel AP, Wenn B, Kockler K, et al.: Solvent effects on kp in organic media?: statement to the response, *Macromol Rapid Commun* 36:1984–1986, 2015.

Hamzehlou S, Ballard N, Reyes Y, et al.: Analyzing the discrepancies in the activation energies of the backbiting and beta-scission reactions in the radical polymerization of n-butyl acrylate, *Polym Chem* 7:2069–2077, 2016.

Harrisson S, Mackenzie SR, Haddleton DM: Unprecedented solvent-induced acceleration of free-radical propagation of methyl methacrylate in ionic liquids, *Chem Commun, 2*,:2850–2851, 2002.

Harrisson S, Mackenzie SR, Haddleton DM: Pulsed laser polymerization in an ionic liquid: strong solvent effects on propagation and termination of methyl methacrylate, *Macromolecules* 36:5072–5075, 2003.

Haven JJ, Zaquen N, Rubens M, et al.: The kinetics of n-butyl acrylate radical polymerization revealed in a single experiment by real time on-line mass spectrometry monitoring, *Macromol React Eng* 11:1700016, 2017.

Heuts JPA, Russell GT: The nature of the chain-length dependence of the propagation rate coefficient and its effect on the kinetics of free-radical polymerization. 1. Small-molecule studies, *Eur Polym J* 42:3–20, 2006.

Houwink R: Zusammenhang zwischen viscosimetrisch und osmotisch bestimmten Polymerisationsgraden bei Hochpolymeren, *J Prakt Chem* 157:15, 1940.

Hutchinson RA, Richards JR, Aronson MT: Determination of propagation rate coefficients by pulsed-laser polymerization for systems with rapid chain growth—vinyl-acetate, *Macromolecules* 27:4530–4537, 1994.

Hutchinson RA, Beuermann S, Paquet DA, et al.: Determination of free-radical propagation rate coefficients for alkyl methacrylates by pulsed-laser polymerization, *Macromolecules* 30:3490–3493, 1997a.

Hutchinson RA, McMinn JH, Paquet DA, et al.: A pulsed-laser study of penultimate copolymerization propagation kinetics for methyl methacrylate n-butyl acrylate, *Ind Eng Chem Res* 36:1103–1113, 1997b.

Hutchinson RA, Beuermann S, Paquet DA, et al.: Determination of free-radical propagation rate coefficients for cycloalkyl and functional methacrylates by pulsed-laser polymerization, *Macromolecules* 31:1542–1547, 1998.

Idowu LA, Hutchinson RA: Solvent effects on radical copolymerization kinetics of 2-hydroxyethyl methacrylate and butyl methacrylate, *Polymers* 11:18, 2019.

Jeličić A, Beuermann S, García N: Influence of ionic liquid structure on the propagation kinetics of methyl methacrylate, *Macromolecules* 42:5062–5072, 2009a.

Jeličić A, García N, Löhmannsröben H-G, et al.: Prediction of the ionic liquid influence on propagation rate coefficients in methyl methacrylate radical polymerizations based on kamlet–taft solvatochromic parameters, *Macromolecules* 42:8801–8808, 2009b.

Johnston-Hall G, Monteiro MJ: Bimolecular radical termination: new perspectives and insights, *J Polym Sci A Polym Chem* 46:3155–3173, 2008.

Johnston-Hall G, Stenzel MH, Davis TP, et al.: Chain length dependent termination rate coefficients of methyl methacrylate (MMA) in the gel regime: accessing k(t)(i,i) using reversible addition-fragmentation chain transfer (RAFT) polymerization, *Macromolecules* 40:2730–2736, 2007.

Johnston-Hall G, Barner-Kowollik C, Monteiro MJ: Design criteria for accurate measurement of bimolecular radical termination rate coefficients via the RAFT-CLD-T method(a), *Macromol Theory Simul* 17:460–469, 2008.

Junkers T, Barner-Kowollik C: The role of mid-chain radicals in acrylate free radical polymerization: branching and scission, *J Polym Sci A Polym Chem* 46:7585–7605, 2008.

Junkers T, Barner-Kowollik C: Optimum reaction conditions for the synthesis of macromonomers via the high-temperature polymerization of acrylates, *Macromol Theory Simul* 18:421–433, 2009.

Junkers T, Voll D, Barner-Kowollik C: Determination of vinyl acetate propagation rate coefficients via high frequency pulsed laser polymerization, *e-Polymers* 9:076, 2009.

Kattner H, Buback M: Chain-length-dependent termination of vinyl acetate and vinyl pivalate bulk homopolymerizations studied by SP-PLP-EPR, *Macromolr Chem Phys* 215:1180–1191, 2014.

Kattner H, Buback M: Chain-length-dependent termination of styrene bulk homopolymerization studied by SP-PLP-EPR, *Macromolecules* 48:309–315, 2015.

Kattner H, Buback M: Chain-length-dependent termination of styrene bulk polymerization up to high degrees of monomer conversion, *Macromolecules* 50:5308–5314, 2017.

Kiparissides C, Macgregor JF, Hamielec AE: Continuous emulsion polymerization of vinyl acetate. Part: II parameter estimation and simulation studies, *Can J Chem Eng* 58:56–64, 1980.

Kockler KB, Haehnel AP, Fleischhaker F, et al.: No apparent correlation of k(p) with steric hindrance for branched acrylates, *Macromol Chem Phys* 216:1573–1582, 2015.

Kockler KB, Fleischhaker F, Barner-Kowollik C: Investigating the propagation kinetics of a novel class of nitrogen-containing methacrylates via PLP-SEC, *Polym Chem* 7:4342–4351, 2016a.

Kockler KB, Haehnel AP, Junkers T, et al.: Determining free-radical propagation rate coefficients with high-frequency lasers: current status and future perspectives, *Macromol Rapid Commun* 37:123–134, 2016b.

Kuchta F-D, van Herk AM, German AL: Propagation kinetics of acrylic and methacrylic acid in water and organic solvents studied by pulsed-laser polymerization, *Macromolecules* 33:3641–3649, 2000.

Lacík I, Beuermann S, Buback M: Aqueous phase size-exclusion-chromatography used for PLP-SEC studies into free-radical propagation rate of acrylic acid in aqueous solution, *Macromolecules* 34:6224–6228, 2001.

Lacík I, Beuermann S, Buback M: PLP-SEC study into free-radical propagation rate of non-ionized acrylic acid in aqueous solution, *Macromolecules* 36:9355–9363, 2003.

Lacík I, Učňová L, Kukučková S, et al.: Propagation rate coefficient of free-radical polymerization of partially and fully ionized methacrylic acid in aqueous solution, *Macromolecules* 42:7753–7761, 2009.

Laki S, Shamsabadi A, Riazi H, et al.: Experimental and mechanistic modeling study of self-initiated high-temperature polymerization of ethyl acrylate, *Ind Eng Chem Res* 59:2621–2630, 2020.

Lena J-B, Deschamps M, Sciortino NF, et al.: Effects of chain transfer agent and temperature on branching and β-scission in radical polymerization of 2-ethylhexyl acrylate, *Macromol Chem Phys* 219:1700579, 2018.

Li D, Li N, Hutchinson RA: High-temperature free radical copolymerization of styrene and butyl methacrylate with depropagation and penultimate kinetic effects, *Macromolecules* 39:4366–4373, 2006.

Liang K, Dossi M, Moscatelli D, et al.: An investigation of free-radical copolymerization propagation kinetics of styrene and 2-hydroxyethyl methacrylate, *Macromolecules* 42:7736–7744, 2009.

Liu S, Srinivasan S, Grady MC, et al.: Backbiting and β-scission reactions in free-radical polymerization of methyl acrylate, *Int J Quantum Chem* 114:345–360, 2014.

Marien YW, Van Steenberge PHM, Kockler KB, et al.: An alternative method to estimate the bulk backbiting rate coefficient in acrylate radical polymerization, *Polym Chem* 7:6521–6528, 2016.

Marien YW, Van Steenberge PHM, Barner-Kowollik C, et al.: Kinetic Monte Carlo modeling extracts information on chain initiation and termination from complete PLP-SEC traces, *Macromolecules* 50:1371–1385, 2017a.

Marien YW, Van Steenberge PHM, Kockler KB, et al.: Estimating the photodissociation quantum yield from PLP-SEC peak heights, *Polym Chem* 8:3124–3128, 2017b.

Mark H: In Sangster R, editor: *Der feste korper*, Leipzig, 1938, Hirzel.

Mark J, NGia K, Graessley W, et al.: *Physical properties of polymers*, 2003, Cambridge University Press.

Mastan E, Li X, Zhu S: Modeling and theoretical development in controlled radical polymerization, *Prog Polym Sci* 45:71–101, 2015.

Matyjaszewski K, Davies TP, editors: *Handbook of radical polymerization*, Hoboken, 2002, John Wiley & Sons, Inc.

Mavroudakis E, Cuccato D, Moscatelli D: On the use of quantum chemistry for the determination of propagation, copolymerization, and secondary reaction kinetics in free radical polymerization, *Polymers* 7:1789–1819, 2015.

Monyatsi O, Nikitin AN, Hutchinson RA: Effect of head-to-head addition on vinyl acetate propagation kinetics in radical polymerization, *Macromolecules* 47:8145–8153, 2014.

Nesvadba P: Radical polymerization in industry, In *Encyclopedia of radicals in chemistry, biology and materials*, 2012, Wiley.

Nikitin AN, Evseev AV: Kinetics of free-radical polymerization with chain-length-dependent constants under initiation by laser pulses: effect of chain length dependence of propagation, *Macromol Theory Simul* 8:296–308, 1999.

Nikitin AN, Hutchinson RA: Determination of the mode of free radical termination from pulsed laser polymerization experiments, *Macromol Theory Simul* 16:29–42, 2007.

Nikitin AN, Castignolles P, Charleux B, et al.: Determination of propagation rate coefficient of acrylates by pulsed-laser polymerization in the presence of intramolecular chain transfer to polymer, *Macromol Rapid Commun* 24:778–782, 2003.

Nikitin AN, Hutchinson RA, Buback M, et al.: Determination of intramolecular chain transfer and midchain radical propagation rate coefficients for butyl acrylate by pulsed laser polymerization, *Macromolecules* 40:8631–8641, 2007.

Nikitin AN, Hutchinson RA, Wang W, et al.: Effect of intramolecular transfer to polymer on stationary free-radical polymerization of alkyl acrylates, 5—consideration of solution polymerization up to high temperatures, *Macromol React Eng* 4:691–706, 2010.

Nikitin AN, Hutchinson RA, Hesse P: Determination of the mode of radical termination from pulsed laser polymerization experiments in the presence of retardation and chain transfer to agent, *Macromol Chem Phys* 214:2670–2682, 2013.

Nikitin AN, Lacík I, Hutchinson RA: A 3D simulation investigation of the influence of temperature increases on the accuracy of propagation rate coefficients determined by pulsed-laser polymerization, *Macromolecules* 49:9320–9335, 2016.

Nikitin AN, Lacık I, Hutchinson RA, et al.: Detection of PLP structure for accurate determination of propagation rate coefficients over an enhanced range of PLP-SEC conditions, *Macromolecules* 52:55–71, 2019.
Nitschke A, Riemann L, Kollenbach L, et al.: Investigation into the kinetics of n-pentyl methacrylate radical polymerization, *Macromol Chem Phys* 221:1900345, 2020.
Noble BB, Coote ML: First principles modelling of free-radical polymerisation kinetics, *Int Rev Phys Chem* 32:467–513, 2013.
Noble BB, Mater AC, Smith LM, et al.: The effects of Lewis acid complexation on type i radical photoinitiators and implications for pulsed laser polymerization, *Polym Chem* 7:6400–6412, 2016.
Odian G: *Principles of polymerization*, 2004, Wiley.
Olaj OF, Schnöll-Bitai I: Laser-flash-initiated polymerization as a tool for evaluating (individual) kinetic constants of free-radical polymerization—5. Complete analysis by means of a single experiment, *Eur Polym J* 25:635–641, 1989.
Olaj OF, Bitai I, Hinkelmann F: The laser-flash-initiated polymerization as a tool of evaluating (individual) kinetic constants of free-radical polymerization, 2. The direct determination of the rate-constant of chain propagation, *Macromol Chem Phys* 188:1689–1702, 1987.
Olaj OF, Vana P, Zoder M, et al.: Is the rate constant of chain propagation kp in radical polymerization really chain-length independent? *Macromol Rapid Commun* 21:913–920, 2000.
Payne KA, D'Hooge DR, van Steenberge PHM, et al.: ARGET ATRP of butyl methacrylate: utilizing kinetic modeling to understand experimental trends, *Macromolecules* 46:3828–3840, 2013.
Peck ANF, Hutchinson RA: Secondary reactions in the high-temperature free radical polymerization of butyl acrylate, *Macromolecules* 37:5944–5951, 2004.
Peklak AD, Butte A, Storti G, et al.: Gel effect in the bulk reversible addition-fragmentation chain transfer polymerization of methyl methacrylate: modeling and experiments, *J Polym Sci A Polym Chem* 44:1071–1085, 2006.
Plessis C, Arzamendi G, Leiza JR, et al.: Modeling of seeded semibatch emulsion polymerization of n-BA, *Ind Eng Chem Res* 40:3883–3894, 2001.
Plessis C, Arzamendi G, Alberdi JM, et al.: Evidence of branching in poly(butyl acrylate) produced in pulsed-laser polymerization experiments, *Macromol Rapid Commun* 24:173–177, 2003.
Purmova J, Pauwels KFD, Agostini M, et al.: Experimental and theoretical evaluation of the reactions leading to formation of internal double bonds in suspension PVC, *Macromolecules* 41:5527–5539, 2008.
Ricci RW, Ditzler MA, Nestor LP: Discovering the beer-lambert law, *J Chem Educ* 71:983–985, 1994.
Rooney TR, Hutchinson RA: Monomer structure and solvent effects on copolymer composition in (meth)acrylate radical copolymerization, *Ind Eng Chem Res* 57: 5215–5227, 2018.
Ruiperez F: Application of quantum chemical methods in polymer chemistry, *Int Rev Phys Chem* 38:343–403, 2019.
Saldívar-Guerra E: Polymer reaction engineering, In *Encyclopedia of polymer science and technology*, 2019, Wiley, pp 1–45.
Sarnecki J, Schweer J: Conditions for the determination of precise and accurate free-radical propagation rate coefficients from pulsed-laser-made polymer, *Macromolecules* 28:4080–4088, 1995.
Schier JES, Cohen-Sacal D, Hutchinson RA: Hydrogen bonding in radical solution copolymerization kinetics of acrylates and methacrylates: a comparison of hydroxy-and methoxy-functionality, *Polym Chem* 8:1943–1952, 2017.
Schweer J: Penultimate model description of the propagation kinetics for the free-radical copolymerization of styrene and methyl-methacrylate, *Makromol Chem Theory Simul* 2:485–502, 1993.

Shipp DA, Solomon DH, Smith TA, et al.: Chain length dependence of radical-radical termination in free radical polymerization: a pulsed laser photolysis investigation, *Macromolecules* 36:2032–2040, 2003.

Smoluchowski M: Versuch einer mathematischen Theorie der Koagulationskinetik kolloider Lösungen, *Z Phys Chem* 92:129, 1917.

Soroush M, Rappe AM: Chapter 5—theoretical insights into chain transfer reactions of acrylates, In Soroush M, editor: *Computational quantum chemistry*, 2019, Elsevier, pp 135–193.

Stach M, Lacík I, Chorvát D, et al.: Propagation rate coefficient for radical polymerization of N-vinyl pyrrolidone in aqueous solution obtained by PLP-SEC, *Macromolecules* 41:5174–5185, 2008.

Stach M, Lacík I, Kasák P, et al.: Free-radical propagation kinetics of N-vinyl formamide in aqueous solution studied by PLP-SEC, *Macromol Chem Phys* 211:580–593, 2010.

Szablan Z, Huaming M, Adler M, et al.: Depolymerization kinetics of di(4-tert-butyl cyclohexyl) itaconate and Mark-Houwink-Kuhn-Sakurada parameters of di(4-tert-butyl cyclohexyl) itaconate and di-n-butyl itaconate, *J Polym Sci A Polym Chem* 45:1931–1943, 2007a.

Szablan Z, Junkers T, Koo SPS, et al.: Mapping photolysis product radical reactivities via soft ionization mass spectrometry in acrylate, methacrylate, and itaconate systems, *Macromolecules* 40:6820–6833, 2007b.

Szablan Z, Lovestead TM, Davis TP, et al.: Mapping free radical reactivity: a high-resolution electrospray ionization-mass spectrometry study of photoinitiation processes in methyl methacrylate free radical polymerization, *Macromolecules* 40:26–39, 2007c.

Tung LH, Runyon JR: Calibration of instrumental spreading for GPC, *J Appl Polym Sci* 13:2397–2409, 1969.

Vachaudez M, D'hooge DR, Socka M, et al.: Inverse dependencies on the polymerization rate in atom transfer radical polymerization of N-isopropylacrylamide in aqueous medium, *React Funct Polym* 73:484–491, 2013.

Van Herk AM: Pulsed initiation polymerization as a means of obtaining propagation rate coefficients in free-radical polymerizations. II review up to 2000, *Macromol Theory Simul* 9:433–441, 2000.

Van Steenberge PHM, Vandenbergh J, D'Hooge DR, et al.: Kinetic Monte Carlo modeling of the sulfinyl precursor route for poly(p-phenylene vinylene) synthesis, *Macromolecules* 44:8716–8726, 2011.

Vir AB, Marien YW, Van Steenberge PHM, et al.: Access to the β-scission rate coefficient in acrylate radical polymerization by careful scanning of pulse laser frequencies at elevated temperature, *React Chem Eng* 3:807–815, 2018.

Vir AB, Marien YW, Van Steenberge PHM, et al.: From n-butyl acrylate Arrhenius parameters for backbiting and tertiary propagation to β-scission via stepwise pulsed laser polymerization, *Polym Chem* 10:4116–4125, 2019.

Wang W, Hutchinson RA: PLP/SEC/NMR study of free radical copolymerization of styrene and glycidyl methacrylate, *Macromolecules* 41:9011–9018, 2008.

Wenn B, Junkers T: Kilohertz pulsed-laser-polymerization: simultaneous determination of backbiting, secondary, and tertiary radical propagation rate coefficients for tert-butyl acrylate, *Macromol Rapid Commun* 37:781–787, 2016.

Willemse RXE, van Herk AM: Determination of propagation rate coefficients of a family of acrylates with PLP-MALDI-ToF-MS, *Macromol Chem Phys* 211:539–545, 2010.

Willemse RXE, van Herk AM, Panchenko E, et al.: PLP-ESR monitoring of midchain radicals in n-butyl acrylate polymerization, *Macromolecules* 38:5098–5103, 2005.

Woecht I, Schmidt-Naake G, Beuermann S, et al.: Propagation kinetics of free-radical polymerizations in ionic liquids, *J Polym Sci A Polym Chem* 46:1460–1469, 2008.

Yu X, Broadbelt LJ: Kinetic study of 1,5-hydrogen transfer reactions of methyl acrylate and butyl acrylate using quantum chemistry, *Macromol Theory Simul* 21:461–469, 2012.

Zammit MD, Coote ML, Davis TP, et al.: Effect of the ester side-chain on the propagation kinetics of alkyl methacrylates—an entropic or enthalpic effect? *Macromolecules* 31:955–963, 1998.

Zetterlund PB, Yamazoe H, Yamada B, et al.: High-conversion free-radical bulk polymerization of styrene: Termination kinetics studied by electron spin resonance, Fourier transform near-infrared spectroscopy, and gel permeation chromatography, *Macromolecules* 34:7686–7691, 2001.

Zhou Y-N, Luo Z-H: State-of-the-art and progress in method of moments for the model-based reversible-deactivation radical polymerization, *Macromol React Eng* 10:516–534, 2016.

CHAPTER FOUR

3D printing as chemical reaction engineering booster

Kristina M. Zentel, Marcel Fassbender, Werner Pauer, and Gerrit A. Luinstra*

Institute of Technical and Macromolecular Chemistry, University of Hamburg, Hamburg, Germany
*Corresponding author: e-mail address: gerrit.luinstra@chemie.uni-hamburg.de

Contents

Abstract

3D printing or additive manufacturing (AM) is a fast expanding technology with manifold possibilities regarding freedom of design and functional prototyping, which is beneficial also in the field of chemical reaction engineering. This review intends to summarize the most frequently applied 3D printing technologies in that regard, the applied printing materials and available equipment, and presents developments in the area. Polymeric materials play a dominant role due to their favorable processing and post-processing options and range of product properties. While the majority of 3D printed products are indeed made of polymeric materials, certain applications, such as heterogeneous catalyst design, require the use of ceramic materials. In chemical engineering, designed and printed objects are successfully applied in form of lab-scale reaction ware, operation units for scale-up approaches as well as to the optimization of reactors and reaction protocols. Significant progress has been made e.g. by coupling chemical functionality with 3D printing, designing smart reactors and exploiting synergetic effects of computational fluid dynamics (CFD) and AM. Current research has demonstrated that developments in chemical engineering can be expanded into new regions as well as accelerated by the use of additive manufacturing technologies.

Advances in Chemical Engineering, Volume 56
ISSN 0065-2377
https://doi.org/10.1016/bs.ache.2020.08.002

Consequently, further applications and developments, especially involving smart materials and (simulation-assisted) innovative design of operation units and reactors can be expected for the future.

1. Introduction

3D printing or additive manufacturing (AM) is receiving increasing attention in many fields: printers can be found at an increasing number of homes (Gibson et al., 2015a), in research institutions (Capel et al., 2018), start-ups as well as established industry (Gunkel, 2020; Stephan, 2019). Applications of the printed objects range from consumer products to research equipment and is particular attractive for remote applications e.g. for space missions (Wong & Pfahnl, 2014). The current Corona Crisis enhanced the visibility of 3D printing and showed one of its inherent merits: its power to produce also complex products, fast and flexible depending on acute demands, especially in the field of plastics was redemonstrated, although the method may be economically not as efficient as e.g. injection molding for mass production of simple parts. At 3D printing farms significant numbers of testing sticks or face shields could be produced easily, without the need of time-consuming preparations or the dependency of supply chains.

The potential of additive manufacturing within chemical engineering is high: AM is expected to lead to a revolution in reaction and separation engineering and thus also may drastically impact chemical industries (Grande et al., 2015). On the one hand heat exchange units as a crucial part of every chemical plant are in focus, as they contain the promise of energy saving (Stark, 2018). Current attention is drawn also by fluidic devices and especially chemical reactors get attention as their geometries and consequently their efficiencies may benefit strongly from the option of additive manufacturing (Capel et al., 2018; Ganneboyina et al., 2018; Guba et al., 2018; Parra-Cabrera et al., 2018). Especially the impact of tailor-made designs (Hein et al., 1997; Taheri et al., 2012) and respective fabrication techniques will change the way reactors and separation units are constructed. At the same time AM offers the option to add chemical functionality to typically inert printing materials and thus give superior performance. It has been summarized how different degrees of freedom, such as matrix materials, blended materials or reactive moieties added were applied successfully after printing with AM or are under investigation in this context (Hartings & Ahmed, 2019). Gradients of chemical composition or mechanical strength were found to be desirable properties in resulting products, as well as materials with high porosity. The items of special interest were heterogeneous catalytic devices, photocatalysts as well as tissue scaffolds. It is referred to adjoining topics, such as flow chemistry (Capel et al., 2013), microfluidics (Au et al., 2016) and engineering (Femmer et al., 2016; Guba et al., 2018) that in combination with the still increasing ease of AM give an additional flavor to the possibilities of speeding up development and comprise a change of paradigm from batch to continuous reaction in research and perhaps more importantly, also in chemical education.

Within the last decade, the interest within industry and science in the topic of additive manufacturing has increased strongly, measured by the number of publications (Fig. 1).

Consequently, a trend within the individual categories relevant for this article is discernible. After expiration of some important patents (Batchelder John, 1995; Deckard Carl, 1986, 1994; Hull Charles, 1984; Smalley Dennis et al., 1994), the development of individual and cost-effective machines (RepRap.org, 2019) and materials could be started. As the industrially available equipment at that time was beyond the usual budget of a university or a start-up company, a trend of publications within the categories 'Instruments & Instrumentations', 'Polymer Science' and 'Materials Science' can be explained. There is the new and ongoing development of methods and technologies for the additive processing of materials beyond the state of the art. This is driven by the demand for materials with an enhanced spectrum of properties (including tailored additives). Since the above mentioned application fields have become a basis within research in recent years, it is to be expected that the trends within the individual possible application sectors such as chemistry and physics will increase faster than before. The protection of new technologies, processes (Fig. 2, Patent Barplot 23%) and materials (Fig. 2, Patent Barplot 17%) will continue to result into more and more specific patent applications. Science communication—also with respect to AM—is mainly provided in form of journal articles and conference contributions. This communication may be reflected in the distribution of publications by document type, where journal articles rank second with 19%. Within the journal articles, the distribution of publications is mainly related to materials science (9%), which is considered an interdisciplinary field and combines chemistry (4%) and physics (6%) as well as life sciences. The authors forecast further protection of

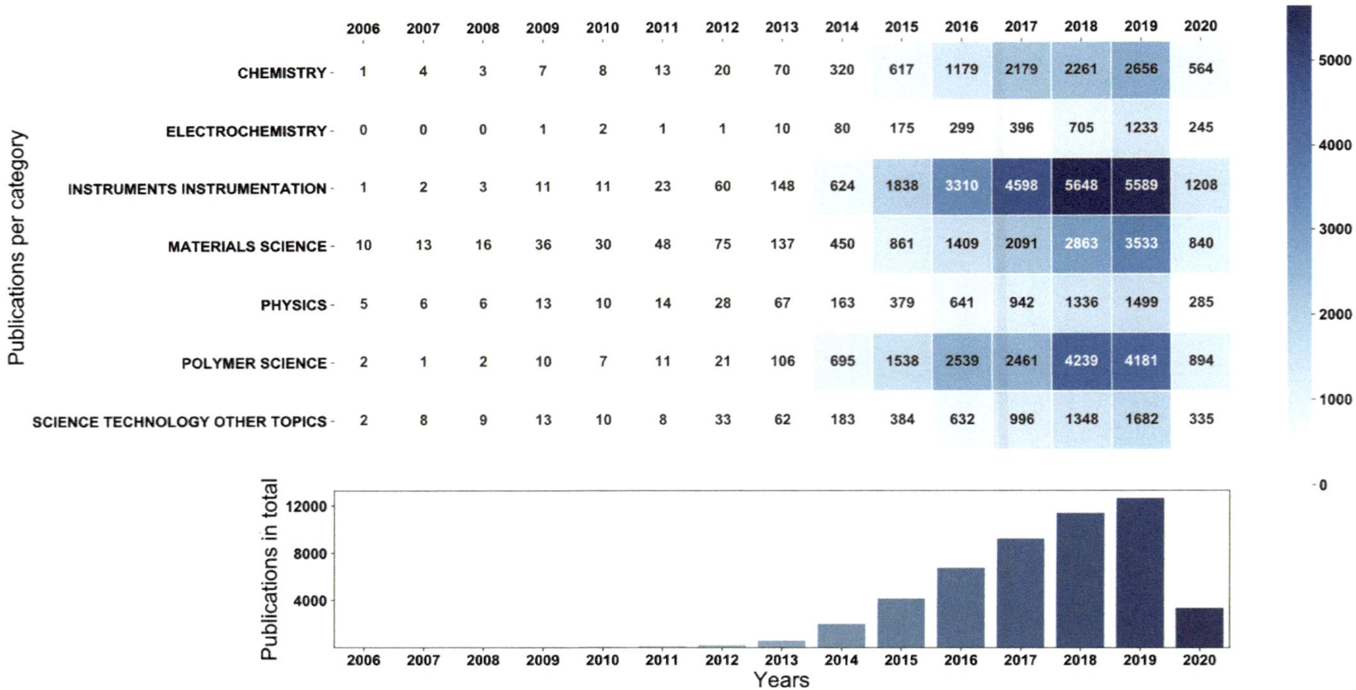

Fig. 1 (Top) Publications per category broken down by year, for the search term "3d print*" on the platform Web of Science (Status as of 28.04.2020), displayed as heatmap. (Bottom) The resulting total numbers of all available categories for the search term "3d print*" on the Platform Web of Science, broken down by year, displayed as barplot.

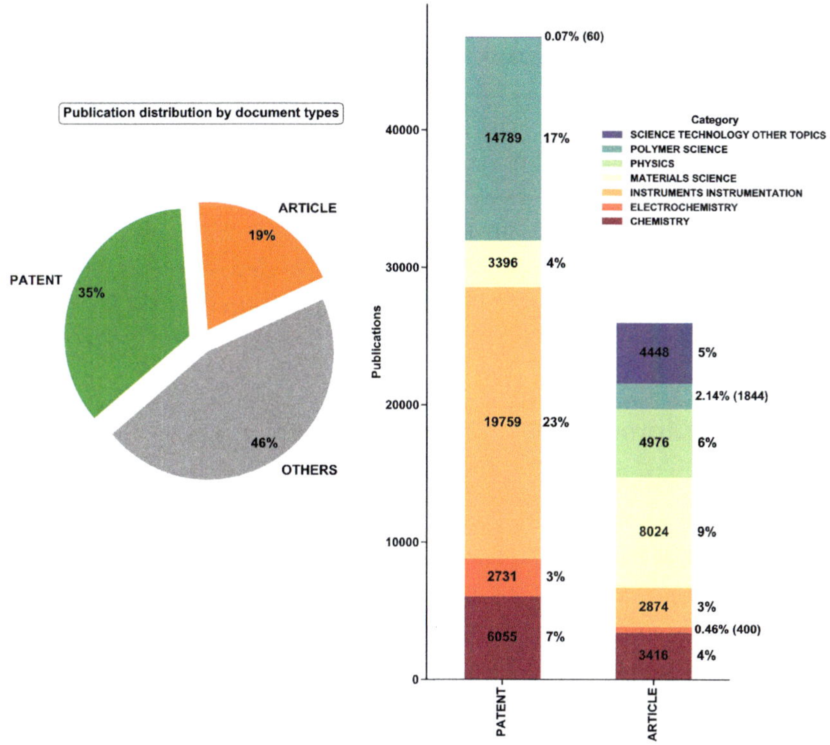

Fig. 2 (Left) Distribution of publications according to document types, for the search term "3d print*" on the platform Web of Science (Status as of 28.04.2020), presented as a pie chart. (Right) The distribution of publications within selected categories of the document types: Patent and journal articles represented as stacked barplot.

technical inventions and utility models leading to patent applications in the coming years, but also an increasing trend within polymer and materials research. The fields of application for the use of additive manufactured products are widening with the growing range of available materials and compounds, as well as the methodological possibilities of 3D and 4D printing.

Polymers play an essential role in 3D printing, as they are the material class applied predominantly as printing material (Ligon et al., 2017). They are majorly used in fused deposition modeling (FDM), selective laser and sintering (SLS), stereolithography (SLA) and direct ink writing (DIW), but can also be found in less widely used printing techniques (Ligon et al., 2017). There are several reasons why polymers play the key role in AM. Besides an easy processing behavior and the availability of affordable printers, the main advantage of polymeric materials is the possibility to tailor properties of the resulting devices. Given the choice of thermoplastics, elastomers, thermosets, polymer blends and composites, functional polymers or biological systems e.g. building speed, resolution, resistance against chemicals and temperature, stability, mechanical properties, surface finish, porosity or even functionality can be tuned. With this toolbox also complex multifunctional and multimaterial devices become available. This review will focus on recent developments in 3D printing with a special focus on polymeric materials and how they enable new developments, find new applications and help overcome challenges especially in the field of chemical reaction engineering.

2. Equipment and technology

At first, a short overview of the most common 3D printing techniques and respective advantages, limitations and applications is given in this section. Then, recent developments, especially in the field of (polymeric) printing materials, are presented.

Fused deposition modeling (FDM) or fused filament fabrication (FFF) is one of the most common and cheapest additive manufacturing processes today (Sathies et al., 2020; Wittbrodt et al., 2013). Scott Crump developed this process in 1988 (Crump, 1992) and patented it in 1992 (Crump, 1989) under the company Stratasys, Ltd. in which he was co-founder. The separation of the two terms has been caused by trademark conflicts. The short form FDM is a trademark of Stratasys, Ltd. (United States Patent and Trademark Office, 2020), whereas FFF has been established in the OpenSource community through the RepRap project (RepRap.org, 2019). The principle of the FFF method is to heat a filament strand consisting of thermoplastic polymer in a nozzle and place it on a printing platform (Fig. 3).

Typical polymers used are standard polymers like polylactic acid (PLA), acrylic butadiene styrene (ABS), polypropylene (PP) and polyvinyl alcohol (PVA). Technical polymers such as polyamide-6 or polyamide-12 (nylon; PA6, PA12), polycarbonate (PC), polyethylene terephthalate (PET), high impact polystyrene (HIPS) next to high-performance polymers polysulfones (PSU, PPSU, PES), polyetherimides (PEI) and polyaryletherketones (PEK, PEEK, PEKK) can be used. Latter usually require machines designed for their processing. The machines work in the Cartesian coordinate system and can be grouped according to their controls. One design style is the delta version (Anzalone et al., 2015; Bell, 2015; RepRap.org, 2014a), where the special feature is the round printing bed combined with an extruder fixed at

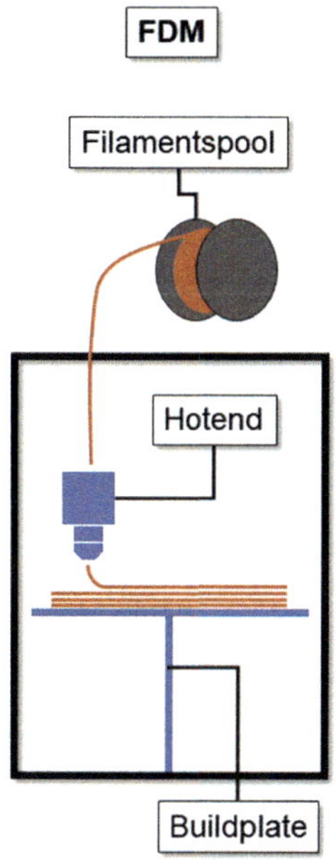

Fig. 3 Representation for an FDM or DIW structure. The material is fed into the hot end by means of a filament roll, heated and deposited layer by layer on the printing plate.

three points in a triangular structure. Each of the three points can move up or down in the z-direction and thus determines the position of the extruder in all three spatial directions (x, y, z). This principle has been taken from the pick & pack industry and is impressive because of its speed in approaching points within space. The second, most common construction style corresponds to the Cartesian 3D printer (RepRap.org, 2014b,c), using x, y, z dimensions of the Cartesian coordinate system. The printing bed can be fixed, or it can move in one of three spatial directions, the z-axis being preferred. The print head is usually guided two-dimensionally on a two-axis plane. The pool of available devices with different equipment is large and ranges from those that have an integrated heated print room to those that do not have a heated print bed. A hot end is located in the print head by melting the thermoplastic polymer and depositing it on the print bed in a strand by moving the print head. The molten filament is conveyed through a nozzle, which has a typical range of 0.25–1 mm in diameter. The nozzle diameter determines the width in which the filament is deposited.

A resolution of 50–150 μm can be achieved (Batista et al., 2019). In the building direction, the layer height is determined beforehand in a software, the slicer. Modern slicers even make it possible to vary the layer height of 3D printing within a print job, so that certain areas receive a higher resolution. Various approaches can be used to create necessary overhangs within an object. If the only possibility is to extrude material from a nozzle, it is possible to create the support structure from the same material. However, in applications such as flow-through reactors, the removal of larger parts of such structures is costly or even physically impossible (/extremely elaborate), since such devices are built as monolithic units and do not consist of several parts. A technical way around this is a 3D printer with a multiextrusion head (Kluczyński et al., 2020; Lopes et al., 2018; Paleń & Mantiuk, 2019), using a second extruder providing an orthogonal material with respect to solubility. Exponents in that regard are PVA (Procopio & Tewari, 2020; Sun & Velasquez-Garcia, 2017) and HIPS. After finishing the production (e.g. in PA-12), these supplementary materials can be removed without complicated procedures by dipping the parts into water (for PVA) or limonene (for HIPS). FDM is especially popular in research, as it is a cheap and easy way to print objects, while at the same time a significant number of polymers are available as extrusion material.

Direct ink writing (DIW) also falls into the category of extrusion-based 3D printing methods that use an ink application nozzle. The applied material is deposited layer by layer under controlled flow conditions. A wide range of materials are already available for use within the DIW. These include highly shear-thinned colloidal suspensions (Morissette et al., 2001; Yun et al., 2019), waxes (Cordonier & Sierros, 2020; Lewis, 2006), molten eutectics (Boley et al., 2017) and colloidal gels (Cheng et al., 2019; Jiang et al., 2019). Solidification to a solid 3D object is achieved by temperature (Boley et al., 2017; Hutmacher et al., 2001), gelation (Cheng et al., 2019; Koffler et al., 2019; Yun et al., 2019), liquid evaporation (Song et al., 1999) or phase change caused by solvents (Gratson et al., 2004; Xu et al., 2006). By combining DIW with the given materials, new fields of application have been opened up, such as 4D printing (Sinha, 2020; Ren et al., 2020) and bioprinting for functional tissues (Cheng et al., 2019; Koffler et al., 2019; Szojka et al., 2017).

Selective laser melting (SLM) and sintering (SLS) are two identical processes, which are only distinguished by the material used (Gibson et al., 2015b). In this process, focused laser beams are used to heat the raw material at designated locations. These laser beams are moved across the construction zone via mirrors in the pressure chamber (Fig. 4). When the production of the current layer is completed, the construction plate is lowered and new powder is scraped up. Yuan et al. present a broad overview of the current

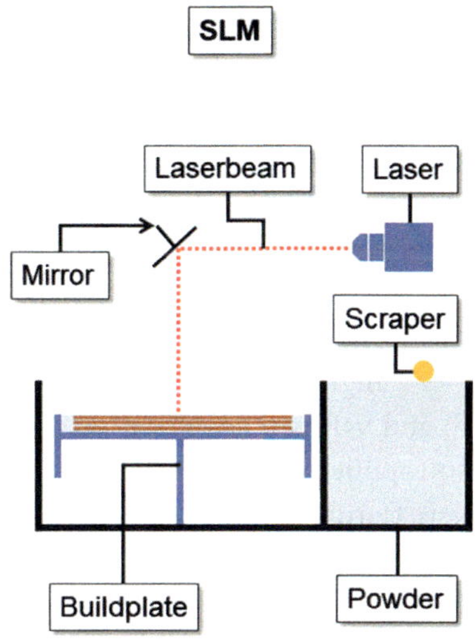

Fig. 4 A laser beam, which is directed onto the printing plate by means of a mirror, melts the presented powder. Once a layer is finished, a new layer of powder is applied to the printing plate by means of a scraper.

state of the art for powder-based additive manufacturing methods, such as selective laser sintering, electron beam melting and selective laser melting (Yuan et al., 2019). If we speak explicitly of SLM, only metal powder is used as pure metal or in an alloy (stainless steel, titanium, aluminum, etc.). The metal is completely melted in an inert gas atmosphere to prevent possible oxidation. This produces a homogeneous mass. With SLS, on the other hand, sintering is the main process, since the powders used (polymers, ceramics) are only heat-treated and not completely melted. The layer height is defined by the particle size of the powder, which is usually over 20 μm (Parra-Cabrera et al., 2018). The use of supporting structures is not necessary here, since the powder bed is arranged all around the objects being built up. As soon as the printing part has been completed, it must be removed from the compacted powder mass and the remaining powder must be removed, usually by means of compressed air. Cavities which are enclosed in themselves cannot be freed from the powder inside.

When developing new materials, different properties must be combined in order to be considered as powder candidates for the sintering or melting process. The prerequisites pertain to two categories: the material properties and the powder properties (Yuan et al., 2019). The material properties can be divided into molecular, thermal and rheological properties. The reusability of the powder is governed by the molecular properties such as the molar mass and functional groups. If the thermal and rheological properties of the material as well as the optical properties of the powder are known, the sinter stable range can be determined. To determine the achievable layer thickness and deposition rate for the powder on the design platform, the flow properties of the powder must be suitable. The current state of the art (Yuan et al., 2019) makes the SLS method a popular technique and the current choice for the production of high-end prototypes in industry. The various materials can be made available in a high purity and quality to produce flow reactors (Gutmann et al., 2017) and valves (Nawada et al., 2019).

Wyn Kelly Swainson applied for the first patent (US 4078229A) in 1975, which describes the basic features of the stereolithography (SLA) process. In the patent, two beams of rays crossing each other to form an object over an active center constitute the claimed technology. A further development followed in 1981 by Hideo Kodama following the concept of photopolymerization by laser (JPS56144478A). The first device for SLA printing (SLA-1) was presented by Charles Hull in 1987 (Snyder et al., 2014). Hull founded 3D Systems in 1983 and is considered the inventor of the SLA process (Bandyopadhyay, n.d.) (US Patent 4575330). Digital light processing technology (DLP) is a similar process and both use a synthetic resin as a base material. The difference between the two technologies is not the material, as with SLM and SLS, but the light source used and how it is projected onto the photopolymer (Fig. 5, left and right). In SLA, the construction of individual layers is carried out using a laser or a UV lamp as the light source, which excites the initiator selectively to start polymerization (Fig. 5, right). The range of polymerization extends from UV light to the visible light range (Au et al., 2016). This is achieved by modern high-intensity lasers and focused LEDs with the appropriate photoinitiators. To stimulate polymerization, the light source is directed across the pressure chamber by means of several mirrors. DLP, on the other hand, uses a projector as the light source and a matrix to shade the area not to be polymerized (Au et al., 2016; Bertsch et al., 1997) (Fig. 5, left). Basically, both techniques start with a resin tank holding the photopolymer, a building platform sinks into this resin and wets it. Then, a laser or projector is used to release the necessary light onto the area where the resin is to be cured. Care must be taken to ensure that each layer is fully cured. After completion, the uncured liquid resin is still suitable for further printing job.

While in stereolithography the choice of resin and the size of the laser spot are partly responsible for the resolution (approximately 0.05–0.15 mm)

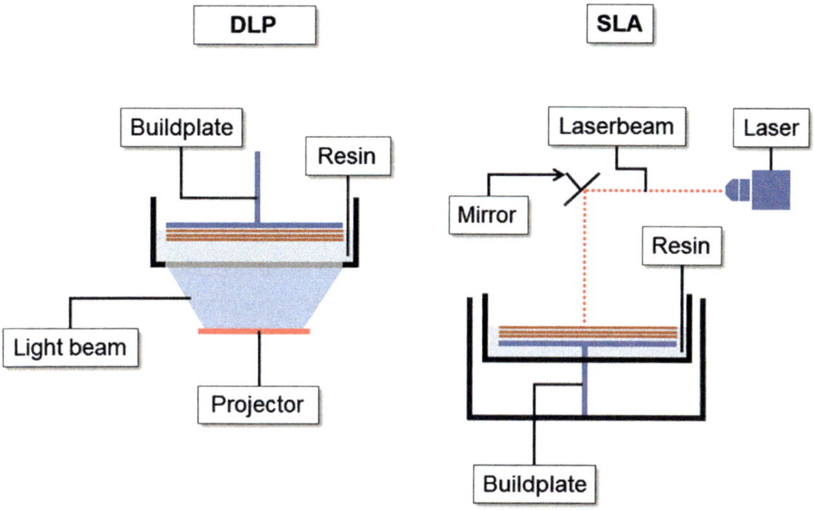

Fig. 5 (Left) In the DLP process, a matrix is projected onto the print platform via a projector. The photoactive polymer is cured by the radiated matrix. (Right) A laser beam, which is directed onto the printing platform by a mirror, is used to polymerize a photoactive resin in the SLA process.

(Au et al., 2016; Sahoo & Das, 2019), in DLP technology it is the pixel resolution of the projector (Au et al., 2016) that can achieve significantly lower resolutions. The advantages of the SLA and DLP process are the chemical bonds that are formed, which provide additional strength within the structure. In addition smooth surface finishes and shear forces are excluded within the printing material (Kontodina et al., 2019). In order to perform SLA printing successfully, it is necessary to adjust various parameters, such as the laser intensity, the speed of the laser as well as the exposure time, the ratio of photoinitiator to monomer and the light sensitivity of the resin used. Mathematical models to determine the curing depth and to avoid over-curing can be applied to better coordinate the interaction of the physical properties.

SLA has a wide field of application, e.g. in biomedicine with significant therapeutic properties such as drug delivery and tissue engineering (Bittner et al., 2018; Oliaei & Nasseri, 2019). Cabin parts and knobs have already been used in aviation (Sahoo & Das, 2019) and also the field of chemical reaction engineering is applying SLA (Figuerola et al., 2019; Zhang et al., 2019a). A 3D printing technology for tiny parts with a surface resolution well into the submicro-meter scale is the two-photon polymerization (TPP). An advanced TPP process enables to print parts with a height of 2.2 mm and with a printing speed of about 0.25 mm^3/h (Yang et al., 2019).

In order to successfully apply 3D printing in the chemical context, robust materials have to be used. Developments in recent years have been reviewed (Kotz et al., 2019). They comprise ceramics, transparent silicate glass, which can be printed both directly and indirectly with a thermal post processing step (Gal-Or et al., 2019), and (fluorinated) polymers. Fluorinated polymers, such as PTFE or perfluoro polyethers (PFPE) are interesting because of their high chemical inertness. Additionally, polymers are of interest for the options of structuring their surfaces on micro or nanoscale and thus desired surface patterns for e.g. chemical arrays can be generated. This was shown for superhydrophobic fluorinated polymer foams (Kotz et al., 2019). 3D printing of PEEK by FDM is also feasible and rewarding in the context of flow reactors. A continuous milli reactor was reported with static mixing elements and an inner free volume of less than or equal to 1 mL (Fig. 6)

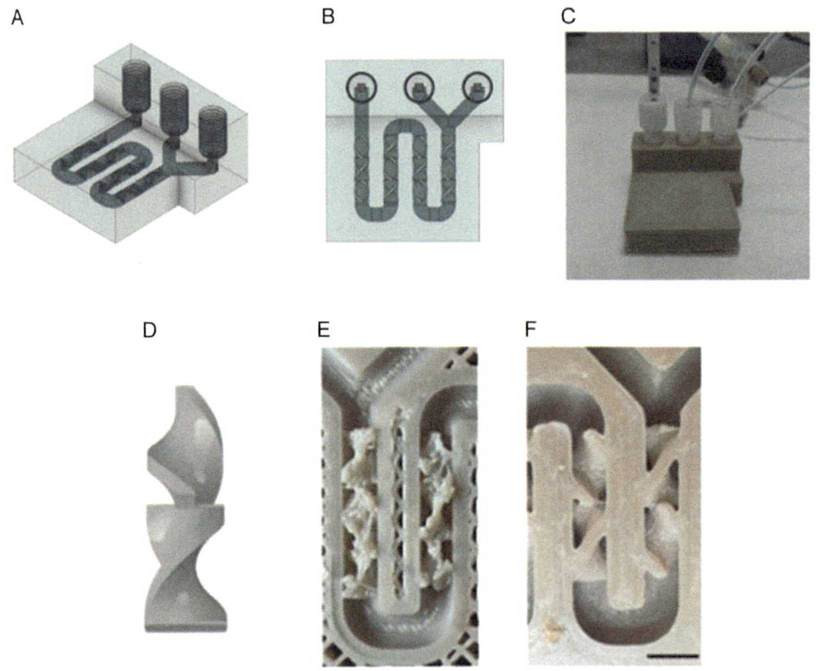

Fig. 6 (Left) CAD image of the designed reactor piece: Orthographic view, top view showing channel and internal mixer geometry printed part; (right) final printed reactor and printed mixer elements with 0.4 mm nozzle, 0.42 mm profile, 3 mm diameter (center), printed mixer elements with 0.4 mm nozzle, 1.2 mm profile, 4 mm diameter (right). *Republished with permission of the Royal Society of Chemistry, from reference Matthew J. Harding, Sarah Brady, Heather O'Connor, et al.: 3D printing of PEEK reactors for flow chemistry and continuous chemical processing,* React Chem Eng *5:728735, 2020. https://doi.org/10.1039/C9RE00408D. Permission conveyed through Copyright Clearance Center, Inc.*

(Harding et al., 2020). The two inlet streams as well as the static mixing geometries in Kenics type design are shown in the CAD drawings (left), as well as the printed reactors (right). They were printed with a 0.4 mm nozzle and 1.2 mm profile, 4 mm diameter (center) or 0.42 mm profile, 3 mm diameter (left). Printing artifacts issues could largely be resolved by using a smaller nozzle diameter. PEEK has a high chemical and thermal resistance, as well as very good mechanical strength. Pressures of up to 60 bar could be hold and the fact that it is printable by FDM makes it a cost effective candidate for continuous manufacturing equipment (Harding et al., 2020).

Besides choosing a neat polymer, fillers or post-processing can be used to enhance the product properties. As early as 2015, a process chain of filament production for the FDM printing process was reported upon. Samples are stiffer in the filament lay-down direction and that the adhesion between the individual layers is weakest in the build-up direction. The filling degree has a linear effect on mechanical properties, whereas the layer thickness has no significant influence. If fibers are incorporated into the filament material, in this work it were glass fibers, they have a positive effect on the final product. If the additively manufactured specimen are compared with those from a compression molding process, the mechanical performance is reduced by 20–30%. However, the use of incorporated glass fibers almost compensates for the loss in mechanical performance. It can be seen here that the FDM process can be used to produce parts that come close to the properties of compression molded ones. There is still a lot of potential in material development for specified material for additive manufacturing by compounding or in-situ synthesis of the raw product with fillers (Carneiro et al., 2015). Efforts using carbon fiber powder in a plastic matrix acrylic butadiene styrene (ABS) were undertaken. In order to determine the effects of the addition of the carbon fibers, tensile and bending tests were carried out. In addition, the fracture surfaces were analyzed using scanning electron microscope (SEM) images. They concluded that by adding carbon fibers to the polymer matrix, the tensile strength and modulus of elasticity could be increased, but with a reduction in toughness, ductility and yield strength. Further investigations have shown that the mechanical properties are dependent on the amount of filler and the length of the individual carbon fibers. It could be shown that a fiber content of 5% by weight showed an increase in tensile strength of 22.5%. In contrast, a 7.5% fiber content by weight resulted in an increase of 30.5% in the modulus of elasticity. There was also an increase in tensile strength and modulus of elasticity for fibers with a length of 150 μm compared to fibers with a length of 100 μm. Samples containing carbon fibers greater than 10% by weight showed an increase in porosity.

This trend could be observed on the fracture surface by SEM images and in the smallest mean values of tensile strength, toughness, yield strength and ductility (Ning et al., 2015). In order to enhance the surface area and yield a highly porous material, devices printed by SLA were treated with hypercrosslinked pyrrole-derived hyperporous carbon before UV post curing. This process proved easy, efficient and gave highly robust results with high surface areas. Corresponding devices are to be applied in extraction and degradation of pollutants from water (Medina et al., 2019).

Ceramic materials are the gold standard for high temperature applications. Typically corresponding precursor materials have a high weight loss and show shrinking during sintering. In the context of chemical processes, a material was presented for non-oxidic ceramic SiCN preparation by digital light processing (DLP) with linear shrinkage in all three dimensions of 20%. The DPLP technique was used to produce ceramic microreactors in centimeter dimension. These were screened for ammonia cracking (Gyak et al., 2019).

The interest in the application of *smart* materials in 3D printed objects, especially reactors and the like, is increasing. These smart materials could act as sensors or control/switching elements. Consequently, the 3D printability of smart materials is investigated and successes have been made (An et al., 2019; Jia et al., 2019; Nguyen et al., 2017; Ren et al., 2020; Wu et al., 2019). (Hydro-)gels are one interesting polymeric material (Chivers & Smith, 2019), which can act stimuli-responsive and thus give interesting options in AM. Consequently, approaches to make them directly printable were made (Jia et al., 2019; Wu et al., 2019). First applications of hydrogels in 3D printed reaction ware are given in Section 4 (Hu et al., 2019; Schmieg et al., 2018). By combining stimuli-responsive materials, mathematics and 3D printing so-called 4D printing is performed (Sadasivuni et al., 2020), which gives encoded structures with programmable self-morphing if an external stimulus is activated. A further 4D printing approach, which is based on the use of a single material and the shape-morphing is regulated by changing printing parameters (e.g. deposition speed) on the fly. Thus, a local control over the nematic arrangement in liquid crystal elastomers gave a prove of concept (Ren et al., 2020). In Fig. 7 several double-layered structures are shown, which were designed to give shape-morphing behavior by varying print path and print speed. In this way a flower shape (a), twisting and bending (b, c) as well as grasping (d) was accomplished. A cellulose-based 3D printing material was developed, which is conductive and thus aimed to be applied in electronic devices, sensors and the like. It can be produced by attaching a conductive polymer to cellulose nanofibers and mixing them

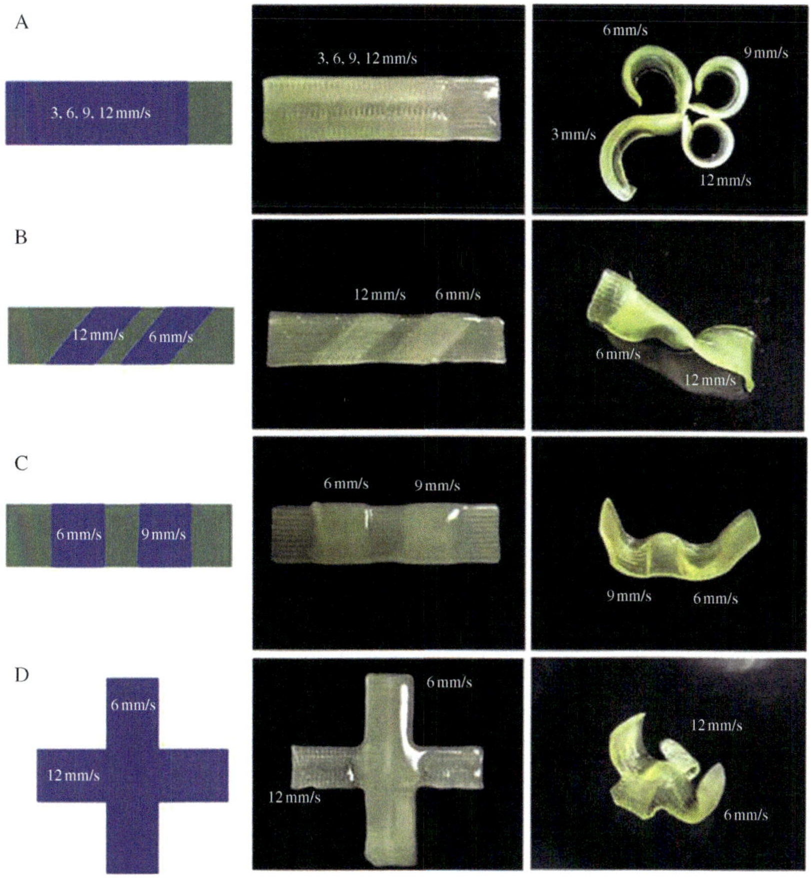

Fig. 7 Shape-morphing patterns enabled by altering the geometry and print speed distribution of each layer. (A) A flower shape assembly with distinct bending curvatures in each petal. (B) A strip with different twisting angles in two joints. (C) A strip with different bending angles in two joints. (D) A gripper with different bending angles in two adjacent arms. (The green area represents the bottom layer with a print speed of $3\,mm\,s^{-1}$, and the blue area represents the top layer with the print speeds marked above.) *Reprinted with permission from Ren L, Li B, He Y, et al.: Programming shape-morphing behavior of liquid crystal elastomers via parameter-encoded 4D printing,* ACS Appl Mater Interfaces *13:15562–15572, 2020. https://doi.org/10.1021/acsami.0c00027. Copyright 2020 American Chemical Society.*

with a crosslinking agent, graphite and a curing agent to yield the printing material (An et al., 2019).

In the context of sustainable 3D printing materials, a biodegradable 3D printing resin out of waste cooking oil has been introduced. The oil was acrylated in a one-step reaction and gives the resin in combination with a photoinitiator. It can be used in commercial LDP printers and achieves high resolutions of down to 100 μm. Thus, a high value 3D printing resin can be directly produced from waste (Wu et al., 2020). With the appropriate treatment, recycled plastics can also be used to give 3D printable material (Huang, 2019), e.g. for the FDM process. Also paper waste has been applied to prepare supporting structures sustainably. Stolz and Mülhaupt used paper waste and an aqueous solution of a polymer binder like polyvinyl alcohol to produce aqueous cellulose pastes suitable for extrusion-based 3D printing. Cellular cellulose composites exhibiting hierarchical pore architectures are fabricated by the printing process, which show high mechanical stability, water resistance and flame retardancy suitable for lightweight construction (Stolz & Mülhaupt, 2020).

3. 3D printed equipment

3D printing has not only made advances in the field of printing technologies and materials, but is also getting more attention with respect to designing and printing useful equipment in the chemistry context, which were traditionally manufactured by top-down methods (Mothes, 2014). Such equipment includes tailor-made labware as well as prototypes for operation units.

3.1 Lab scale

Using AM to build configurable reaction ware, e.g. for chemical synthesis was started as early as 2012 (Kitson et al., 2012). A review, covering efforts from 2012 to 17 was focused on laboratory hardware, analytical equipment, teaching aids, micro and milli fluidics, catalysis and many biological and pharmaceutical aspects (Capel et al., 2018). Consecutive developments have been made since then encompassing small parts, e.g. mixing elements (Cronin, 2013; Hnát et al., 2019; Huang et al., 2020; Stiefel et al., 2015), process units, such as separators (Pu & Su, 2018) and complete reaction set ups (Kitson et al., 2016; Neumaier et al., 2019; Penny et al., 2019). Polymeric materials, especially PP, are mostly the material of choice (Dragone et al., 2013), as they are inexpensive, printable via FDM, inert to most chemicals and can be used over a relevant temperature range. Low cost lab equipment for the use in continuous flow chemistry is accessible for under 300€. It consisted of four Arduino controlled syringe pumps and microreactors made of PP. With this equipment they successfully performed and optimized the preparation of glycosyl donors and subsequently glycosylation reactions in a second reactor in cascade flow (Neumaier et al., 2019).

A detailed protocol was published explaining how basically anyone with appropriate printer access can print versatile reaction ware. Polypropylene was the recommended inert material, being resistant to most solvents, acids, bases etc. and at the same time reasonably printable via the FDM method. Moreover, the incorporation of a PP composite was reported of in form of a viscous catalyst-silicone material. This allowed the fabrication of sealed, multi-component reactors based on specific requirements for complex and/or multiple experimental steps (Kitson et al., 2016). Also requirements for the FDM manufacturing process to produce gas tight laboratory equipment was been put forward. The requirements are both on software and hardware side. On the software side, the challenge lies in the settings within the selected slicer software. It turns out that a layer height greater than 75 μm leads to the formation of air channels within the layers. With larger layer heights, the filament is extruded from the nozzle in a rather cylindrical shape and deposited on top of each other. The wall thickness should be set to a multiple of the nozzle diameter, but at least three times the diameter. If an improper temperature is used during the extrusion of the given thermoplastic, the individual layers will not fuse together and no homogeneous surfaces will be formed. The tolerances of the filament strand should be reduced to a minimum, as it is not usual up to now to carry out compensation on the printer itself. To check the gas tightness, channels with a length of 4 cm, 0.5 cm inner diameter and a wall thickness of 1.2 cm were printed. In order to connect these with already existing screw connections, threads were made additively at the respective pipe ends. The materials used were PLA, ABS, TPU, PP and PVDF. The overpressure tests were carried out according to the test methods C1, C2 and D1 of DIN EN 1779. An overpressure test of 600 kPa, with minimal pressure loss, was successfully carried out on the materials mentioned above (Gelhausen et al., 2018).

A further 3D printed reactor for continuous synthesis in laboratory scale was presented (Penny et al., 2019). The entire plant consists of simple, inexpensive and at the same time robust components. The reactor used is a printed tube reactor on a millimeter scale from PP, which is tempered by means of a standard laboratory heating plate. While the mixing device is printed from PP, for the printing of all other necessary devices PLA was used. The mean residence time can easily be adapted to the synthesis. The concept is tested by means of organic coupling reactions (Penny et al., 2019).

A flow-through reaction system with freeze-thaw valves was manufactured from Ti6Al-4V by selective laser melting. This chip consists of a T-junction with two separate circulation jackets for cooling and heating.

This microreactor can withstand a pressure of at least 25 MPa. The respective switching times between frozen and thus closed channel to open channel were 1–2.5 min with a passive switching approach. It could be shown that numbering up is possible, toward more complex switching networks. The authors hope that micro-selective laser melting will lead to a significantly better resolution of their T-junction, which promises improved phase definition, reduced dead volumes and switching times (Nawada et al., 2019). In the context of performing challenging heterogeneous catalysis in microreactors, additional labware was readily prepared via AM to facilitate the process: a heterogeneous benzylation reaction of toluene with benzyl chloride was performed with silica-immobilized $FeCl_3$ nanocatalysts. A gas-liquid separator and a buffer vessel, which were 3D printed from a photopolymer resin (acrylated monomers, acrylated oligomers and photoinitiators) were applied during the experiment (Pu & Su, 2018). Also, an inline shower-head mixer, which was printed by binder jetting and DMLS. Intensified mixing of two streams was achieved while maintaining minimal pressure drop. CFD simulations as well as a diazo coupling as model reaction were performed and showed that mixing behavior was improved compared to commercial micro T-mixers, especially for low Reynolds numbers (Zhang et al., 2019b). Another example of the implementation of 3D printing in experimental set-ups was a membrane reactor to control the electrochemical depolymerization of lignin into its monomeric subunits as valuable chemicals. As the reaction is unselective, overoxidation has to be prevented, which could be realized in a continuous process with in situ nanoporous filtration. In order to promote turbulence and minimize fouling, mixing elements were 3D printed and successfully incorporated into the reactor (Stiefel et al., 2015).

Another important aspect of 3D printing in the lab environment are individually embedded sensors (Dai et al., 2020). An early example for 3D printed micro reaction ware from polypropylene (PP) coupled with inline ATR and/or UV–Vis flow-cells demonstrate the usability of the reaction ware by an sequential imine formation and reduction as well as an alkylation reaction (Kitson et al., 2012). An easy and inexpensive way to fabricate embedded electrochemical sensors, which in the chemical engineering background will be specially interesting with respect to redox-active analytes, could be the single-step simultaneous printing of PLA and a commercially available graphene-PLA composite with a dual extruder system. Such sensors can be produced and successfully applied for hydrodynamic electrochemistry, which they tested by the oxidation of catechol. There is room for improvement with the help of AM, e.g. by increasing the surface area or flow path (O'Neil et al., 2019).

In another study, 3D printed graphene electrodes modified with Prussian blue were developed and tested as a promising platform for sensing molecular targets. The 3D printed electrodes were subjected to a electrochemical treatment in order to create oxygen functionalities, which could then be used as anchors for Prussian blue. It was possible to apply them in sensing hydrogen peroxide in commercial milk and mouth wash and they exhibited similar behavior to conventional electrodes, thus posing an easily modifiable new type of electrode (Katic et al., 2019). A microflow cytometer as a lab-on-chip device (LoC) out of polycarbonate (PC, Accura 60) can count particles without optical marking via embedded optical fibers. The shading of the light by the particles determines its size. Particles up to a size of 5.5×10^4 particles per mL and particle sizes in a mixture could be counted. With the help of an evolutionary multi-objective optimization algorithm the particle counting could be optimized. Square channels in the size range of $250\,\mu m \times 250\,\mu m$ were used. The total time to produce the lab-on-chip device was 5 h and the total cost was about 13€ (Hampson et al., 2018)

3.2 Operation units

Also chemical engineering is expected to strongly benefit from AM techniques, especially in the context of design optimization (Stark, 2018). Thus, research is increasingly focusing on 3D printing of operation units. Interestingly, polymeric materials are also the material of choice for those applications. In this context, a methodical procedure for the design and production of additively manufactured packing structures in lab scale has been put forward. Crystal lattices were the starting point for CFD simulations in order to investigate and optimize dry specific pressure drop and liquid distribution of the structures. The hexagonal closest packing showed the lowest pressure drop, while structures based on it showed higher pressure drop but improved liquid distribution. The SLS technique with PA-12 was finally used to print the optimized structure and experimental validation is on-going (Neukäufer et al., 2019).

Cylindrical monoliths exhibiting different geometries of transport channels were prepared by 3D printing of thermoplastic polymers (PLA, PVA, PS and mixtures). The monoliths were designed with CAD (Fig. 8). Structure complexity increased from monolith I to IV: while structure I has straight 1D pore channels, monolith II shows 1D axial pore channels

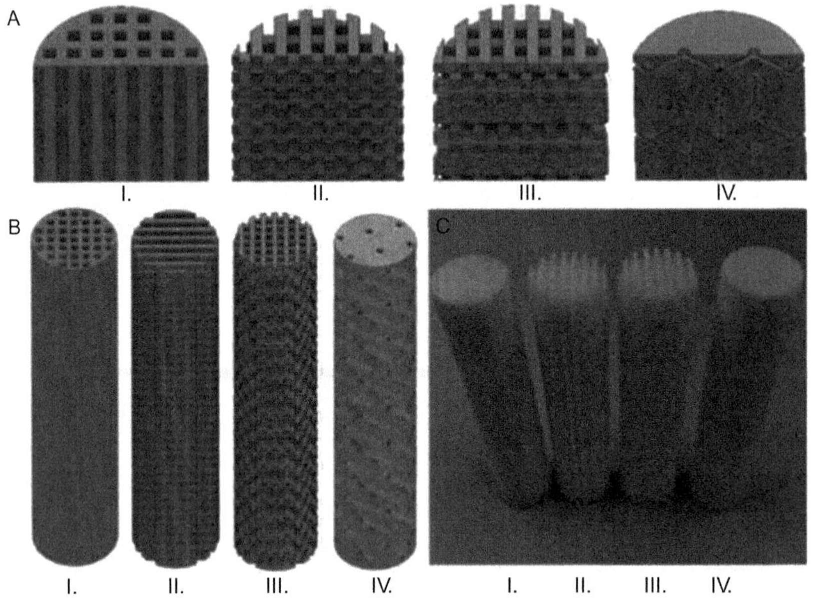

Fig. 8 CAD designed monoliths: I with 1D pore channels; II with AB-type stacking and 1D pore channels with radial connections; III with ABCD-type stacking giving a higher tortuosity of the channels; and IV with pore channels that connect in tetrahedral nodes in a diamondoid structure. *Reprinted with permission from reference Sebastian Hock, Marcus Rose: 3D-structured monoliths of nanoporous polymers by additive manufacturing,* Chem Ing Tech *92:525–531, 2020. https://doi.org/10.1002/cite.201900149. Published by Wiley.*

with radial connections, monolith II has a 3D pore system with increased tortuosity and the channels of monolith IV exhibit a diamondoid structure. Printing was successful with a maximum deviation of ± 0.2 mm over the full length and no obvious defects. The hydrodynamic characteristics and printed composite materials of microporous hyper-crosslinked polymer with an specific surface area of up to 171 $m^2 g^{-1}$ are promising in the catalytic context (Hock & Rose, 2020).

Crystallization in a 3D printed device in continuous flow was studied using a vat polymerization to build up a 2 mL straight tubular micro reactor with some bows for the crystallization of paracetamol. Heat transfer characteristics were determined in a very detailed way. It could be shown that such a device allows to crystallize a significant part as metastable orthorhombic crystals. However, it is emphasized that reactor fouling is a complicating factor (Okafor et al., 2019).

DLP (direct light processing) was used to produce packings of polyamide for rotating packed beds (RPBs). The polymeric replica exhibited sufficient structural and chemical stability for the experimental investigation of the structures. Slide discrepancies in the pressure drop characteristics were observed, which are likely caused by the different material properties. However, comparably effective interfacial areas were found in the mass transfer experiments and material as well as production costs could be significantly reduced by application of AM: thus it poses a technology for the development of improved RPB packing designs and opens the door to a wide range of new packing morphologies (Gładyszewski & Skiborowski, 2018).

AM has successfully been applied to produce small-scale gas-solid fluidized beds, which cannot only be used to perform fast and easy screening experiments but also translate lab scale results reliably to larger scales. A series of geometries was printed by SLA and characterized with respect to pressure drop, flow regime and gas velocities. The 3D printed devices usually exhibit higher surface roughness relative to conventional materials, but this study showed that does not effect fluidization characteristics or modify wall effects (McDonough et al., 2019).

Stirring tools can be costly even on a laboratory scale. These reactor parts are essential to heat supply and removal and also indispensable for particle generation and dispersion (Römpp et al., 1998). A cost-effective approach is to manufacture polyamide stirrers by means of FFF with the option to optimization. This was shown for a batch emulsion copolymerization of styrene and butyl acrylate. The reaction temperature window was in the range of 60–85 °C. A turbidity probe was used to monitor the course of the reaction. This allowed the determination of a mixing quality ratio. In addition, a scale-up from a 0.5 L to a 10 L batch reactor was carried out with adapted stirring tools. A comparable droplet dispersion between the reactors could be determined (Bettermann et al., 2019a).

4. Reactors and reactions

A special focus of both chemists and engineers are 3D printed reactors: the geometry can be designed in order to optimize e.g. flow conditions, and sensors or catalytically active species can be embedded in the reactor. While some research concerned reactions in batch, increasing research effort is focused on continuous flow systems. In micro and milli fluidics the benefits of small scales and high surfaces can be used (Nielsen et al., 2020). At the same time studies with increasing reactor volumes are published, indicating that also larger scale can benefit from AM. Irrespective of application (biological sector, water treatment, typical chemical reactions), reactor operation mode or scale, polymeric 3D printed devices dominate research in this area.

4.1 Batch reactors

An interesting example of computer-aided reactor design and subsequent application was given by Zhao et al. (2019). They designed a fractal tree-like photobioreactor on the basis of a bifurcation tree algorithm and fabricated it via SLA-3D printing. The reactors in operation are shown in Fig. 9 (bottom). The reactor has a high surface to volume ratio and shows good flow

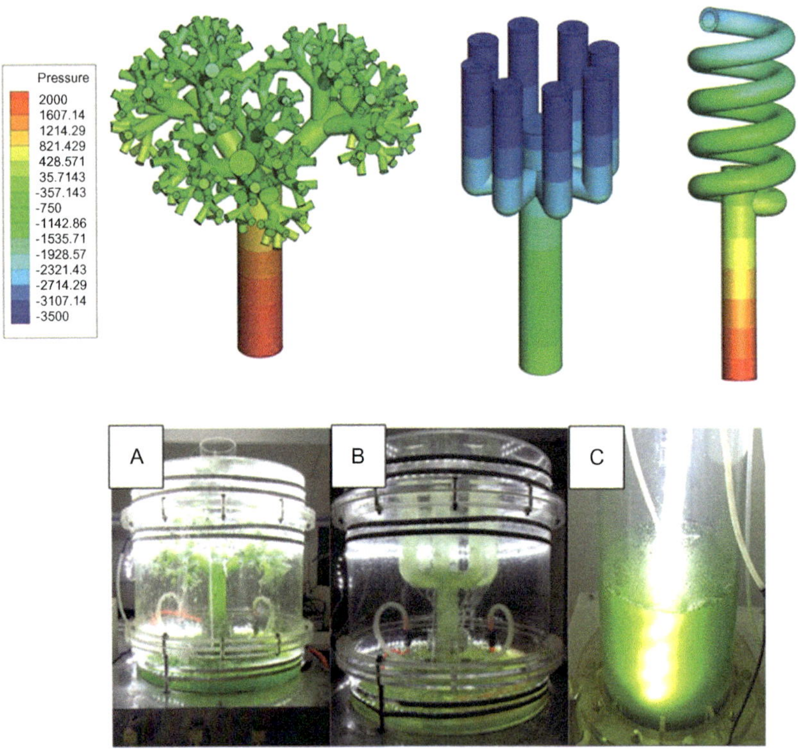

Fig. 9 (Top) Time averaged relative pressure in three-types PBRs as calculated by CFD for three different reactor designs(Unit: Pascal); (Bottom) Chlorella culture in the different PBRs at the 4th day in (A) the fractal tree-like PBR, (B)multitubular PBR and (C) column PBR. *Reprinted from Publication Nielsen JB, Hanson RL, Almughamsi HM, Pang C, Fish TR, Woolley AT: Microfluidics: innovations in materials and their fabrication and functionalization,* Anal Chem *92:150–168, 2020 with permission from Elsevier, Copyright 2019.*

performance, low pressure drop (Fig. 9 top) and good diffusion characteristics in CFD simulations. Moreover, it was tested for the bioreaction of Chlorella culture in comparison to multi-tubular and conventional designs of photobioreactors. The new design enhanced turbulence and produced smaller bubble sizes, so that it could improve mass transfer and give a high photosynthetic conversion efficiency, high CO_2 capture efficiency and high growth rate of biomass. Consequently, this is a nice example of process intensification using 3D printing technology (Zhao et al., 2019).

A counter-currently (CC) operated additively manufactured reactor was used for the enzymatic decarboxylation of ferulic acid (FA) to 2-methoxy-4-vinylphenol (MVP) together with an *in situ* extraction with *n*-heptane. Periodic open-cell structures (POCS) give the carrier for the enzyme phenolic acid decarboxylase, which could be optimized regarding flow characteristics via rapid prototyping. The POCS were printed using SLS and polyamide as material. Also a distributor for the extraction phase was targeted and optimized, showing how AM can be enabled in enzyme catalysis (Büscher et al., 2020). 3D printing of reactors and embedding sensors proved to be a powerful technique for enzymatic reactions. Highly porous monolithic enzyme carriers, which could be used for subsequent enzyme immobilization, were connected in series as well as to a pumping and analyzing system, including a temperature and pH control. A prove of concept with Glucose Oxidase and Horseradish Peroxidase as model enzymes was also given (Kazenwadel et al., 2016).

In order to produce materials such as metal-organic framework compounds or polyoxometalates, a hydrothermal reaction in a closed system is usually applied, at standard pressure and temperatures above the boiling point of the solvent. The problem with this methodology is that it is a one-step process (batch reaction) and does not allow to study multi-step or more complex reaction sequences. Deeper insights could accelerate the discovery and optimization of new materials. An approach to more complex reaction conditions various templates were prepared using the FDM technique and PP as material. Within these templates, different reactant concentrations can be present to control the composition parameters. These mixtures are then mixed together at a chosen time. Thus, it was possibly performing more complex reactions and to achieve structures that could not be achieved previously (Lin et al., 2018).

3D printing by stereolithography was used to print and modify four different geometries of chitosan-coated monolithic porous filters for the application in heavy metal ion removal from waste water. The acrylate-based polymer substrates produced using 3D printing were functionalized first by silane coupling and subsequent chitosan growth, starting from the coupled outmost epoxide group via epoxide–amine reaction. The product showed comparable absorption capacities to commercially available filters, but high reusability, simplified manufacturing and customized design could be achieved additionally (Zhang et al., 2019a). Another example of improving waste water was treatment in a hybrid composite of surface-modified 3D-printed ABS polymer, on which a lacunary polyoxometallate (POM) anion has been deposited. Latter remove transition metal pollutants from water. The surface modification was carried out by hydrolysis in a basic medium. Scanning electron microscopy images showed a morphological transformation of the surface towards macroporous channels. By Raman spectroscopy carboxylate and amide groups could be detected, the later POM anchor points. A very good metal removal efficiency of up to 67% for Cu^{2+}, 44% for Ni^{2+} and 41% for Co^{2+} was found (Ji et al., 2018).

Selective laser sintering was applied successfully to produce a catalyst for hydrogenation consisting of a porous yet solid matrix of PP in which the catalytically active material Pd/SiO_2 is incorporated. The material exhibited good mechanical and thermal stability as well as good resistance to leaching. Although the printing process itself did not seem to have a major effect on the catalytic performance, the hydrogenation experiments were successful. They showed that an efficient, easy-to-use and reusable heterogeneous catalyst was developed, which should be suitable for further tailor-made applications, e.g. in flow reactors (Lahtinen et al., 2019).

The recovery of catalysts has a high priority in the chemical industry. Catalysts are usually expensive and should not be detectable in the final product. The immobilization of catalysts on the surface is a good approach here. Adhesion of the catalyst to the surface of the support plays a major role here, as otherwise the catalyst is rinsed out again with the solvent phase. Sanchez et al. show a solution approach by tricatalysis over Cu(II)/Cu(I)/Pd(0). Different sizes and shapes for the immobilization of the catalysts were realized by Fused Deposition Modeling (FDM) and Direct Ink Writing (DIW). By using Direct Ink Writing a silica carrier could be produced additively. It was then coated with polyimide/palladium and was thus a monolithic catalyst support. Using the FDM method, a polypropylene capsule containing copper(II) was produced, which was loaded onto polystyrol-supported 1,5,7-triazabicyclo[4.4.0]dec-5-ene. A great potential is assigned to the combination of both catalyst supports with other catalytic materials, such as magnetic catalysts, which thus allow complex multistage one-pot syntheses without contamination of the reaction solution. The result of the combination of the supports are separated catalyst supports which can be separated from each other again in order to recycle and reuse the catalysts contained in them individually (Sanchez Diaz-Marta et al., 2019).

4.2 Continuous flow reactors

4.2.1 Polymeric reactors

The spread of additive manufacturing within organic synthesis can make good progress if individual flow-through reactors tailored to the reaction are used. Materials such as nylon, HIPS (high impact polystyrene) and PLA (polylactic acid) were applied to produce catalytic flow reactors. This shows the advantage of "functional" rapid prototyping over classical reactor construction. Within a very short time, reactors differing in size, shape and channel dimensions, could be prepared and screened. This led to an optimized reaction control for the two-stage Henry reaction in which a nitroaldol condensation with subsequent hydrogenation was carried out. Target products were 1,2-amino alcohols such as norephedrine, metaraminol and methoxamine. Under the optimized conditions a change of solvent and an intermediate cleaning were not necessary. A diastereoselectivity of 90% ee could be obtained. For industrial application an upscaling of the optimized reactor to 10 mL was performed (Rossi et al., 2017).

A continuous production of hydroxyapatite nanoparticles was developed using Y-shaped 3D printed reactors. The nanoparticles show similar properties to natural bone and are typically used in scaffolds, coating and as carriers for drugs or proteins. Two different reactors were printed using the FDM technique and ABS as polymeric material. ABS is stable under the process conditions and gives durability for the pressure used in the system. The reactors were successfully applied and mapped with respect to process parameters and product properties, demonstrating the advantages of continuous synthesis also in biomedical field (Latocha et al., 2018).

Oscillating flow-through reactors in the microscale range have the advantage over normal tube reactors that the built-in baffles lead to increased mixing. The low-cost stereolithography method produced a miniaturized oscillating flow-through sonic plate reactor that could produce monodisperse silver nanoparticles. These silver nanoparticles were more stable in time and had a narrower particle size distribution compared to the particles in a conventional tube reactor. The frequency and amplitude applied to the flow frequency also had a positive effect on the residence time distribution of the reactor (Okafor et al., 2017).

Another approach to couple chemical functionality with 3D printed devices was in the creation of coatings on respective devices in a post-printing approach: by SLA printed devices were treated with metal-organic framework (MOF)/polymer mixed-matrix coating (MMC). Submicrometric crystals of a zeolitic imidazolate framework (ZIF-67) were dispersed in a binary mixture comprising a polymer and an organic solvent, applied to the high-surface 3D printed device and finally heated up gently (60 °C). Those devices were successfully tested for water pollutant degradation: potassium peroxymonosulfate was activated to form sulfate radicals and enhance the degradation of organic dyes (in this case rhodamine B) in water. Those model tests were successful and showed high reproducibility and reusability in both stirred-tank and flow-through column formats (Figuerola et al., 2019).

A miniaturized 3D printed reactor with an integrated mass spectrometric ionization component, a nanoelectrospray ionization (nano-ESI) capillary, allowing direct sampling and fast MS analysis of the reaction solution was developed. The microreactor was tested successfully for the online analysis of a Diels-Alder reaction and the subsequent retro Diels-Alder reaction. It is fabricated from PP, which guarantees reusability and shows a high resistance towards most chemicals, so that possible future applications are wide (Scotti et al., 2017).

As polymers are a dominant part of the chemical industry production, polymerization reactions are also of special interest in the context of AM. The production of parallelized microfluidic droplet maker devices was reported using rapid AM (Femmer et al., 2015). The microreactors developed were used for the production of small droplet sizes. These droplets in turn serve as containers for an emulsion polymerization of microgels. The homogenized particle size of the microgels is determined by the droplet size in the microreactor. The droplet size presented here, with a diameter of approx. 500 μm and the monodisperse microgels produced in it, cannot be produced by standard emulsion or precipitation polymerization. The microgels are crosslinked photochemically. With the reactor presented here, a droplet emulsion at a rate of 3 L/h can be achieved (Femmer et al., 2015).

A continuously operated PLA reactor was used in two cases for continuous emulsion copolymerization. It was also shown that a polymer product change is possible during operation. The conversion were in the range of 81–99% and had stable zeta potential values. A stable operating condition was shown by a thermal imaging camera. This was easy to show, as the reactor itself was only air-cooled (Bettermann et al., 2018). Development cycles in which iteration steps are performed to implement improvements in a chemical reactor are usually lengthy processes. The coupling of modeling by means of CAD with subsequent CFD simulation is already an established procedure (Gutmann et al., 2017; Neukäufer et al., 2019; Papetti et al., 2018; Zhao et al., 2019) and can now take up thrust again. The coupling between CAD, CFD and rapid prototyping promises short and fast iteration steps for individual development stages. Short development cycles were demonstrated (Bettermann et al., 2019b) and the results taken for developing specific requirements of the already known TBR reactor along CFD simulations (Bettermann et al., 2018). Dead zones were eliminated and the progressive formation of Dean vortices were taken advantage of. Additionally, static mixing elements adopted from Kenics at defined positions along the reactor axis. They are shown in orange in Fig. 10 (left). Corresponding CFD simulations are given in Fig. 10 (right) and demonstrated good radial mixing due to the mixing elements, which lasts also for 30–20 mm (for a mean residence time of 5 or 15 mins, respectively) after the mixing elements due to memory effects. This demonstrates the fruitful symbiosis of the three different fields CAD, CFD and AM, which can be exploited intensively in the future.

An advanced reactor concept incorporates a reactor module and implements the autonomous behavior in a flow reactor that has been manufactured additively. This module, based on poly(*N*-isopropylacrylamide) (pNIPAM), reacts as an actuator to an exothermic emulsion polymerization of vinyl acetate and Versa(R) 10 and manipulates the flow regime of the reactor. The pNIPAM hydrogel serves as a temperature responsive valve and thus acts as an interface to the first level of quality control by separating the reactant flow from the product flow. The smart module is shown in Fig. 11: Depending on the temperature of the flowing medium that reaches the hydrogel chamber, the pNIPAM is either in its swollen state and leads the medium into the waste line ($T > LCST$, green). As soon as the temperature reaches the LCST of pNIPAM (orange) due to the exothermic reaction, the gel collapses and the medium is transferred to the product tank. The reaction time of the hydrogel for the respective circuit is approx. 10 min and can be repeated several times. Hydrogels themselves are known to have stimuli not only on temperature, but also on pH and solvent composition. An adaptation to another reaction system, like the one shown here, is therefore possible. This opportunity leads to the fact that online analytics within a flow-through reactor can be controlled by integrated intelligent hydrogels and that the usual measuring sensors can only be operated as fail-safe (Hu et al., 2019).

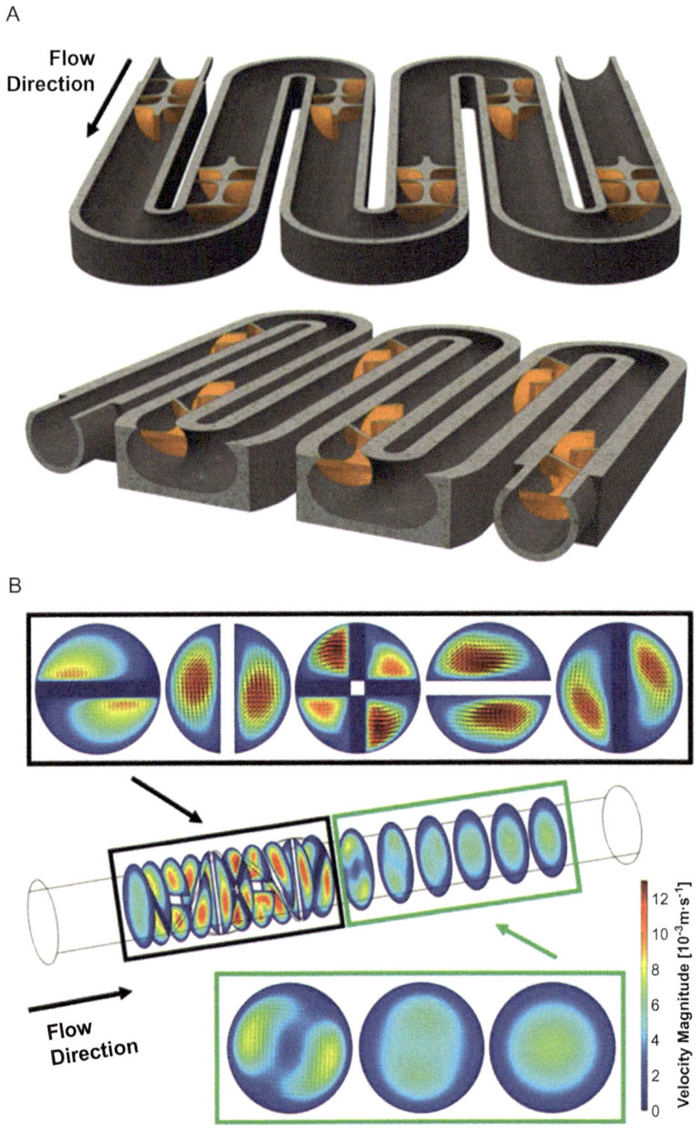

Fig. 10 (Left) CAD drawing of the final iteratively optimized reactor design obtained during the digital and lean development method by implementing 12 static mixing elements (SME) adopted from Kenics at targeted positions into the original reactor design. Perspective representations showing cross sectional views; z-axis: −12 mm (top) and z-axis: −5 mm. y-axis: −12 mm (bottom); (right) Velocity magnitude overview along the first straight tube of the optimized reactor design for a mean residence time of 5 min showing every second cutplane of the 3D-plot in more detail. The enlarged representation of the mixing segment (black box, top) visualized the generation of radial flow due to SMEs by cutplanes every 5 mm. The following part of the straight tube (green box, bottom) showed the abatement of radial mixing by cutplanes every 10 mm. *Reprinted from Publication Bettermann S, Kandelhard F, Moritz H-U, Pauer W, Digital and lean development method for 3D-printed reactors based on CAD modeling and CFD simulation,* Chem Eng Res Des *152:71–84, 2019. https://doi.org/10.1016/j.cherd.2019.09.024 with permission from Elsevier, Copyright 2020.*

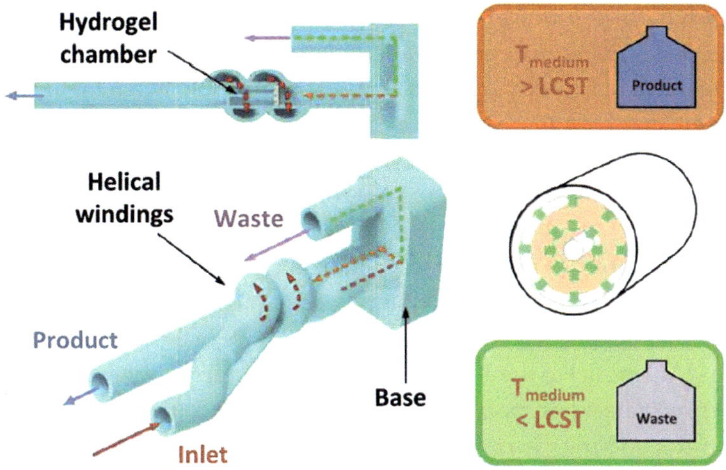

Fig. 11 Smart reactor module concept with self-actuated outlet control based on temperature-responsive hydrogel for a continuous process. Product of exothermic reaction is intended to trigger gel collapse (T > lower critical solution temperature (LCST)) by heat transfer in helical windings, which leads to opened product outlet. During interrupted reaction with resulting drop in temperature (T < LCST) gel is intended to swell in order to block passage through product outlet. *Reprinted from Publication Hu X, Karnetzke J, Fassbender M, et al.: Smart reactors—combining stimuli-responsive hydrogels and 3D printing,* Chem Eng J *387:123413, 2019. https://doi.org/10.1016/j.cej.2019.123413 with permission from Elsevier, Copyright 2020.*

Flow-through reactors were developed with hydrogel grids, each of which was produced additively. Three different enzyme reactors with alcohol hydrogenase, benzoyl formate decarboxylase, β-galactosidase were operated, screening four different enzymatic reactions within the reactors. An effectiveness factor of 6–14% could be achieved in batch experiments. A trend could be observed in the transfer from the batch to the flow reactor. The higher the effectiveness factor within the batch reactor, the better the performance in the flow reactor. However, non-optimal flow profiles within the flow-through reactor reduced the product yield. 3D-printed hydrogels usually have smaller structural channels at the outer than the inner channels. By introducing the hydrogels into the reactor and thus causing irregular deformation, they influence the flow profile, which has a negative effect on the obtained yields. However, it is still possible to generate an optimized reactor with the optimal amount of enzymes to maximize the yield (Schmieg et al., 2018).

4.2.2 Metal and inorganic reactors

Besides printing soft and even stimuli responsive materials, it is also possible to prepare metal reactors by AM. This might be favorable when highly reactive species are involved, more harsh reaction conditions are required or improved heat transfer is unavoidable. In this regard a flow-through reactor was fabricated by selective laser melting (SLM) for a two-stage difluoro methylation reaction with gaseous fluoroform. This reactor was made of stainless steel 316 L. Due to the fast exothermic reaction, a yield of 81% could be obtained after recrystallization. First test prints of 10 mm long tubes and different inner diameters, showed good reproducibility from a size of 0.6 mm on, depending on the CAD model to the actual additive manufactured inner diameter of the tube. A short heat transport path was achieved by placing the reaction channels on the cooling channel. A computer-aided fluid dynamics simulation (Computational Fluid Dynamics; CFD) was carried out here to evaluate the mixing geometry. The mentioned boundary conditions were chosen equal to the expectation of the experimental conditions. To characterize the reactor, a residence time distribution was recorded by a step experiment. The volume determined was 1.85 mL, which corresponds well with the design volume of 1.89 mL. This underlines the accuracy of the manufacturing method (Gutmann et al., 2017).

Implementation on inline sensors is fast and easy for 3D printed reactors and their geometries can be adjusted simultaneously or in a next step (Capel et al., 2017; Maier et al., 2019). This proved to allow fast and easy process investigation and optimization. Two different flow-through reactors were designed and printed with the aim to integrate inline and online analysis via a HPLC pump. This pump was coupled with a UV–Vis spectrometer. This set-up was then used to determine the optimum reaction conditions for two synthesis (a ketone functional group interconversion and a fused polycyclic heterocycle formation), via spectroscopic and chromatographic analysis with respect to conversion. The reactors used were manufactured using stereolithography methods and selective laser melting. These reactors are the first metallic multifunctional 3D printed metal fluidic devices known from the SLM process (Capel et al., 2017).

An actively stirred lab-scale tank cascade and a split-and-recombine stainless-steel reactor designed and printed were used for an oxidation reaction of a Grignard in homogeneous phase. The CAD drawing of the reactor as well as the respective printed reactor are shown in Fig. 12. The reactor consists of a pre-cooling section and a series of mixing elements. The sections are scalable, which enables easy numbering up. Moreover, this

A

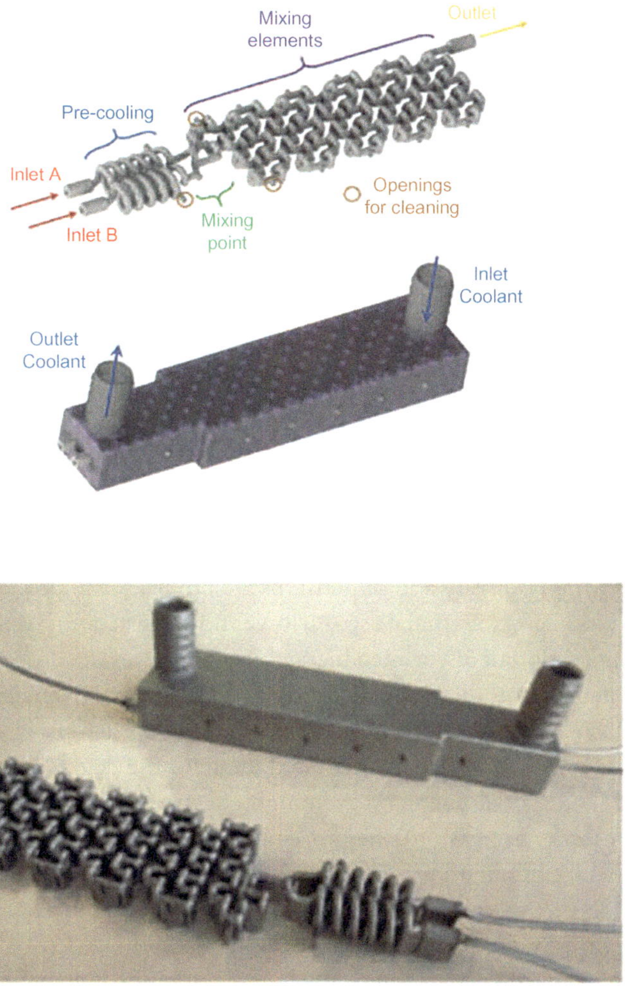

B

Fig. 12 (Left) CAD drawings of the split and recombine reactor: Overview of the reactor sections. By repeating the structure elements, the reactor can be scaled arbitrarily. The lower image shows a 3D printing support, which was generated through a Boolean operation to incorporate the reaction channel into the cooling shell. (Right) 3D printed split and recombine reactor for the oxidation of Grignard reagents with and without a cooling shell. After postprocessing, 1/16″ capillaries were welded to the in- and outlets for connection with standard HPLC equipment. *Reprinted with permission from reference Maier MC, Lebl R, Sulzer P, et al.: Development of customized 3D printed stainless steel reactors with inline oxygen sensors for aerobic oxidation of Grignard reagents in continuous flow,* React Chem Eng *4:393–401, 2019. https://doi.org/10.1039/c8re00278a published by The Royal Society of Chemistry.*

set-up is incorporated into a cooling shell. Multiple fiber optic oxygen sensors were integrated in order to observe the reaction course inline. The stainless-steel reactors were compared with a coiled micro PFA reactor with respect to performance. All reactors showed similar conversions, however the split-and-recombine stainless steel micro reactor designed by the authors had a significant higher selectivity. The concept is applicable in the hundred's milliliter range per day (Maier et al., 2019).

Inorganic materials are typical for heterogeneous catalysts in chemical industry. Consequently, approaches to print such catalysts or post-process 3D printed objects into the desired shape are being made. The oxidation of C_3H_6 was carried out mediated by designated printed catalyst structures in Al_2O_3 implied by stereolithography. Polyhedral open cell lattice catalyst substrates were optimized with respect to mass transfer and flow-through resistance. Systematic variation of the geometrical characteristics with the help of numerical simulations was powerful taking into account the current status of additive manufacturing (AM) techniques. Simulated results could be confirmed by experiments, and it was pointed out that choosing and controlling coating, porosity and dimensions are crucial for improved conversions, may reduce precious metal use and lower flow resistance (Papetti et al., 2018).

Application of 3D printed hierarchical structures of entirely TiO_2 for heterogeneous photocatalytic processes was demonstrated (Mendez-Arriaga et al., 2019). A solid square-lattice double-diagonal structures was obtained by robocasting and deposition and tested them for the degradation of acesulfame (ACE) as a model pollutant. Although effectivity was reduced compared to nano powder suspensions, they are easier to handle in application than suspensions and exhibited very good long-term stability (Mendez-Arriaga et al., 2019). Nickel-alumina based monolithic catalysts for the CO_2 hydrogenation to methane were prepared by 3D printing. Optimized structures regarding functionality and design could be achieved. Additionally, tomographic 'chemical imaging' (X-Ray Diffraction Computed Technology) was used in operando studies to follow the activation of active Ni-based species and their distribution under methanation conditions. Here, it could be shown that Ni species were distributed more homogeneously if 3D printed compared to other catalyst forms, such as beads or pellets (Middelkoop et al., 2019).

4.3 Additional and possible future applications

Given the importance of continuous processes in chemical industry, concepts for the development of chemical synthesis under continuous conditions should be taught early in undergraduate or graduate level depending on the studies subjects. Millireactors made of PP were proposed as useful items in teaching fluid dynamics, chemical kinetics, and reaction engineering. Examples of T and Y mixers using colored aqueous solutions are of course persuasive in the context of teaching (Tabassum et al., 2018). This might not only ease teaching for instructors, but also enhance understanding and fascination amongst students.

5. Conclusion and outlook

3D printing is a technology that has developed from a niche market to broad applications in the consumer market as well as research and development institutions. This is also mirrored by a steady increase of published articles and patents. While the steep increase in the field of instrumentation started in 2014, the research area, namely material science, chemistry and polymer science are following with a short delay of only 1–2 years. Here, especially polymer science has to be pointed to as polymers are the material of choice when it comes to AM. The advantages of polymers, especially with respect to processing and their wide range of achievable properties, are highly beneficial in the context of 3D printing. Various examples are reported that deal with the development of 3D printing equipment and technology, especially printing material. Besides using various types of well-known polymers, such as PLA, PET, PA, PEEK and more, fillers or post-processing techniques are studied in order to obtain the desired product properties, e.g., mechanical stability or biodegradability. Moreover, smart materials such as hydrogels or liquid crystal elastomers, which can act as sensors or controlling/switching elements are investigated.

AM technology is exploited to give freely designable objects, which are useful either in lab-scale applications, such as tailor-made and cheap reaction ware, lab equipment with optionally embedded sensors, or which are used within the chemical engineering context such as up-scaling approaches. Applications range from free catalyst design, packing structures to stirrer geometries and fluidized beds. Reactors are playing an outstanding role in chemical engineering. 3D printing enables free reactor design in the sense of functional prototyping, which opens up a wide field of reactor optimizations and also for reaction protocols. Extremely high surface-to-volume ratios can be reached, which is beneficial for example in the context of cooling or photoreactions. Coupling chemical functionality with 3D printing e.g. via surface modification or catalyst immobilization also gives exciting options, especially when coupled to high surface areas. In other approaches, smart reactors have been combined either with each other or with sensing applications and successfully tested to control reactions. All experimental approaches and developments can be supported and accelerated via the performance of corresponding CFD simulations. The strong combination of CFD and AM is increasingly adopted, especially if flow behavior is to be optimized.

Although mainly polymeric materials are used within 3D printing, metals or inorganic materials prove to be the better choice in some cases, especially if high chemical or thermal resistance, good (thermal) conductivity or special mechanical or catalytic activities are required. Additionally, reprinting of polymer devices after experience with them in metal or ceramics may be attractive in long term production. All those manifold possibilities and applications demonstrate that AM has the potential to shape the future in chemistry, material science and (chemical) engineering. Further developments are expected especially in the area of smart materials—either by printing them directly or creating them via post-processing—and the innovative design and optimization of operation units and reactors. In this context the smart combination of simulation-based design and fast experimental feedback loops by rapid prototyping can generate synergetic effects.

References

X. An, S. Tang, H. Du 2019, CN110845822 A.

Anzalone GC, Wijnen B, Pearce JM: Multi-material additive and subtractive prosumer digital fabrication with a free and open-source convertible delta RepRap 3-D printer, *Rapid Prototyp J* 21:506–519, 2015. https://doi.org/10.1108/RPJ-09-2014-0113.

Au AK, Huynh W, Horowitz LF, Folch A: Mikrofluidik aus dem 3D-Drucker, *Angew Chem* 128:3926–3946, 2016. https://doi.org/10.1002/ange.201504382.

A. Bandyopadhyay, S. Bose (editors), Additive manufacturing, CRC Press, Boca Raton, 2020.

Batchelder John S: *US 5653925 A,* 1995.

Batista M, Valerga AP, Salguero J, Fernandez-Vidal SR, Girot F: 2. State of the art of the fused deposition modeling using PLA: improving the performance. In Davim JP, editor: *Additive and subtractive manufacturing*, Berlin, Boston, 2019, De Gruyter, pp 59–112.

Bell CA: *3D printing with delta printers,* New York, NY, 2015, Apress.

Bertsch A, Zissi S, Jézéquel JY, Corbel S, André JC: Microstereophotolithography using a liquid crystal display as dynamic mask-generator, *Microsyst Technol* 3:42–47, 1997. https://doi.org/10.1007/s005420050053.

Bettermann S, Schroeter B, Moritz H-U, Pauer W, Fassbender M, Luinstra GA: Continuous emulsion copolymerization processes at mild conditions in a 3D-printed tubular bended reactor, *Chem Eng J* 338:311–322, 2018. https://doi.org/10.1016/j.cej.2018.01.038.

Bettermann S, Stuhr R, Moritz H-U, Pauer W: Customizable 3D-printed stirrers for investigation, optimization and scale-up processes of batch emulsion copolymerizations, *Chem Eng Sci* 206:50–62, 2019a. https://doi.org/10.1016/j.ces.2019.05.026.

Bettermann S, Kandelhard F, Moritz H-U, Pauer W: Digital and lean development method for 3D-printed reactors based on CAD modeling and CFD simulation, *Chem Eng Res Des* 152:71–84, 2019b. https://doi.org/10.1016/j.cherd.2019.09.024.

Bittner SM, Guo JL, Melchiorri A, Mikos AG: Three-dimensional printing of multilayered tissue engineering scaffolds, *Mater Today (Kidlington)* 21:861–874, 2018. https://doi.org/10.1016/j.mattod.2018.02.006.

Boley JW, Chaudhary K, Ober TJ, et al: High-operating-temperature direct ink writing of mesoscale eutectic architectures, *Adv Mater (Weinheim, Ger)* 29:, 2017, 1604778. https://doi.org/10.1002/adma.201604778.

Büscher N, Spille C, Kracht J, et al: Counter-currently operated reactive extractor with additively manufactured enzyme carrier structure, *Org Process Res Dev* , 2020. https://doi.org/10.1021/acs.oprd.0c00205.

Capel AJ, Edmondson S, Christie SDR, Goodridge RD, Bibb RJ, Thurstans M: Design and additive manufacture for flow chemistry, *Lab Chip* 13:4583–4590, 2013. https://doi.org/10.1039/c3lc50844g.

Capel AJ, Wright A, Harding MJ, et al: 3D printed fluidics with embedded analytic functionality for automated reaction optimisation, *Beilstein J Org Chem* 13:111–119, 2017. https://doi.org/10.3762/bjoc.13.14.

Capel AJ, Rimington RP, Lewis MP, Christie SDR: 3D printing for chemical, pharmaceutical and biological applications, *Nat Rev Chem* 2:422–436, 2018. https://doi.org/10.1038/s41570-018-0058-y.

Carneiro OS, Silva AF, Gomes R: Fused deposition modeling with polypropylene, *Mater Des* 83:768–776, 2015. https://doi.org/10.1016/j.matdes.2015.06.053.

Cheng Y, Chan KH, Wang X-Q, et al: Direct-ink-write 3D printing of hydrogels into biomimetic soft robots, *ACS Nano* 13:13176–13184, 2019. https://doi.org/10.1021/acsnano.9b06144.

Chivers PRA, Smith DK: Shaping and structuring supramolecular gels, *Nat Rev Mater* 4:463–478, 2019. https://doi.org/10.1038/s41578-019-0111-6.

Cordonier GJ, Sierros KA: Unconventional application of direct ink writing: surface force-driven patterning of low viscosity inks, *ACS Appl Mater Interfaces* 12:15875–15884, 2020. https://doi.org/10.1021/acsami.9b22976.

Cronin L: *US20150010461A1*, 2013.

Crump S S: *(Crump, S. Scott) US 5121329 A*, 1989.

Cump S.S., (Crump, S. Scott) US 5340433 A, 1992.

Dai J, Ogbeide O, Macadam N, et al: Printed gas sensors, *Chem Soc Rev* 49:1756–1789, 2020. https://doi.org/10.1039/c9cs00459a.

Deckard Carl R: *US 4863538 A,* 1986.

Deckard Carl R: *US 5597589 A,* 1994.

Dragone V, Sans V, Rosnes MH, Kitson PJ, Cronin L: 3D-printed devices for continuous-flow organic chemistry, *Beilstein J Org Chem* 9:951–959, 2013. https://doi.org/10.3762/bjoc.9.109.

Femmer T, Jans A, Eswein R, et al: High-throughput generation of emulsions and microgels in parallelized microfluidic drop-makers prepared by rapid prototyping, *ACS Appl Mater Interfaces* 7:12635–12638, 2015. https://doi.org/10.1021/acsami.5b03969.

Femmer T, Flack I, Wessling M: Additive manufacturing in fluid process engineering, *Chem Ing Tech* 88:535–552, 2016. https://doi.org/10.1002/cite.201500086.

Figuerola A, Medina DAV, Santos-Neto AJ, et al: Metal–organic framework mixed-matrix coatings on 3D printed devices, *Appl Mater Today* 16:21–27, 2019. https://doi.org/10.1016/j.apmt.2019.04.011.

Gal-Or E, Gershoni Y, Scotti G, et al: Chemical analysis using 3D printed glass microfluidics, *Anal Methods* 11:1802–1810, 2019. https://doi.org/10.1039/c8ay01934g.

S.R. Ganneboyina, N. Jada, S. Kali, S. Abdul Rauf, K. Sarangi 2018, IN2018-11030292.

Gelhausen MG, Feuerbach T, Schubert A, Agar DW: 3D printing for chemical process laboratories I: materials and connection principles, *Chem Eng Technol* 41:618–627, 2018. https://doi.org/10.1002/ceat.201700294.

Gibson I, Rosen D, Stucker B: The impact of low-cost AM systems. In Gibson I, Rosen D, Stucker B, editors: *Additive manufacturing technologies: 3D printing, rapid prototyping and direct digital manufacturing*, ed 2, New York, Heidelberg, Dodrecht, London, 2015a, Springer, pp 293–301.

Gibson I, Rosen D, Stucker B, editors: *Additive manufacturing technologies: 3D printing, rapid prototyping and direct digital manufacturing*, New York, Heidelberg, Dodrecht, London, 2015b, Springer.

Gładyszewski K, Skiborowski M: Additive manufacturing of packings for rotating packed beds, *Chem Eng Process Process Intensif* 127:1–9, 2018. https://doi.org/10.1016/j.cep.2018.02.024.

Carlos A. Grande, Jean-Luc Dubois, Juan C. Piquero Camblor, Ørnulv Vistad, Terje Didriksen, Richard Blom, Aud I. Spjelkavik and Duncan Akporiaye, 2015 Additive Manufacturing: Haute Couture for Chemical Industries, https://doi.org/10.13140/RG.2.1.1356.5520.

Gratson GM, Xu M, Lewis JA: Microperiodic structures: direct writing of three-dimensional webs, *Nature* 428:386, 2004. https://doi.org/10.1038/428386a.

Guba F, Tastan Ü, Gugeler K, Buntrock M, Rommel T, Ziegenbalg D: Rapid prototyping for photochemical reaction engineering, *Chem Ing Tech* 7:2155, 2018. https://doi.org/10.1002/cite.201800035.

Gunkel GP: *Betriebswirtschaftliche Perspektiven des 3D-Drucks,* Wiesbaden, 2020, Springer Fachmedien Wiesbaden.

Gutmann B, Köckinger M, Glotz G, et al: Design and 3D printing of a stainless steel reactor for continuous difluoromethylations using fluoroform, *React Chem Eng* 2:919–927, 2017. https://doi.org/10.1039/c7re00176b.

Gyak K-W, Vishwakarma NK, Hwang Y-H, Kim J, Yun H-s, Kim D-P: 3D-printed monolithic SiCN ceramic microreactors from a photocurable preceramic resin for the high temperature ammonia cracking process, *React Chem Eng* 4:1393–1399, 2019. https://doi.org/10.1039/c9re00201d.

Hampson SM, Rowe W, Christie SDR, Platt M: 3D printed microfluidic device with integrated optical sensing for particle analysis, *Sens Actuators B* 256:1030–1037, 2018. https://doi.org/10.1016/j.snb.2017.10.041.

Harding MJ, Brady S, O'Connor H, et al: 3D printing of PEEK reactors for flow chemistry and continuous chemical processing, *React Chem Eng* 5:, 2020, 728735. https://doi.org/10.1039/C9RE00408D.

Hartings MR, Ahmed Z: Chemistry from 3D printed objects, *Nat Rev Chem* 3:305–314, 2019. https://doi.org/10.1038/s41570-019-0097-z.

Hein P, Lauschke G, Ott S, Schierholz W, Schmidt U: 1998 *EP0996498B1*, 1997.

Hnát J, Kodým R, Denk K, Paidar M, Žitka J, Bouzek K: Design of a zero-gap laboratory-scale polymer electrolyte membrane alkaline water electrolysis stack, *Chem Ing Tech* 91:821–832, 2019.

Hock S, Rose M: 3D-structured monoliths of nanoporous polymers by additive manufacturing, *Chem Ing Tech* 92:525–531, 2020. https://doi.org/10.1002/cite.201900149.

Hu X, Karnetzke J, Fassbender M, et al: Smart reactors—combining stimuli-responsive hydrogels and 3D printing, *Chem Eng J* 387:, 2019, 123413. https://doi.org/10.1016/j.cej.2019.123413.

B. Huang 2019. CN110698807.

Huang X, Wang D, He B, Liu Q, Hu L, Jiang G: A 3D-printed modularized purification system for rapid, high-throughput MALDI-MS analysis of small-volume biological samples, *Chem Commun* 56:1637–1640, 2020.

Hull Charles W: *US 4575330 A,* 1984.

Hutmacher DW, Schantz T, Zein I, Ng KW, Teoh SH, Tan KC: Mechanical properties and cell cultural response of polycaprolactone scaffolds designed and fabricated via fused deposition modeling, *J Biomed Mater Res* 55:203–216, 2001. https://doi.org/10.1002/1097-4636(200105)55:2<203:aid-jbm1007>3.0.co;2-7.

Ji Y, Ma Y, Ma Y, Asenbauer J, Passerini S, Streb C: Water decontamination by polyoxometalate-functionalized 3D-printed hierarchical porous devices, *Chem Commun (Cambridge, UK)* 54:3018–3021, 2018. https://doi.org/10.1039/c8cc00821c.

R. Jia, X. Ding, X. Xu, Z. Gan, Z. Hui, C. Zhao, R. Liu, X. He 2019. CN110669305.

Jiang P, Yan C, Guo Y, et al: Direct ink writing with high-strength and swelling-resistant biocompatible physically crosslinked hydrogels, *Biomater Sci* 7:1805–1814, 2019. https://doi.org/10.1039/c9bm00081j.

Katic V, Dos Santos PL, Dos Santos MF, et al: 3D printed graphene electrodes modified with Prussian blue: emerging electrochemical sensing platform for peroxide detection, *ACS Appl Mater Interfaces* 11:35068–35078, 2019. https://doi.org/10.1021/acsami.9b09305.

Kazenwadel F, Biegert E, Wohlgemuth J, Wagner H, Franzreb M: A 3D-printed modular reactor setup including temperature and pH control for the compartmentalized implementation of enzyme cascades, *Eng Life Sci* 16:560–567, 2016. https://doi.org/10.1002/elsc.201600007.

Kitson PJ, Rosnes MH, Sans V, Dragone V, Cronin L: Configurable 3D-printed millifluidic and microfluidic 'lab on a chip' reactionware devices, *Lab Chip* 12:3267–3271, 2012. https://doi.org/10.1039/c2lc40761b.

Kitson PJ, Glatzel S, Chen W, Lin C-G, Song Y-F, Cronin L: 3D printing of versatile reactionware for chemical synthesis, *Nat Protoc* 11:920–936, 2016. https://doi.org/10.1038/nprot.2016.041.

Kluczyński J, Śnieżek L, Kravcov A, et al: The examination of restrained joints created in the process of multi-material FFF additive manufacturing technology, *Materials (Basel)* 13:903, 2020. https://doi.org/10.3390/ma13040903.

Koffler J, Zhu W, Qu X, et al: Biomimetic 3D-printed scaffolds for spinal cord injury repair, *Nat Med* 25:263–269, 2019. https://doi.org/10.1038/s41591-018-0296-z.

Kontodina T, Tzetzis D, Davim JP, Kyratsis P: 5. Additive manufacturing for patient-specific medical use. In Davim JP, editor: *Additive and subtractive manufacturing*, Berlin, Boston, 2019, De Gruyter, pp 199–228.

Kotz F, Risch P, Helmer D, Rapp BE: High-performance materials for 3D printing in chemical synthesis applications, *Adv Mater (Weinheim, Ger)* 31, 2019, e1805982. https://doi.org/10.1002/adma.201805982.

Lahtinen E, Turunen L, Hänninen MM, Kolari K, Tuononen HM, Haukka M: Fabrication of porous hydrogenation catalysts by a selective laser sintering 3D printing technique, *ACS Omega* 4:12012–12017, 2019. https://doi.org/10.1021/acsomega.9b00711.

Latocha J, Wojasiński M, Jurczak K, Gierlotka S, Sobieszuk P, Ciach T: Precipitation of hydroxyapatite nanoparticles in 3D-printed reactors, *Chem Eng Process Process Intensif* 133:221–233, 2018. https://doi.org/10.1016/j.cep.2018.10.001.

Lewis JA: Direct ink writing of 3D functional materials, *Adv Funct Mater* 16:2193–2204, 2006. https://doi.org/10.1002/adfm.200600434.

Ligon SC, Liska R, Stampfl J, Gurr M, Mülhaupt R: Polymers for 3D printing and customized additive manufacturing, *Chem Rev* 117:10212–10290, 2017. https://doi.org/10.1021/acs.chemrev.7b00074.

Lin C-G, Zhou W, Xiong X-T, et al: Digital control of multistep hydrothermal synthesis by using 3D printed reactionware for the synthesis of metal-organic frameworks, *Angew Chem Int Ed* 57:16716–16720, 2018. https://doi.org/10.1002/anie.201810095.

Lopes LR, Silva AF, Carneiro OS: Multi-material 3D printing: the relevance of materials affinity on the boundary interface performance, *Addit Manuf* 23:45–52, 2018. https://doi.org/10.1016/j.addma.2018.06.027.

Maier MC, Lebl R, Sulzer P, et al: Development of customized 3D printed stainless steel reactors with inline oxygen sensors for aerobic oxidation of Grignard reagents in continuous flow, *React Chem Eng* 4:393–401, 2019. https://doi.org/10.1039/c8re00278a.

McDonough JR, Law R, Reay DA, Zivkovic V: Fluidization in small-scale gas-solid 3D-printed fluidized beds, *Chem Eng Sci* 200:294–309, 2019. https://doi.org/10.1016/j.ces.2019.01.048.

Medina DAV, Figuerola A, Rodriguez F, et al: Hyperporous carbon-coated 3D printed devices, *Appl Mater Today* 14:29–34, 2019. https://doi.org/10.1016/j.apmt.2018.11.001.

Mendez-Arriaga F, La Calleja Ed, Ruiz-Huerta L, Caballero-Ruiz A, Almanza R: TiO_2 3D structures for environmental purposes by additive manufacturing: photoactivity test and reuse, *Mater Sci Semicond Process* 100:35–41, 2019. https://doi.org/10.1016/j.mssp.2019.04.034.

Middelkoop V, Vamvakeros A, de Wit D, et al: 3D printed Ni/Al_2O_3 based catalysts for CO_2 methanation—a comparative and operando XRD-CT study, *J CO_2 Util* 33:478–487, 2019. https://doi.org/10.1016/j.jcou.2019.07.013.

Morissette SL, Lewis JA, Clem PG, Cesarano J, Dimos DB: Direct-write fabrication of $Pb(Nb,Zr,Ti)O_3$ devices: influence of paste rheology on print morphology and component properties, *J Am Ceram Soc* 84:2462–2468, 2001. https://doi.org/10.1111/j.1151-2916.2001.tb01036.x.

Mothes H: *WO2016026475A1,* 2014.

Nawada SH, Aalbers T, Schoenmakers PJ: Freeze-thaw valves as a flow control mechanism in spatially complex 3D-printed fluidic devices, *Chem Eng Sci* 207:1040–1048, 2019. https://doi.org/10.1016/j.ces.2019.07.036.

Neukäufer J, Hanusch F, Kutscherauer M, Rehfeldt S, Klein H, Grützner T: Methodology for the development of additively manufactured packings in thermal separation technology, *Chem Eng Technol* 42:1970–1977, 2019. https://doi.org/10.1002/ceat.201900220.

Neumaier JM, Madani A, Klein T, Ziegler T: Low-budget 3D-printed equipment for continuous flow reactions, *Beilstein J Org Chem* 15:558–566, 2019. https://doi.org/10.3762/bjoc.15.50.

Nguyen DT, Aines RD, Baker SE, et al: *WO2018175812A3,* 2017.

Nielsen JB, Hanson RL, Almughamsi HM, Pang C, Fish TR, Woolley AT: Microfluidics: innovations in materials and their fabrication and functionalization, *Anal Chem* 92:150–168, 2020.

Ning F, Cong W, Qiu J, Wei J, Wang S: Additive manufacturing of carbon fiber reinforced thermoplastic composites using fused deposition modeling, *Compos Part B Eng* 80:369–378, 2015. https://doi.org/10.1016/j.compositesb.2015.06.013.

Okafor O, Weilhard A, Fernandes JA, Karjalainen E, Goodridge R, Sans V: Advanced reactor engineering with 3D printing for the continuous-flow synthesis of silver nanoparticles, *React Chem Eng* 2:129–136, 2017. https://doi.org/10.1039/c6re00210b.

Okafor O, Robertson K, Goodridge R, Sans V: Continuous-flow crystallisation in 3D-printed compact devices, *React Chem Eng* 4:1682–1688, 2019. https://doi.org/10.1039/c9re00188c.

Oliaei SNB, Nasseri B: 6. Stereolithography and its applications. In Davim JP, editor: *Additive and subtractive manufacturing*, Berlin, Boston, 2019, De Gruyter, pp 229–250.

O'Neil GD, Ahmed S, Halloran K, Janusz JN, Rodríguez A, Terrero Rodríguez IM: Single-step fabrication of electrochemical flow cells utilizing multi-material 3D printing, *Electrochem Commun* 99:56–60, 2019. https://doi.org/10.1016/j.elecom.2018.12.006.

Paleń D, Mantiuk R: Do-it-yourself multi-material 3d printer for rapid manufacturing of complex luminaries. In Kacprzyk J, editor: *Advances in soft and hard computing*, Cham, 2019, Springer International Publishing, pp 469–480.

Papetti V, Dimopoulos Eggenschwiler P, Della Torre A, Lucci F, Ortona A, Montenegro G: Additive manufactured open cell polyhedral structures as substrates for automotive catalysts, *Int J Heat Mass Transf* 126:1035–1047, 2018. https://doi.org/10.1016/j.ijheatmasstransfer.2018.06.061.

Parra-Cabrera C, Achille C, Kuhn S, Ameloot R: 3D printing in chemical engineering and catalytic technology: structured catalysts, mixers and reactors, *Chem Soc Rev* 47:209–230, 2018. https://doi.org/10.1039/c7cs00631d.

Penny MR, Rao ZX, Peniche BF, Hilton ST: Modular 3D printed compressed air driven continuous-flow systems for chemical synthesis, *Eur J Org Chem* 2019:3783–3787, 2019. https://doi.org/10.1002/ejoc.201900423.

Procopio A, Tewari D: Opportunities and challenges of 3D-printed pharmaceutical dosage forms. In Shegokar R, editor: *Drug delivery trends: expectations and realities of multifunctional drug delivery systems*, San Diego, 2020, Elsevier, pp 15–44.

Pu X, Su Y: Heterogeneous catalysis in microreactors with nanofluids for fine chemicals syntheses: benzylation of toluene with benzyl chloride over silica-immobilized $FeCl_3$ catalyst, *Chem Eng Sci* 184:200–208, 2018. https://doi.org/10.1016/j.ces.2018.03.049.

Ren L, Li B, He Y, et al: Programming shape-morphing behavior of liquid crystal elastomers via parameter-encoded 4D printing, *ACS Appl Mater Interfaces* 13:15562–15572, 2020. https://doi.org/10.1021/acsami.0c00027.

RepRap.org: Category:Delta: All Projects With a Symmetrical Delta Configuration. Linear or Angular Motion With a Moving Extruder, 2014a. https://reprap.org/wiki/Category:Delta. accessed 22 April 2020.

RepRap.org: Category:Cartesian-XY-head: all projects with the extruder head on an XY translation system and an independent Z axis, 2014b. https://reprap.org/wiki/Category:Cartesian-XY-head. accessed 22 April 2020.

RepRap.org: Category:Cartesian-XZ-head: all projects with the head moving in the X and Z plane with the build platform moving in the Y direction, 2014c. https://reprap.org/wiki/Category:Cartesian-XY-head. accessed 22 April 2020.

RepRap.org: Fused Filament Fabrication—RepRap, 2019. https://reprap.org/wiki/Fused_filament_fabrication.

Römpp H, Zorll U, Adler H-JP, editors: *Römpp-Lexikon Lacke und Druckfarben*, Stuttgart, 1998, Thieme.

Rossi S, Porta R, Brenna D, Puglisi A, Benaglia M: Stereoselective catalytic synthesis of active pharmaceutical ingredients in homemade 3D-printed mesoreactors, *Angew Chem Int Ed Engl* 56:4290–4294, 2017. https://doi.org/10.1002/anie.201612192.

Sadasivuni KK, Deshmukh K, Al-Maadeed MA-A, editors: *3D and 4D printing of polymer nanocomposite materials: processes, applications, and challenges*, Amsterdam, 2020, Elsevier.

Sahoo P, Das SK: Emerging trends in additive and subtractive manufacturing. In Davim JP, editor: *Additive and subtractive manufacturing*, 1:Berlin, Boston, 2019, De Gruyter, pp 1–58.

Sanchez Diaz-Marta A, Yanez S, Tubio CR, et al: Multicatalysis combining 3D-printed devices and magnetic nanoparticles in one-pot reactions: steps forward in compartmentation and recyclability of catalysts, *ACS Appl Mater Interfaces* 11:25283–25294, 2019. https://doi.org/10.1021/acsami.9b08119.

Sathies T, Senthil P, Anoop MS: A review on advancements in applications of fused deposition modelling process, *Rapid Prototyp J* 26(4):669–687, 2020. https://doi.org/10.1108/RPJ-08-2018-0199.

Schmieg B, Döbber J, Kirschhöfer F, Pohl M, Franzreb M: Advantages of hydrogel-based 3d-printed enzyme reactors and their limitations for biocatalysis, *Front Bioeng Biotechnol* 6:211, 2018. https://doi.org/10.3389/fbioe.2018.00211.

Scotti G, Nilsson SME, Haapala M, et al: A miniaturised 3D printed polypropylene reactor for online reaction analysis by mass spectrometry, *React Chem Eng* 2:299–303, 2017. https://doi.org/10.1039/c7re00015d.

Sinha SK: Additive manufacturing (AM) of medical devices and scaffolds for tissue engineering based on 3D and 4D printing. In Sadasivuni KK, Deshmukh K, Al-Maadeed MA-A, editors: *3D and 4D printing of polymer nanocomposite materials: processes, applications, and challenges*, Amsterdam, 2020, Elsevier, pp 119–160.

Smalley Dennis R, Vorgitch Thomas J, Manners Chris R, Hull Charles W, Vandorin Stacie L: *US 5597520 A,* 1994.

Snyder TJ, Andrews M, Weislogel M, et al: 3D systems' technology overview and new applications in manufacturing, engineering, science, and education, 3D print, *Addit Manuf* 1:169–176, 2014. https://doi.org/10.1089/3dp.2014.1502.

Song JH, Edirisinghe MJ, Evans JRG: Formulation and multilayer jet printing of ceramic inks, *J Am Ceram Soc* 82:3374–3380, 1999. https://doi.org/10.1111/j.1151-2916.1999.tb02253.x.

Stark AK: Manufactured chemistry: rethinking unit operation design in the age of additive manufacturing, *AIChE J* 64:1162–1173, 2018. https://doi.org/10.1002/aic.16118.

Stephan D: Chemie unter Druck: Was bedeutet der Siegeszug der additiven Fertigung?, 2019. https://www.mission-additive.de/amp/chemie-unter-druck-was-bedeutet-der-siegeszug-der-additiven-fertigung-a-846702/.

Stiefel S, Lölsberg J, Kipshagen L, Möller-Gulland R, Wessling M: Controlled depolymerization of lignin in an electrochemical membrane reactor, *Electrochem Commun* 61:49–52, 2015. https://doi.org/10.1016/j.elecom.2015.09.028.

Stolz B, Mülhaupt R: Cellular, mineralized, and programmable cellulose composites fabricated by 3D printing of aqueous pastes derived from paper wastes and microfibrillated cellulose, *Macromol Mater Eng* 305, 2020, 1900740. https://doi.org/10.1002/mame.201900740.

Sun Z, Velasquez-Garcia LF: Monolithic FFF-printed, biocompatible, biodegradable, dielectric-conductive microsystems, *J Microelectromech Syst* 26:1356–1370, 2017. https://doi.org/10.1109/JMEMS.2017.2746627.

Szojka A, Lalh K, Andrews SHJ, Jomha NM, Osswald M, Adesida AB: Biomimetic 3D printed scaffolds for meniscus tissue engineering, *Bioprinting* 8:1–7, 2017. https://doi.org/10.1016/j.bprint.2017.08.001.

Tabassum T, Iloska M, Scuereb D, et al: Development and application of 3D printed mesoreactors in chemical engineering education, *J Chem Educ* 95:783–790, 2018. https://doi.org/10.1021/acs.jchemed.7b00663.

Taheri H, Sarin Y, Ozero B: *US9340470B2,* 2012.

United States Patent and Trademark Office: Type Drawing—FDM: Serial Number: 74133656. Registration Number: 1663961, 2020. http://tmsearch.uspto.gov/bin/showfield?f=doc&state=4801:nhukl2.2.45.

Wittbrodt BT, Glover AG, Laureto J, et al: Life-cycle economic analysis of distributed manufacturing with open-source 3-D printers, *Mec Dent* 23:713–726, 2013. https://doi.org/10.1016/j.mechatronics.2013.06.002.

Wong JY, Pfahnl AC: 3D printing of surgical instruments for long-duration space missions, *Aviat Space Environ Med* 85:758–763, 2014. https://doi.org/10.3357/ASEM.3898.2014.

Wu X., Tu Y., Li X. and Chen X., 2019. CN110804193 A.

Wu B, Sufi A, Ghosh Biswas R, et al: Direct conversion of McDonald's waste cooking oil into a biodegradable high-resolution 3D-printing resin, *ACS Sustain Chem Eng* 8(2):1171–1177, 2020. https://doi.org/10.1021/acssuschemeng.9b06281.

Xu M, Gratson GM, Duoss EB, Shepherd RF, Lewis JA: Biomimetic silicification of 3D polyamine-rich scaffolds assembled by direct ink writing, *Soft Matter* 2:205, 2006. https://doi.org/10.1039/b517278k.

Yang H, Zhao Y-Y, Zheng M-L, et al: Stepwise optimized 3D printing of arbitrary 3D structures at millimeter scale with high precision surface, *Macromol Mater Eng* 304(11), 2019, 1900400. https://doi.org/10.1002/mame.201900400.

Yuan S, Shen F, Chua CK, Zhou K: Polymeric composites for powder-based additive manufacturing: materials and applications, *Prog Polym Sci* 91:141–168, 2019. https://doi.org/10.1016/j.progpolymsci.2018.11.001.

Yun X, Lu B, Xiong Z, et al: Direct 3D printing of a graphene oxide hydrogel for fabrication of a high areal specific capacitance microsupercapacitor, *RSC Adv* 9:29384–29395, 2019. https://doi.org/10.1039/C9RA04882K.

Zhang D, Xiao J, Guo Q, Yang J: 3D-printed highly porous and reusable chitosan monoliths for Cu(II) removal, *J Mater Sci* 54:6728–6741, 2019a. https://doi.org/10.1007/s10853-019-03332-y.

Zhang H, Kopfmüller T, Achermann R, et al: Accessing multidimensional mixing via 3D printing and showerhead micromixer design, *AICHE J* 66(4), 2019, e16873. https://doi.org/10.1002/aic.16873.

Zhao L, Zeng G, Gu Y, et al: Nature inspired fractal tree-like photobioreactor via 3D printing for CO_2 capture by microaglae, *Chem Eng Sci* 193:6–14, 2019. https://doi.org/10.1016/j.ces.2018.08.057.

CHAPTER FIVE

Renewable feedstocks in emulsion polymerization: Coating and adhesive applications

Miren Aguirre, Shaghayegh Hamzehlou, Edurne González, and Jose R. Leiza*

POLYMAT and Kimika Aplikatua Saila, Kimika Fakultatea, University of the Basque Country UPV-EHU, Joxe Mari Korta Zentroa, Donostia-San Sebastian, Spain

*Corresponding author: e-mail address: jrleiza@ehu.eus

Contents

Abstract

Industrialization has remarkably improved our life quality at the cost of an increase in the environmental pollution. Social awareness about environmental impact of our actions and the climate change is significantly increasing and environmental regulations are pushing the industry toward the use of non-toxic and environmentally friendly products. Consequently, the number of scientific works dealing with the use of renewable reagents has increased substantially in the last decade. In this context, this chapter highlights the green and sustainable alternatives that have been used so far in the open literature for the production of waterborne polymeric dispersions. A special focus on coatings and adhesives applications has been done. The most relevant renewable resource reagents such as (macro)monomers, surfactants, crosslinkers, chain transfer agents and nanofillers are summarized, paying special attention to emulsion and miniemulsion polymerization processes. In addition, some guidelines for the sustainability assessment of new reagents and processes are presented. Therefore, this chapter serves as guide for scientist interested in the development of sustainable waterborne coatings and adhesives.

Advances in Chemical Engineering, Volume 56
ISSN 0065-2377
https://doi.org/10.1016/bs.ache.2020.07.004

1. Introduction

Polymer materials emerged at the beginning of the XIX century as materials to revolutionize the market, improving human's standard of living, offering enhanced properties and cheap prices. Since then, the annual polymer production has been increasing and it is expected to continue growing in the upcoming years. The worldwide polymer production during 2020 is expected to be around 400 million of tones, but this number will grow up to 1000 million of tones by 2050 (McKinsey, 2019). Due to the vast production of polymers, regulations are necessary to avoid dramatic environmental consequences such as the acceleration of the climate change or the accumulation of plastics in the oceans and landfills. In this line, during the last years, several scientist have focused their research to find strategies to decrease the dependence on raw materials derived from fossil fuels, the improvement of the polymer waste treatment as well as the reduction of greenhouse gases emissions during their production. Waterborne polymer dispersions are good candidates to produce more sustainable polymer products and they are mainly produced by emulsion polymerization even if miniemulsion, microemulsion, suspension and dispersion polymerization are also used (Asua, 2007). They offer some advantages with respect to solvent borne products; since water is used as solvent and the amount of volatile organic compounds (VOCs) emitted into the atmosphere is decreased considerably. Moreover, as the polymer particles are dispersed in the aqueous phase, for the same polymer content as the solvent-based counterpart, the viscosity of the medium is much lower improving the processability of the final product. Besides, the viscosity of the medium can also be controlled with the particle size distribution obtaining high solids content latexes with low viscosity (Mariz et al., 2011). On the other hand, the use of water as the continuous phase improves the heat removal capacity during the polymerization process enabling a better temperature control. Additionally, emulsion polymerization offers a unique feature due to the compartmentalization of the growing radicals in the polymer particles, thus both, high polymerization rates and molar masses can be obtained simultaneously (Asua, 2004).

Waterborne dispersions are widely used for many applications like coatings, adhesives, drug delivery systems, textiles, cosmetics, automotive industry and so on (Asua, 1997; Urban and Takamura, 2002a). More recently novel ways of processing polymer latexes to produce nanofibers (by electrospinning) (Agarwal and Greiner, 2011; De San Luis et al., 2018) and additive manufacturing (Vat photopolymerization (VP) or stereolithography) to produce objects in 3D with light resolution (Scott et al., 2020) have emerged spanning the applications of these waterborne polymeric dispersions. According to Grand View Research, emulsion polymerization product market is dominated by coatings and adhesives, being their manufacture more than half of the total waterborne polymer production in 2019 (Grand View Research, 2019).

A typical formulation with the most common reagent used in emulsion polymerization is presented in Table 1 with a solids content around 55%.

Although water is used as continuous phase, the other half of the recipe, around the 50%–55%, consists of raw materials which predominantly come from fossil fuel resources. In the last years scientists, either from academia or industry, are directing their research efforts toward the use of less hazardous and more sustainable chemicals. Not only monomers have been investigated, which are the major components of a polymer latex, but also conventional and reactive emulsifiers, crosslinkers, chain transfer agents (CTA) as well as fillers. These novel materials mainly come from biomass feedstock (see Fig. 1) (Cummings et al., 2019a; Molina-Gutiérrez et al., 2019; Murawski and Quirino, 2018; Petri, 2015).

Biobased polymers are increasing in popularity as it can be observed from the constantly increasing number of publications and reviews (see Fig. 2). It is important to note that the use of renewable feedstock as raw materials for the production of new polymers is not always a sustainable process. In some cases, the harsh conditions needed to extract the targeted product, the required modifications that imply long synthesis and purification processes, and their low performance compared to their oil-based counterparts

Table 1 Typical formulation for emulsion polymerization.

Reagent	**wt (%)**
Monomer	50–55
Functional monomer	0–5
Surfactant	0.5–3
CTA	0–10
Crosslinker	0–5
Initiator	0.5
Water	45

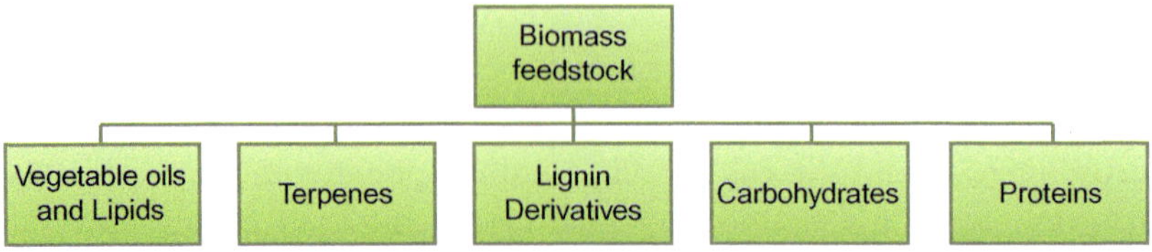

Fig. 1 Biomass classification. *Reprinted with permission from Molina-Gutiérrez S, Ladmiral V, Bongiovanni R, et al: Radical polymerization of biobased monomers in aqueous dispersed media,* Green Chem *21:36–53, 2019. doi: 10.1039/c8gc02277a.*

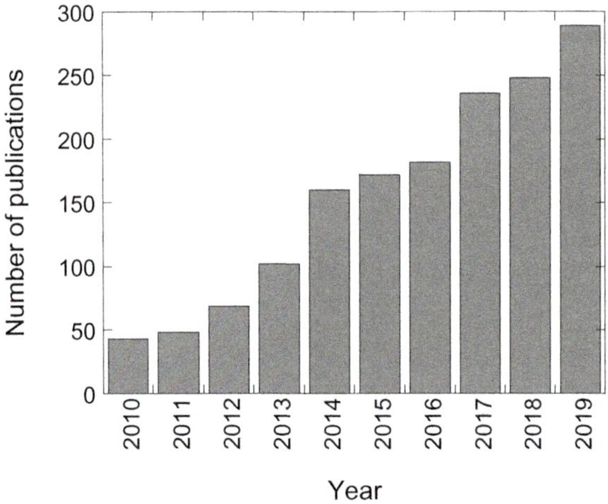

Fig. 2 Number of publications per year using the term "biobased polymers." *Source: Web of Science.*

(with very cheap prices) make these biobased materials not that environmentally friendly, increasing the final product's carbon footprint for instance.

Sustainable design is guided by principles such as the "*12 principles of Green Chemistry*" (Anastas and Warner, 1998) and the "*12 principles of Green Engineering*" (Anastas and Zimmerman, 2003). These principles serve as guide for scientists to minimize or eliminate the environmental impact of chemical products and processes based on fundamental science and engineering aspects, respectively.

More recently, Dubé and coworkers applied these principles to polymer production processes (Dubé and Salehpour, 2014) and more precisely to emulsion polymerization (Zhang and Dube, 2017). Zhang and Dube (2017) claimed that the sustainability of a process goes well beyond the replacement of toxic and/or nonrenewable materials. Thus, they explored other engineering issues in emulsion polymerization such as the waste prevention, the energy efficiency maximization and the minimization of potential accidents.

In order to evaluate the environmental impact of an entire chemical process, tools such as Life Cycle Assessment (LCA) should be considered. LCA is an assessment method which consists of quantitatively evaluating all the potential environmental impacts of a product, considering its whole lifecycle (from design through manufacture and use to recycling or reuse) (Caillol, 2014; Guinée, 2001; Tabone et al., 2010).

In 2015, McElroy et al. (2015) developed a unified CHEM21 metrics toolkit to evaluate sustainability of new chemical reactions as a quantitative extension of "*The 12 Principles of Green Chemistry*." This toolkit provides a comprehensive and holistic range of criteria for measuring how green a reaction is, giving the possibility to monitor, measure, and compare different methodologies.

The toolkit consists of four levels with increasing complexity toward the design of an optimized process. These levels are: "Zero Pass," "First Pass," "Second Pass" and "Third Pass" (Fig. 3). The "Zero Pass" toolkit level is

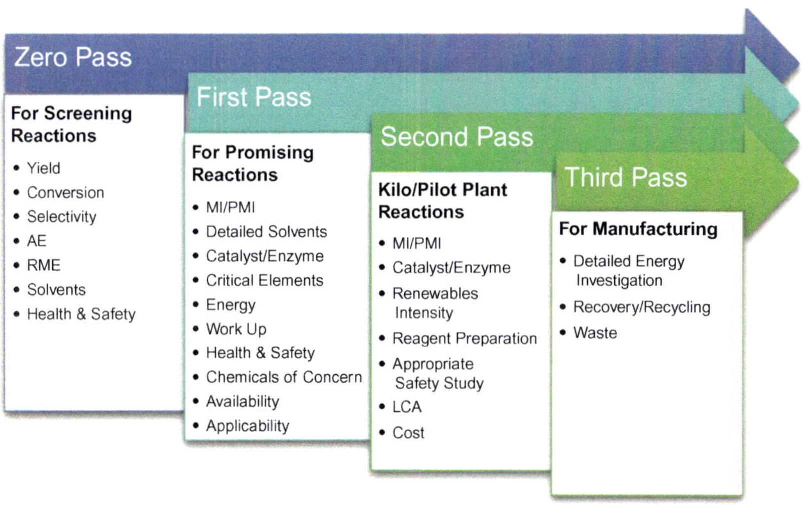

Fig. 3 Structure of the CHEM21 metrics toolkit levels showing the parameters covered at each Pass. AE: atom economy, RME: reaction mass efficiency, MI: mass intensity, PMI: process mass intensity and LCA: life cycle assessment. *Reprinted with permission from McElroy CR, Constantinou A, Jones LC, et al: Towards a holistic approach to metrics for the 21st century pharmaceutical industry,* Green Chem *17:3111–3121, 2015. doi: 10.1039/c5gc00340g.*

focused on identifying issues in the use of highly hazardous solvents and chemicals as well as providing an initial assessment of the reaction efficiency. Reactions that show potential at the "Zero Pass" should be further evaluated by the "First Pass," which is also designed for laboratory research at the discovery stage. If the new synthesis procedure passes the "First Pass," the "Second Pass" and "Third Pass" evaluation levels should be assessed. These two last passes are considered "industrial" toolkits and are design for reactions at pilot scale and beyond. The key parameters that are evaluated in the "Second Pass" and "Third Pass" are considered too time consuming in earlier stages or not appropriate on a smaller scale (below 2 L), such as, a detailed energy investigation, the costs, the renewable intensity, LCA and the resulting waste.

Although during the last decade, several review articles have been published about the production, performance and applications of polymers derived from renewable sources (Cummings et al., 2019a; Imam et al., 2013; Kobayashi, 2013; Magalhães et al., 2019; Molina-Gutiérrez et al., 2019; Murawski and Quirino, 2018; Petri, 2015; Smeets et al., 2017), only few of these works have been focused on the use of renewable resources for the production of waterborne dispersed polymers (Cummings et al., 2019a; Molina-Gutiérrez et al., 2019; Smeets et al., 2017). Smeets et al. (2017) focused their review on the use of carbohydrates as surfactant, initiator, grafting site or macromonomer for the synthesis of carbohydrate functionalized hybrid latex particles. Two years later, Cummings et al. (2019a) published a review about the attempts to modify and incorporate various types of starch in emulsion polymerizations. Lately, Molina-Gutiérrez et al. (2019) published a review paper on the development of novel biobased monomers specially designed for free radical polymerization reactions that could be used in (mini)emulsion and suspension polymerizations. The use of biobased surfactants, crosslinking agents or fillers, however, was out of the scope of their work. In addition, it should be mentioned that none of these reviews was centered on the final applications of the biobased latexes.

This chapter is focused on reviewing green and sustainable alternatives to the use of fossil fuel derivatives as raw materials in the production of waterborne polymeric dispersions for coating and adhesive applications. Therefore, the work presented herein serves as guide for scientist that are interested in the use of renewable feedstock's for the development of waterborne coatings and adhesives.

2. Renewable resource reagents in waterborne polymeric dispersions for coatings and adhesives applications

The global coatings industry has an estimated production of 40 million tons per year. Even though traditionally solvent-based coatings have been the market leaders, the increasingly strict environmental regulations have pushed coating industry to move toward more environmentally friendly technologies, such as waterborne products. Waterborne disperse polymers, are being increasingly used for textile coatings, paper coating, as binder in household and industrial paints, as well as for various types of protective coatings for automobiles, aircraft, ships, and more (Asua, 1997; Keddie and Joseph, 2010; Urban and Takamura, 2002a).

On the other hand, waterborne adhesives and sealants are also gaining importance, more precisely pressure sensitive adhesives (PSA). The main feature of the PSAs is that the bonding between different surfaces is established by applying a light pressure on them (Benedek and Heymans, 1997; Creton, 2003; Gutsche, 2005). They have been widely used since the 19th century for many applications. Three main categories of PSA can be found in the market, which are directly related to their polymerization process: solvent-borne, water-borne, hot-melt and radiation cured. The general production of PSA is expected to grow in the upcoming years, however, due to the environmental regulations the waterborne PSAs are the ones that are gaining more interest (Tellica, 2019).

As it has been mentioned in the introduction of this chapter, in the recipe of waterborne polymeric dispersions monomers, functional monomers, surfactants, crosslinkers, CTA and initiators are used. These formulations are designed and tuned depending on the final application of the dispersion. In this section of the chapter, the different renewable reagents that have been used for coatings and adhesives applications will be described. The main difference between a coating and adhesive formulation remains in the glass transition temperature (Tg) of the polymer. In a coating, the Tg should be below the application temperature (normally room temperature), in order to form a coherent film, but high enough to give good mechanical properties and to avoid problems such as dirt pick up and blocking. These counteracting requirements can be overcome using structured polymer particles or latexes able to self-crosslink during film formation

(Barquero et al., 2019, 2020; González et al., 2006a,b; Limousin et al., 2019a,b; Min et al., 1983; Stubbs and Sundberg, 2008). The Tg of the polymers used for coatings is usually around 10–15 °C. On the contrary, adhesives should have a Tg below −20 °C to obtain a good adhesion (Benedek and Heymans, 1997; Creton, 2003; Urban and Takamura, 2002b).

The principal characteristic of a waterborne PSA product is that it offers a good balance of tack, peel and shear resistances (Benedek and Heymans, 1997). The former, give the material the ability to adhere quickly to the surface. The second is the capacity to resist the removal by peeling, whereas the last one is responsible to resist flow when shear forces are applied. The balance between these three properties will determine the end use properties of a PSA. In a common PSA formulation, monomers, initiator, emulsifier, crosslinkers and CTA can be found. The type of components used in the formulation, their amounts, as well as the process parameters, will determine the microstructure of the final polymer chains which, at the same time, will directly affect to the final adhesive properties. The most critical parameter to define the end use properties of the PSAs is the molar mass distribution, determined by the soluble and gel (the insoluble fraction in a good solvent, which normally presents molar masses above 10^7 g/mol) fractions of the polymer chains, which will be mainly controlled by the amount of crosslinkers and CTAs (Jovanović and Dubé, 2004). Both, crosslinkers and CTAs can be used to fine tune the microstructure of the polymer and hence to balance the adhesive and cohesive forces that are required in the multiple applications of these adhesives found in the market (Chauvet et al., 2005).

It is worth mentioning that the renewable resource reagents will be described in the context of the application where they were used (e.g., coating), but this does not imply that its use is limited only to this type of application.

2.1 Monomers

The monomers are the main components of either coatings or adhesives formulation, and this is why many researches focused their studies in the total or partial replacement of petroleum based monomers. In the open literature, renewable monomers have been used in a broad range of applications (Cummings et al., 2019a; Dubé and Salehpour, 2014; Magalhães et al., 2019; Molina-Gutiérrez et al., 2019; Petri, 2015; Smeets et al., 2017; Zeeb et al., 2014); however, as it has been already highlighted in the introduction, in this chapter only renewable source monomers that have been used for the production of coatings and adhesives will be considered. In Table 2 a summary of the monomers that will be considered within this

Table 2 Summary of the renewable (macro)monomers used in coating and adhesives applications.

Biobased source	(Macro)monomer structure	Polymerization method	References
Soybean oil	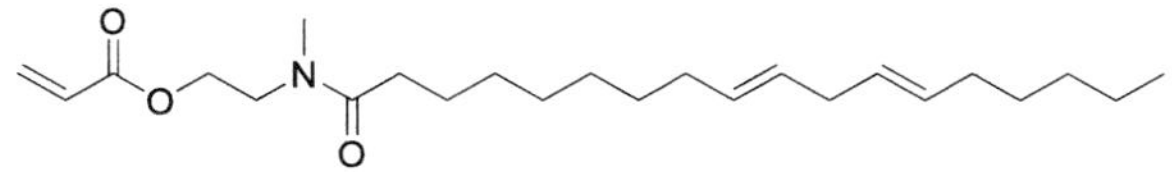	Emulsion	Delatte et al. (2014), Kaya et al. (2013)
Soybean oil	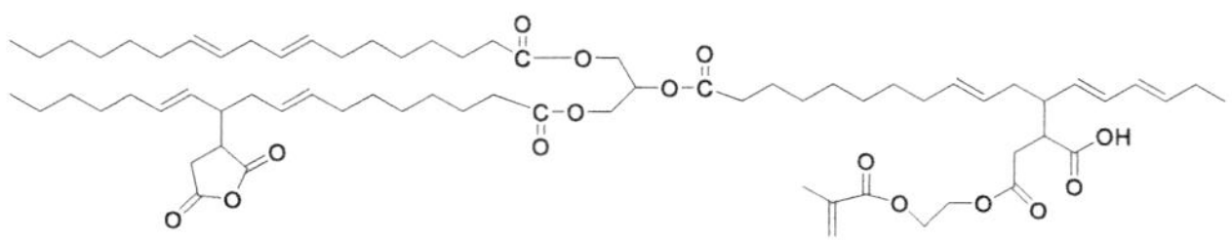	Miniemulsion	Quintero et al. (2006)
Sunflower, olive oil	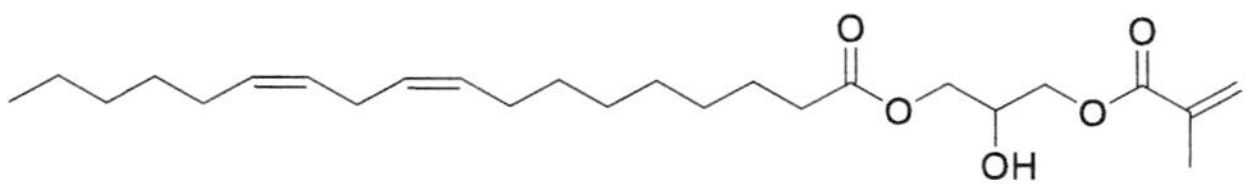	Miniemulsion	Moreno et al. (2014a)
Sunflower, olive oil	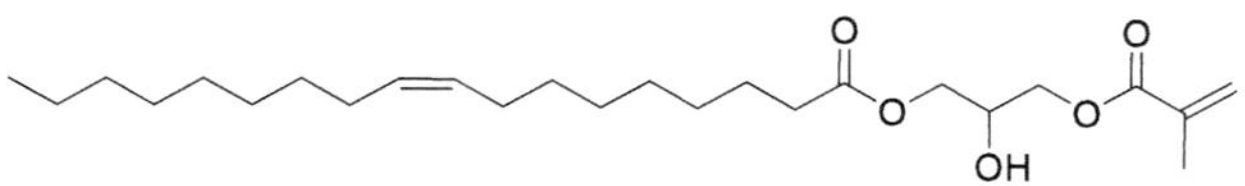	Miniemulsion	Moreno et al. (2012)
Soybean oil	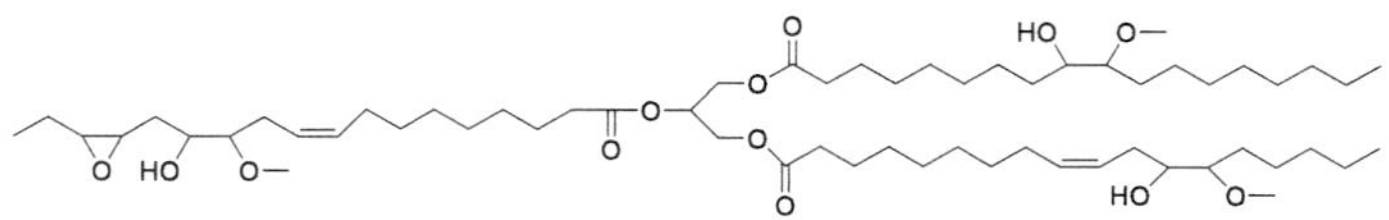	Used in the synthesis of PU seeds	Lu et al. (2011)

Continued

Table 2 Summary of the renewable (macro)monomers used in coating and adhesives applications.—cont'd

Biobased source	(Macro)monomer structure	Polymerization method	References
Soybean oil	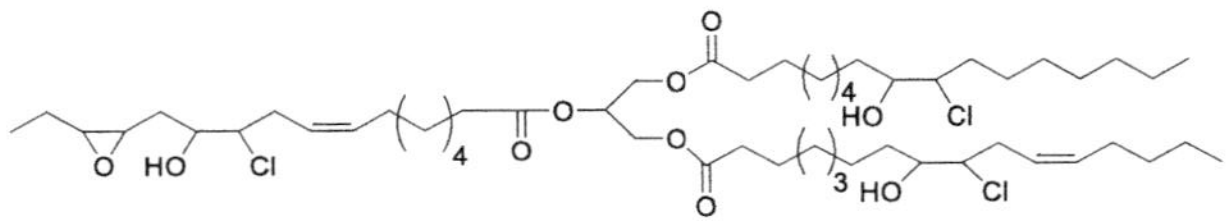	Used in the synthesis of PU seeds	Lu and Larock (2007), Lu and Larock (2011)
Soybean oil	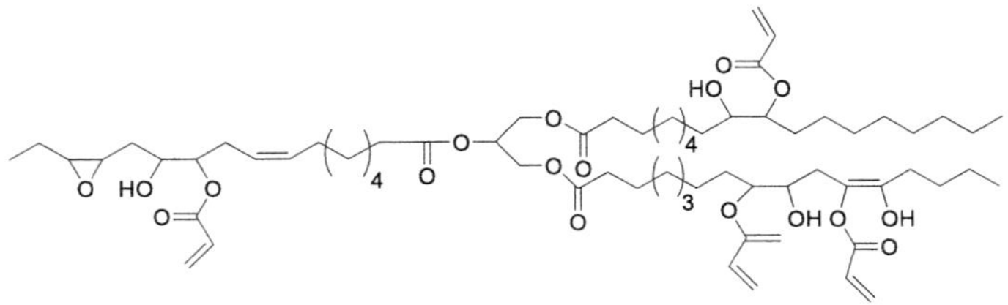	Used in the synthesis of PU seeds	Lu and Larock (2011)
Soybean oil	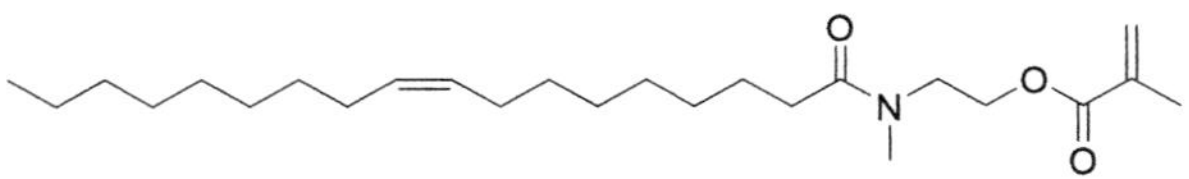	Emulsion	Lamm et al. (2019)
Olive/ soybean oil	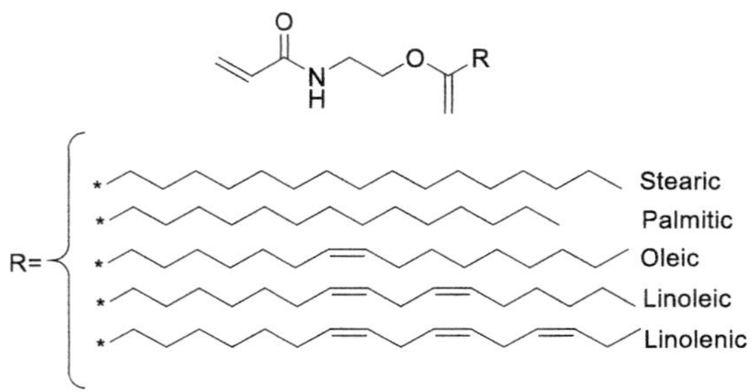	Emulsion	Kingsley et al. (2017)

Rapeseed oil	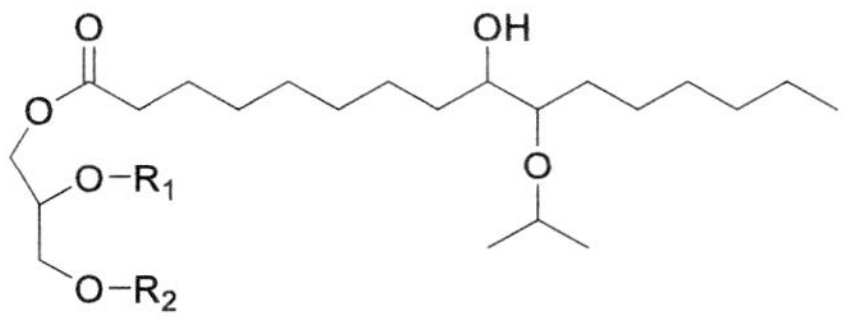	Used in the synthesis of PU as polyol	Bullermann et al. (2014)
Fructose	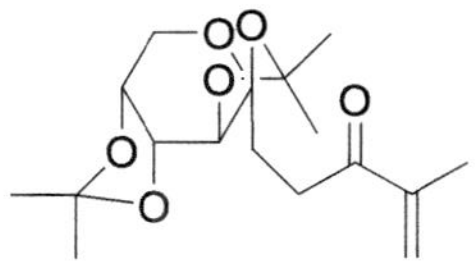	Emulsion	Desport et al. (2018)
Plant oil	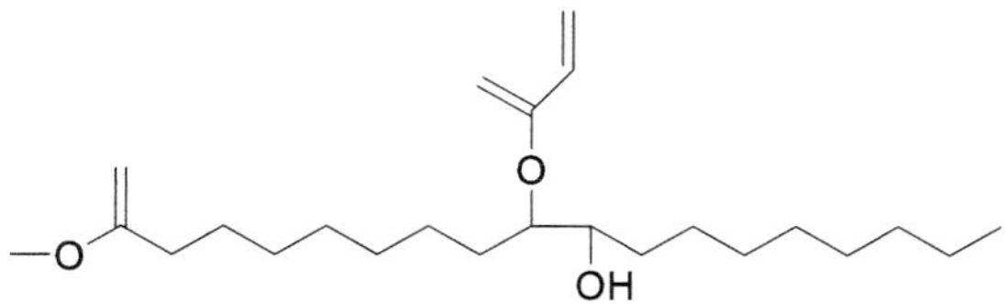 Acrylated Methyl Oleate	Emulsion	Bunker and Wool (2002)
Plant oil	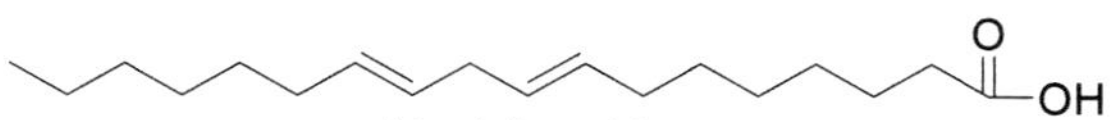 Linoleic acid	Emulsion	Roberge and Dubé (2016)
Pine resin	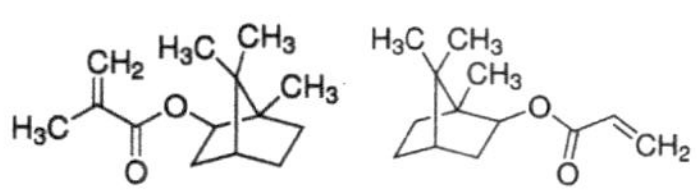 IBOMA IBOA	Miniemulsion	Badía et al. (2018), Zhang et al. (2016)

Continued

Table 2 Summary of the renewable (macro)monomers used in coating and adhesives applications.—cont'd

Biobased source	(Macro)monomer structure	Polymerization method	References
Castor oil	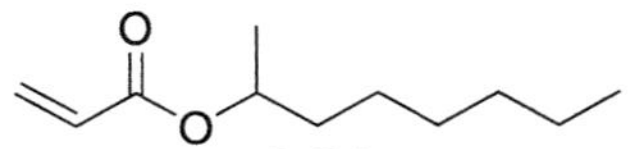 2-OA	Emulsion	Badía et al. (2018, 2020)
Black pepper	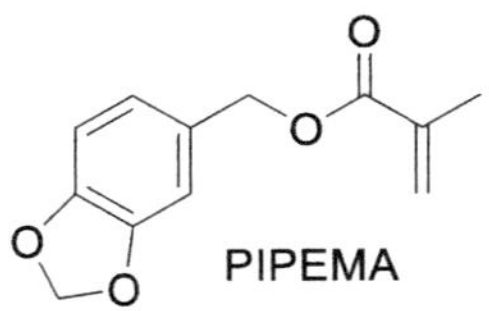 PIPEMA	Emulsion	Badía et al. (2019)
Dextrose	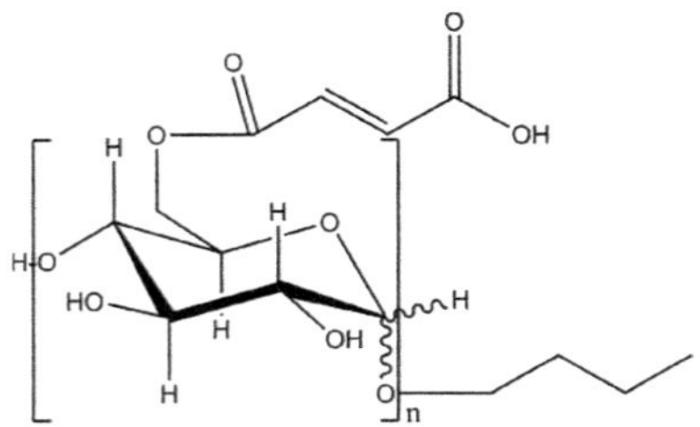 Ecomer®	Emulsion	Badía et al. (2020), Pokeržnik and Krajnc (2015)

chapter is presented. In this first section, monomers and macromonomers will be described. It is important to mention that, as IUPAC suggests, the term *monomer* will refer to molecules that can undergo polymerization to form macromolecules (IUPAC, 1996) and the term *macromonomer* will be used in cases in which a molecule that can undergo polymerization has more than one repeating unit (IUPAC, 1997).

Vegetable oil derived monomers have gained plenty of attention in waterborne coatings as they help to reduce the amount of VOCs. Linoleic acid (LA) and oleic acid (OlA) are the main components of sunflower, olive and soybean oil as it is shown in Table 2. Both are unsaturated fatty acids with 18 carbon atoms, but whereas OlA has only one double bond in the aliphatic chain, LA bears two (cis-9 and cis-12). The unsaturation of the fatty acids barely polymerizes by free radical polymerization.

Incorporation of these macromonomers in conventional emulsion polymerization reactions is challenging due to their low water solubility. To overcome this diffusion limitation, miniemulsion polymerization has been used to (co)polymerize vegetable oil derived macromonomers in order to obtain waterborne coatings (Moreno et al., 2012, 2014a,b; Quintero et al., 2006).

In an attempt to synthesize fully renewable source coatings, monomers derived from high oleic soybean oil and cardanol were synthesized and copolymerized (Demchuk et al., 2019). Fig. 4 shows the chemical structure of both monomers. The soybean oil monomer was synthesized by esterification of the oil with *N*-hydroxyethyl acrylamide in the presence of a catalyst

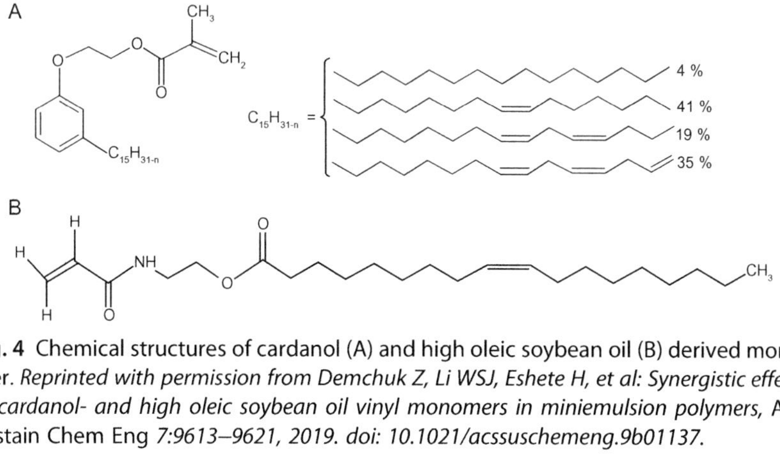

Fig. 4 Chemical structures of cardanol (A) and high oleic soybean oil (B) derived monomer. *Reprinted with permission from Demchuk Z, Li WSJ, Eshete H, et al: Synergistic effects of cardanol- and high oleic soybean oil vinyl monomers in miniemulsion polymers,* ACS Sustain Chem Eng *7:9613–9621, 2019. doi: 10.1021/acssuschemeng.9b01137.*

(sodium hydroxide, NaOH). The methacrylate cardanol monomer synthesis involved a two-step reaction including hydroxyethylation of the phenolic hydroxyl group and its later methacrylation. Note that this last monomer was earlier successfully copolymerized by miniemulsion polymerization with methyl methacrylate (MMA) (Li et al., 2018). In spite of the successful synthesis of fully renewable latexes, the molar mass obtained were quite low (maximum of 26,000 g/mol in the case of the soybean oil monomer homopolymerization) and they reduced by increasing the amount of cardanol-based monomer as it is prone to undergo chain transfer to monomer reactions during the polymerization. However, the unsaturation of the monomers opens the possibility of crosslinking the final polymer films improving the mechanical properties of the coatings. The crosslinking density of the film, and hence its mechanical properties, was controlled by adjusting the monomer ratio used during the synthesis.

Moreno et al. (2014a) explored the use of LA in waterborne coatings as a renewable source. The authors functionalized the LA by using glycidyl methacrylate (GMA) and a chromium (Cr III) based catalyst (AMC-2) to synthesize methacrylated LA (MLA). Latexes with fully renewable source monomers were successfully synthesized. Nevertheless, low molar masses were obtained due to the chain transfer reactions provoked by the labile allylic hydrogens of the LA backbone during the polymerization. In addition, no gel was formed, when potassium persulfate (KPS) was used as initiator, indicating that the unsaturation of LA was not participating in the free radical polymerization. On the contrary, the LA double bonds were prone to undergo auto-oxidative crosslinking during annealing of the films. At 120 °C the polymer film was completely cured after 1 h. This would give the opportunity to tune the mechanical properties of the film during film formation and overcome low molar mass issue of the polymer. Analogous methacrylate OlA (MOlA) monomer (Fig. 5) was also synthesized and used by the same authors to create latexes by free radical miniemulsion polymerization with potential application on coatings and films (Moreno et al., 2012, 2014b). It was found that unlike with MLA, significant amount of gel was formed in this case through the crosslinking of the OlA double bonds (Moreno et al., 2012). The binders synthesized by MOlA and MLA homopolymerization and copolymerization with MMA were formulated as waterborne paints and their performance was compared with a commercial decorative waterborne paint. The formulated paints with renewable monomers showed similar performance in terms of hardness, gloss, rheological behavior and open time (Moreno et al., 2015).

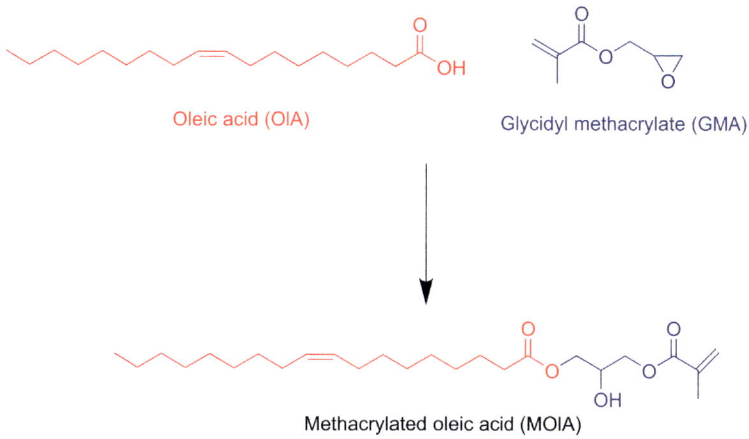

Fig. 5 Scheme of the synthesis of the methacrylated oleic acid (MOlA).

Notably during the synthesis of the monomer, a significant amount of the OlA remained unreacted that after the polymerization acted as plasticizer in the resulting coating. The authors addressed this issue by using the free OlA as precursor for *in situ* surfactant generation. This method was based on the formation of surfactant at the oil–water interface by mixing the organic phase (containing OlA), and the aqueous phase (containing potassium hydroxide, KOH). The neutralization reaction lowers the interfacial tension of the oil–water interface leading to instantaneous phase separation on a local scale and to the formation of droplets without the need to apply high energy (Prokopov and Gritskova, 2001). The acid/base ratio and type of counterion are key parameters in the formation of surfactant. Even though stable latexes were formed when appropriate OlA/base ratio was used, clear films were only formed when the miniemulsion was sonicated (Moreno et al., 2014b). In a following work, Moreno el al. copolymerized MOlA with α-methylene-γ-butyrolactone (α-MBL) by miniemulsion polymerization (Moreno et al., 2014c). Incorporating different α-MBL amounts and maintaining the double bonds of the OlA, it was possible to obtain polymers with different T_g and mechanical properties, giving the possibility to follow this route for the synthesis of either coatings or adhesives with high (29 wt%) renewable materials content.

In spite of the diffusion problems of these vegetable oil monomers through the water phase, some other derivatives showed sufficient water solubility to be copolymerized by conventional emulsion polymerization. More hydrophilic methacrylated (Fig. 6) (Lamm et al., 2019) and acrylated

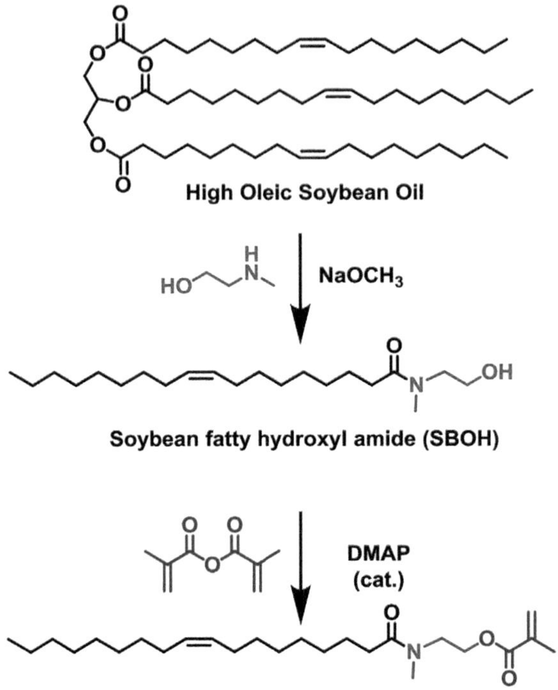

Fig. 6 Synthesis of soybean methacrylate starting with high oleic soybean oil via a fatty amide alcohol intermediate. *Reprinted with permission from Lamm ME, Li P, Hankinson S, et al: Plant oil-derived copolymers with remarkable post-polymerization induced mechanical enhancement for high performance coating applications,* Polymer *174:170–177, 2019. doi: 10.1016/j.polymer.2019.04.072.*

soybean oil (Delatte et al., 2014; Kaya et al., 2013) monomers were copolymerized with other common monomer such as (meth)acrylates and styrene (St) by semibatch emulsion polymerization showing the feasibility of using these monomers in industrial processes. Jensen and coworkers also used MOlA, but they copolymerized it with vinyl pivalate (VPi) (Jensen et al., 2016). VPi is commonly polymerized by heterogeneous suspension polymerization, however, in this work, they were able to copolymerize it with MOlA by emulsion polymerization. High incorporation of MOlA content to the polymeric backbone was achieved, obtaining latexes with high molar masses and tunability in terms of T_g by varying the MOlA content enabling these materials to be used either as coatings or adhesives.

Voronov et al. (Demchuk et al., 2016; Kingsley et al., 2017) also proved the feasibility of copolymerizing plant oil-based vinyl monomers with St, MMA and vinyl acetate (VAc) by emulsion polymerization. The monomers derived from soybean and olive oil contained from none to three double bonds in the aliphatic chain of the fatty acid (see Fig. 6). As expected, the molar mass of the resulting polymers decreased by increasing the unsaturation in the fatty acid due to chain transfer reactions. Additionally, the toughness of rigid PSt and PMMA films increased by lowering the Tg. Also the water resistance of PVAc films was improved.

Biobased aromatic containing monomers polymerizable by free radical polymerization are of great interest to replace St, which is present in the majority of synthetic latex formulations (Van Herk, 2013). Cardanol derived monomers are potential candidates for this replacement (Table 2). In the reference mentioned above (Demchuk et al., 2019), a cardanol derived monomer was introduced which was used in miniemulsion polymerization. Caillol et al. synthesized a cardanol derived methacrylic monomer (by epoxidation of cardanol followed by methacrylation) with sufficient hydrophilicity to be copolymerized in conventional emulsion polymerization (Ladmiral et al., 2017). Fig. 7 shows the synthesis scheme of this monomer. Homopolymerization and copolymerization reactions with MMA were performed in a batch process. The polymerization kinetics of the cardanol-based monomer was found to be comparable to the one of MMA. The authors also addressed the film-forming capacity of the latex showing its potential application in coatings. The films were photo-crosslinked using thiol-ene chemistry in which a small amount of dithiol were added to the films. The thiol-ene reaction happened between dithiol and the unsaturations available in the cardanol-based monomer increasing substantially the gel fraction of the formed films.

Hybrids of polyurethane (PU) and acrylic latexes have gained much interest as they combine the advantages of acrylic polymers (hardness, gloss, weatherability and chemical resistance) with the advantages of PUs (excellent adhesion and toughness) (Daniloska et al., 2014; Lopez et al., 2011, 2013).

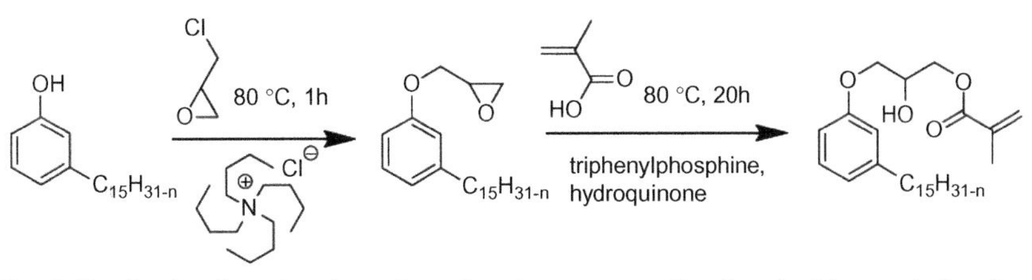

Fig. 7 Synthesis of cardanol methacrylated monomer. *Reprinted with permission from Ladmiral V, Jeannin R, Fernandes Lizarazu K, et al: Aromatic biobased polymer latex from cardanol,* Eur Polym J *93:785–794, 2017. doi: 10.1016/j.eurpolymj.2017.04.003.*

These latexes can be used as binders of decorative and protective coatings and the potential monomers are presented in Table 2. Moreover, in an attempt to produce more environmentally friendly hybrid latexes, methoxylated soybean oil polyols (MSOL) with OH functionality ranging from approximately 2.4 to as high as 4.0 were used to prepare the renewable PU dispersions (Lu and Larock, 2008). This PU dispersions were used as seed particles in emulsion polymerization of acrylates to prepare core-shell hybrid particles (Lu et al., 2011) where the hydrophilic parts of the PU acted as surfactant making possible the synthesis of surfactant-free PU/acrylic hybrids. The mechanical properties of these hybrids could be easily tuned by controlling the crosslinking degree in both the PU shell and the acrylate core, as well as the grafting between both phases. The films obtained covered a wide range of mechanical properties from elastomeric polymers to tough and hard plastics. A chlorinated soybean oil-based polyol and an acrylated epoxidized one were also successfully used to synthesize more environmentally friendly PU/acrylate hybrids in the presence of sodium dodecyl sulfate (SDS) as surfactant (Lu and Larock, 2007, 2011). Stable hybrid latexes were synthesized which contained up to 60 wt% soybean oil-based polyol as a renewable resource. In these cases, the high grafting degree of the acrylics onto the PU improved the compatibility between phases leading to a high interpenetration between them. The thermal and mechanical properties of these films were comparable to polypropylene glycol-based PU/acrylic hybrid latex films.

Bullerman et al. replaced petroleum-based polyols, such as polycaprolactone polyol and polypropylene glycol, with another one coming from rapeseed oil (Bullermann et al., 2014). Since the OH number of the rapeseed-based polyol was higher than the petroleum-based ones, hydroxyl equivalents of the polyols were fixed at the same level. PU dispersions with up to 42% renewable material content were produced. Even though the work was more focused on studying the self-emulsifying of the PU dispersions using trimellitic anhydride (TMA) and dimethylol propionic acid (DMPA), they concluded saying that varying the TMA and DMPA content as well as shifting from petroleum-based polyols to renewable ones, resulted in a broad portfolio of different PU dispersions with huge versatility in mechanical properties.

Fructose base monosaccharides can also be used to synthesize renewable methacrylated monomers (see Table 2). Desport et al. (2018) synthesized a fructose base methacrylate (MF) and copolymerized it with butyl acrylate (BA) by emulsion polymerization. The methacrylate polymerizable group

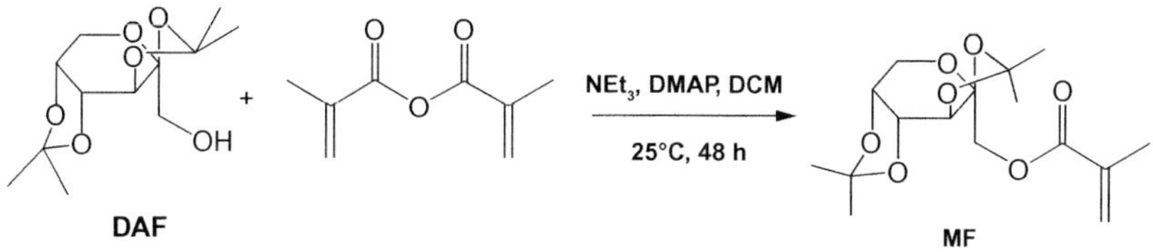

Fig. 8 Functionalization of protected fructose in solvent at ambient temperature. *Reprinted with permission from Desport JS, Moreno M, Barandiaran MJ: Fructose-based acrylic copolymers by emulsion polymerization,* Polymers (Basel) *10:1–11, 2018. doi: 10.3390/polym10050488.*

was incorporated to the fructose through the primary alcohol present in the monosaccharide as shown in Fig. 8. High conversions were achieved and solids content up to 45% were reached. The high Tg values measured confirmed the ability of this MF monomer to substitute conventional hard monomers in coatings and adhesive formulations such as MMA.

Up to now, mostly, the (macro)monomers that haven been used for the production of waterborne coatings have been presented. However, as it has been shown in Table 2, some of the (macro)monomers have been also used for the production of PSAs. The first paper published using renewable resources in a PSA was published in 2002 by Bunker and Wool (2002) and in this pioneer work, they synthesized acrylated methyl oleate (AMO) monomer (see Fig. 9), starting from methyl oleate, which can be extracted from plant oils. The polymerization ability of AMO was demonstrated in free radical emulsion polymerization yielding high molar masses, high monomer conversions and crosslinking. However, due to the low solubility of this monomer in the aqueous phase, long polymerization times and high surfactant concentrations were necessary.

The authors overcome this problem by shifting the polymerization process to miniemulsion polymerization (Bunker et al., 2003). They synthesized different PSAs, not only copolymerizing the AMO with MMA but also varying the crosslinker content. The adhesive properties were tested in terms of rheological behavior, tack, peel and shear strength. The polymers derived from the AMO renewable monomer showed similar properties compared to the petroleum based ones in most of the cases, although they showed lower peel resistance. They concluded that this problem could be solved by optimizing the crosslinker amount and type even though this was not proved.

In order to synthesize more sustainable PSAs by miniemulsion polymerization, Pu et al. substituted half of the petroleum-based monomers (BA/VAc/MMA) by macromonomers synthesized using lactide and

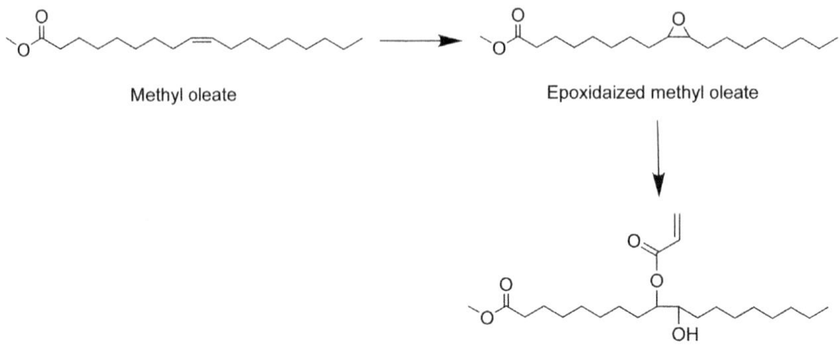

Fig. 9 Functionalization of methyl oleate to obtain the acrylated methyl oleate monomer.

ε-caprolactone (Pu et al., 2012). They were able to increase the amount of biobased content up to 50 wt% in the final latex, maintaining the adhesive properties.

In this line, Roberge and Dubé synthesized PSA changing partially the amount of petroleum based monomers (Roberge and Dubé, 2016). They proposed to use conjugated linoleic acid (CLA), which comes from plant oils, by emulsion copolymerization with St and BA. Due to the high price of CLA, they used the cheap version of it, which is a mixture between CLA and oleic acid. This last compound reacts with the St and BA forming oligomers and interfering dramatically in the terpolymer composition and hence, in the adhesive properties. They faced this problem varying the composition mixture and adding crosslinkers and CTAs to control the final polymer molar mass.

Isobornyl methacrylate (IBOMA) is a monomer with relatively high T_g that can be obtained from a renewable resource (see Table 2), more precisely, camphene is obtained by catalytic isomerization of alpha-pinene, which is one of the main components of pine resin (Wisser et al., 1997). Then, camphene can be reacted with (meth)acrylic acid in the presence of aqueous sulfuric acid to yield isobornyl methacrylate. Zhang et al. (2016) copolymerized IBOMA (see Fig. 10) with BA, 2-hydroxyethyl acrylate (2-HEA) and acrylic acid (AA).

The adhesive properties such as, peel and shear resistance were better when IBOMA was used instead of MMA in the formulation, even if tack was maintained, due to the interactions of the IBOMA with the treated and untreated polypropylene (PP) substrates used. Moreover, IBOMA as

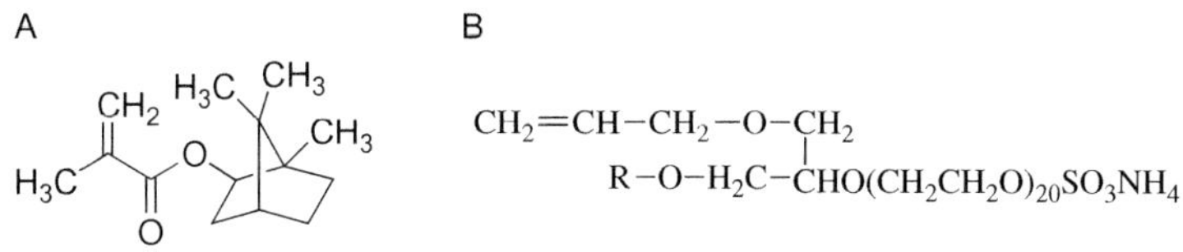

Fig. 10 Renewable materials used in the work of Zhang et al. (2016). (A) Isobornyl methacrylate (IBOMA) and (B) reactive emulsifier SR-20.

well as isobornyl acrylate (IBOA) was also used by Badía and coworkers to synthesize PSAs with biobased contents up to 72% (Badía et al., 2018). They copolymerized IBOMA with 2-octyl acrylate (2-OA), a monomer coming from castor oil, one of the most important renewable resources among vegetable oils. Precisely, 2-octanol is produced through a cracking process of ricinoleic acid (the major component of castor oil attached to glycerol) obtaining sebacic acid as side product, which is a diacid commonly employed in the Nylon industry (Mubofu, 2016). Badía and coworkers found that the soft/hard comonomer composition of commercial PSA formulations (e.g., BA/MMA or 2-HEA/MMA) did not yield the same performance because of the substantial differences on the microstructures (molar mass distribution) and the resulting T_g of the new copolymers that made the PSAs too cohesive. Fine-tuning of the IBOMA and CTA content used in the formulations was necessary to obtain competitive PSAs with at most 70 wt% of biobased content. Lower biobased contents (by replacing 2-OA by 2-HEA) allowed slightly improving performance at the price of lowering biobased content.

The same authors produced another set of PSAs but this time with piperonyl methacrylate (PIPEMA) which gave to the latex UV-light-tunable features (Badía et al., 2019). The PIPEMA was synthesized from piperonyl alcohol, which comes from black pepper and it was homopolymerized by emulsion polymerization as well as copolymerized with 2-OA and MAA. The effect of the hard monomer PIPEMA was compared synthesizing the same latexes but using another hard monomer, IBOMA. The benzodioxol structures presented in the PIPEMA's structure (see Fig. 11), allowed the tuning of the adhesive properties by UV radiation increasing the shear resistance and SAFT without damaging substantially the adhesion. This was due to the formation of a crosslinked network presented in Fig. 11. The photoreaction mechanism was proved by rheological and nuclear magnetic resonance studies.

In the latest publication of these authors, Ecomer® was copolymerized with different acrylic (2-OA, BA), methacrylic (MMA, IBOMA) and

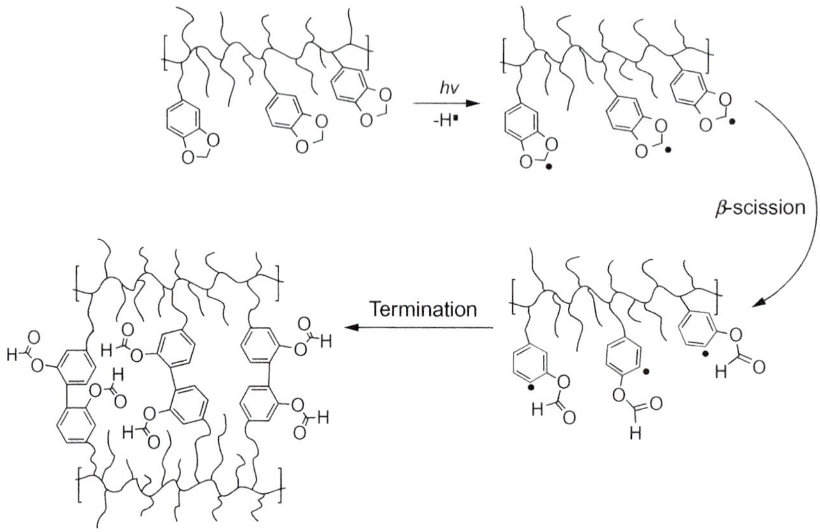

Fig. 11 The proposed photoreaction mechanism for adhesive films containing PIPEMA by Badia and coworkers. *Reprinted with permission from Badía A, Santos JI, Agirre A, et al: UV-tunable biobased pressure-sensitive adhesives containing piperonyl methacrylate,* ACS Sustain Chem Eng *7:19122–19130, 2019. doi: 10.1021/acssuschemeng.9b05067.*

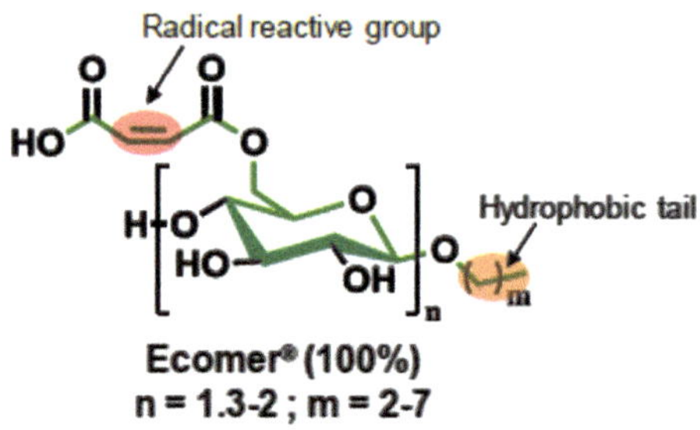

Fig. 12 Ecomomer® macromonomer's structure. *Reprinted with permission from Badía A, Barandiaran MJ, Leiza JR: Development of biobased waterborne coatings containing Ecomer®: an alkyl polyglucoside maleic acid ester monomer,* Prog Org Coat *147:105708, 2020.*

St monomers, but this time to obtain a coating formulation (Badía et al., 2020). As presented in Table 2, Ecomer® is a sugar-based macromonomer that has been developed by Ecosynthetix. It is an alkyl polyglucoside (APG) maleic acid ester coming mainly from dextrose with an average oligomerization degree between 1.3 and 2 and having a hydrophobic tail (see structure in Fig. 12).

Badia et al. showed that the incorporation of the Ecomer® monomer was not complete because of the substantial water solubility of Ecomer® and the unfavorable reactivity ratios with common monomers. Thus, the authors found that the incorporation of the Ecomer® was the lowest for the MMA (65%), then IBOMA (71%) and the highest for St (85%) in the semibatch emulsion copolymerizations using 2OA/Ecomer® and the mentioned monomers. Interestingly, the authors found that the unreacted Ecomer® present in the films could be further incorporated by a caramelization process (annealing the films at 120 °C). This process consisted on condensation reactions between the different hydroxyl groups of the macromonomer (see Fig. 13). This process served not only to incorporate the free Ecomer®, but also to enhance the final mechanical and coating properties of the polymeric film.

Pokeržnik and Krajnc used the same APG macromonomer and copolymerized it with BA by semibatch emulsion polymerization processes (Pokeržnik and Krajnc, 2015). They claimed that the APG macromonomer acted as a polymerizable surfactant, also known as surfmer. However, even if they were able to measure the critical micellar concentration (cmc) of the APG macromonomer which resulted very high (~0.1 mol/L), they used additional surfactants (anionic, non-ionic and polyvinyl alcohol (PVA)) in

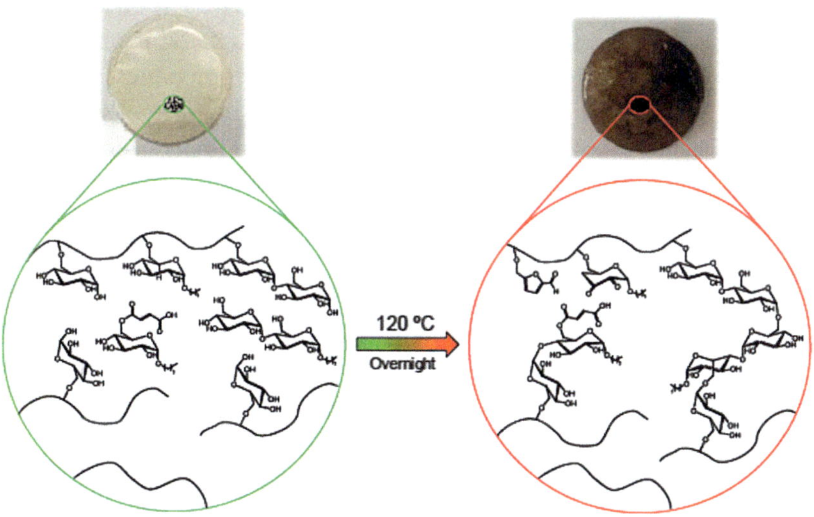

Fig. 13 Caramelization process to incorporate the unreacted Ecomer® to the final polymer. *Reprinted with permission from Badía A, Barandiaran MJ, Leiza JR: Development of biobased waterborne coatings containing Ecomer®: an alkyl polyglucoside maleic acid ester monomer,* Prog Org Coat *147:105708, 2020.*

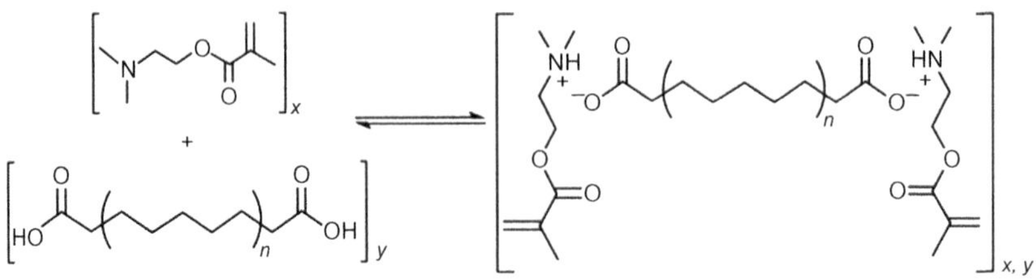

Fig. 14 Protic ionic liquid monomer, synthesized using one aliphatic dicarboxylic group and two DMAEMA. *Reprinted with permission from Fernandes AM, Moreno M, Adboudzadeh MA, et al: Odorless polymer latexes based on renewable protic ionic liquids for pressure-sensitive adhesives,* Green Mater *2:24–30, 2014. doi: 10.1680/gmat.13.00010.*

the formulation. They did not perform any reaction using the APG as the unique stabilizing agent of the emulsion which make us consider that this butyl polyglucoside maleic acid ester was acting as macromonomer, copolymerizing with BA, as it was explained in the work of Badia and coworkers, and not as surfmer as Pokeržnik and Krajnc claimed. Nevertheless, they observed improved adhesive properties when increasing the APG content in the formulation, more precisely, tack was slightly enhanced while peel and shear resistance showed a bigger improvement.

Ionic liquid monomers could be another green monomer alternative, they are synthesized by neutralization and a subsequent proton transfer between a Brønsted acid and a base (Mecerreyes, 2011). They became popular due to their ionic conductivity, thermal and chemical stability that together with the polymer properties make these compounds unique for some specific applications (Yuan et al., 2013). Fernandes et al. synthesized odorless polymer latexes using ionic liquid monomers and non-ionic surfactant in dispersion polymerization (Fernandes et al., 2014). The ionic liquids were synthesized by a proton-transfer reaction (see Fig. 14) between renewable aliphatic dicarboxylic acids and 2-dimethylaminoethyl methacrylate (DMAEMA). Different carboxylic acid and acid/base ratios were tried. The adhesive properties of the PSAs were promising particularly in terms of peel resistance.

2.2 Surfactants

Renewable resources were also used to replace some typical surfactants used in emulsion polymerization. Although the most common surfactants are low molar mass synthetic molecules with a hydrophobic tail and charged hydrophilic head, reactive surfactants (surfmers), polymeric stabilizers and solid spherical particles (Pickering emulsions) can also be used.

Xu et al. used a cationic starch with low viscosity (industrial product CS-8) as surfactant in a soap-free emulsion polymerization to prepare a styrene-acrylic latex for paper sizing and paper coating applications. To provide sufficient electrostatic stabilization, 5–10 wt% cationic starch and a cationic dimethyl amino ethyl methacrylate *co*-monomer were used in the polymerization. The synthesized core-shell latex showed good performance as a binder for paper coating (Xu and Hu, 2012). The same authors, in a similar work, used CS-8 as sole emulsifier in the synthesis of styrene-acrylate latex for the surface treatment of decorative base papers. It was possible to obtain stable latexes with particle size around 160 nm using 10 wt% of starch. The surface treatment performance of this latex was compared with the one of a commercial polyurethane. A significant improvement was observed on smoothness of the paper (printability) while maintaining the absorbability (Xu et al., 2013a).

Another surfactant from a renewable source that has been used in emulsion polymerization is asolectin, which comes from soybeans. Asolectin (Fig. 15) is a mixture of phospholipids with 76% of polyunsaturated fatty acids. The amphiphilic nature of asolectin, due to its fatty acid chains and phosphate and ammonium groups, make this compound a good candidate to be used as emulsifier in emulsion polymerization. In addition, due to the unsaturation of the fatty acid, asolectin can crosslink with polymer chains upon curing. Asolectin was used successfully as surfactant in emulsion polymerization of divinyl benzene (DVB)/BA/tung oil. Stable latexes were synthesized using 10 wt% of asolectin. (Johns et al., 2016).

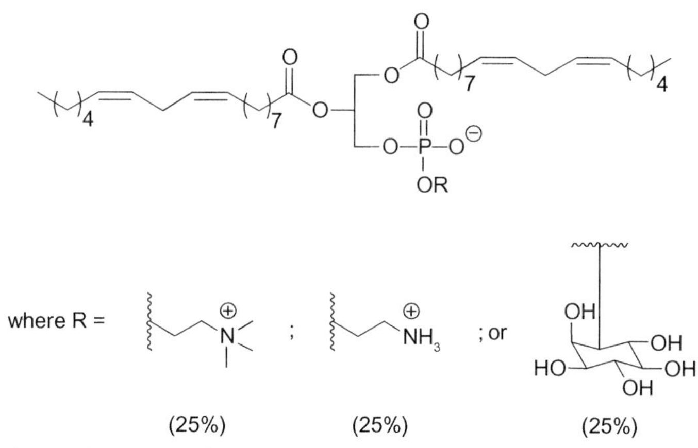

Fig. 15 Chemical structure of asolectin. *Reprinted with permission from Johns A, Edwards K, Inglesby S, et al: Emulsion polymerization of tung oil-based latexes with asolectin as a biorenewable surfactant,* Coatings *6:56, 2016. doi: 10.3390/coatings6040056.*

Alkyl polyglycosides (APG), are glucose based nonionic surfactants. Their structure is similar to the macromonomer mentioned in the previous section, Ecomer® but they do not contain neither double reactive bonds nor carboxyl groups. This type of surfactants are synthesized using fatty acids or sugars as raw materials. The different sugar units give the hydrophilic character, while the hydrophobicity will be given by alkyl chains. Lijun Chen used this type of surfactants in combination with an anionic emulsifier, sodium dodecyl benzene sulfonate (SDSB), to copolymerize MMA/BA/MAA in a seeded semi-batch emulsion polymerization (Chen, 2013). It was confirmed that increasing the APG amount in the formulation the particle size decreased and the ionic stability improved. However, the water sensitivity increased. The main drawback of APGs is that as they are non-reactive surfactants, they might migrate to the film/air interface dissolving quite easily in water due to its hydrophilic character. Nevertheless, this type of surfactant could be used both in adhesive or coatings formulation to replace nonylphenol ethoxylates (NPEOs).

Lazaridis et al. (2001) investigated the effect of the APG surfactant molecular structure and concentration on the polymerization kinetics, latex stability and particle size distribution of semi-batch emulsion copolymerization of VAc/BA (80:20). They observed that the final average particle size exhibited a "U-shape" behavior with respect to the APG surfactant concentration and that the optimum particle stabilization was achieved at a specific hydrophilic/hydrophobic APG chain length ratio.

The use of renewable surfactant in adhesives formulation has also been addressed in the literature. For instance, Lim et al. synthesized biodegradable nonionic surfactants using decanoic acid, ethylene oxide and the renewable resource sorbitol (Lim et al., 2014). Different surfactants were prepared varying the ethylene oxide units (see Fig. 16).

These nonionic surfactants were incorporated in the emulsion polymerization of 2-ethylhexyl acrylate (2-EHA), BA and AA. The adhesive properties measured in terms of peel and shear resistance were better than the ones obtained when commercial nonionic surfactants (Tween 40) were used. This was attributed to the better interfacial properties of the new surfactants.

Pickering stabilization can also be achieved by the use of renewable materials, such as starch nanoparticles (SNPs), starch nanocrystals (SNC) and cellulose nanocrystals (CNC). It is important to underline that SNPs, SNCs and CNCs can also be used as reinforcing nanomaterials. In this section, only the works that used them as surfactant will be described (without

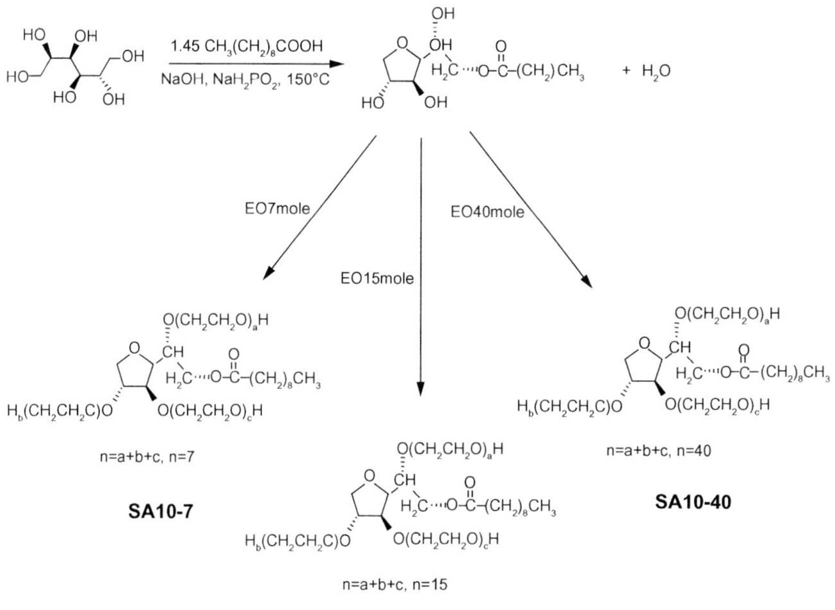

Fig. 16 Synthesis route to obtain SA10 surfactants. *Reprinted with permission from Lim JC, Lee MC, Lim T, et al: Synthesis of sorbital based nonionic surfactants and characterization of interfacial and adhesive properties for waterborne pressure sensitive adhesives,* Colloids Surf A Physicochem Eng Asp *446:80–89, 2014. doi: 10.1016/j.colsurfa.2014.01.018.*

the addition of any extra conventional surfactant in the polymerization recipe). The works in which these nanomaterials were used as additives are summarized in the section of nanofillers (Section 2.5). In order to use SNPs, CNCs and SNCs as stabilizer, their size should be smaller than the one of the latex particles. In an attempt to use SNPs to stabilize a miniemulsion polymerization of butyl methacrylate (BMA) using a hydrophobic initiator, it was found that the SNPs by themselves were not sufficient to stabilize the monomer droplets. However, their combination with a cationic surfactant was effective and reduced the required surfactant amount by a factor of 4 (BelHaaj et al., 2013). Later, the same authors were able to use SNC as sole stabilizer in one-pot aqueous emulsion polymerization of BMA at 70 °C using a water-soluble radical initiator (Haaj et al., 2014a,b). Stable latex particles as small as 121 nm at 12 wt% concentration of SNC were obtained. Films of the Pickering latexes showed better mechanical properties and optical transparency compared to a blend made of latex and SNC (Haaj et al., 2014b). In another work, amphiphilic SNPs, synthesized from starch octenyl succinic ester through a nanocoprecipitation process, were used as Pickering stabilizer in emulsion polymerization of St (Pei et al., 2016). The authors synthesized SNPs with sizes between 70 and 190 nm. Smaller SNPs have larger specific area and hence, higher coverage ability, leading to the formation of smaller latex particles. A linear correlation between the SNPs diameter with the one of the final latex particles was found.

Cellulose nanocrystals (CNCs) are rigid rod-shape nanoparticles which are extracted from wood pulp, cotton and other cellulose based biosources (Habibi et al., 2010) and have been used as Pickering stabilizers in different emulsion and suspension polymerizations due to their amphiphilic nature (Ben Mabrouk et al., 2019; Errezma et al., 2018; Jiménez Saelices et al., 2019; Limousin et al., 2019c). Errezma et al. used aldehyde-functionalized CNC as stabilizer in the Pickering emulsion polymerization of BMA. CNCs with the highest aldehyde content imparted better colloidal stability and therefore, latex particles with lower size were obtained. The optical and mechanical properties of the nanocomposites were dependent on aldehyde and CNC content. Latexes stabilized by CNCs with the highest aldehyde content showed better transparency, but worse mechanical properties (Errezma et al., 2018). Limousin et al. (2019c) polymerized MMA/BA using a cationic initiator in the presence of CNCs. The electrostatic interactions between the anionic CNCs and the cationic initiator drove the adsorption of the CNCs to the polymer/water interface leading to the formation of CNC armored latex particles. High temperatures were needed to obtain homogeneous and transparent, but brittle films. In order to solve this issue, the authors blended the CNC armored latex with a conventional, cationically stabilized latex yielding films with outstanding mechanical properties and capable of forming a film at room temperature. Ben Mabrouk et al. (2019) synthesized poly(vinyl acetate) (PVAc) latexes by emulsion polymerization using CNCs as stabilizer. They saw that the addition of polyethylene glycol methacrylate (MPEG) as co-monomer, favored the binding of CNCs to the surface of the polymer particles improving the colloidal stability. The film transparency of the Pickering latexes was higher than the one produced by mixing PVAc and CNCs. Furthermore, they saw a higher reinforcing effect when MPEG was used as co-monomer.

2.2.1 Polymeric surfactants

Amphiphilic polymer chains, composed by hydrophobic and hydrophilic chains, can act as stabilizer in *surfactant-free emulsion* polymerization.

The stability of the latex in this approach of stabilization depends on the characteristics of the amphiphilic chains such as chain length of the hydrophobic and hydrophilic part.

Starch is by far the least expensive co-binder which blends with poly(meth)acrylate latexes used in paper coating to modify the rheology and water absorbance of the coating. However, it was shown that the blend of starch and acrylate latexes suffers binder migration to the surface of the film, causing multiple problems such as dusting and picking during printing (Du et al., 2011). The migration problem was addressed by surfactant-free emulsion polymerization in which starch is grafted to polymer chains and acts as surfactant. Enzymatically hydrolyzed starch (ES) or oxidized starch was grafted with the polymer latex (Cheng and Zhao, 2017; Cheng et al., 2014a,b, 2015, 2018). Enzymatically hydrolyzed starch-g-P(St-co-BA) latex was synthesized by Cheng et al. through the reaction between the hydroxyl groups of the anhydroglucose units within the starch backbone and the double bonds of the vinyl monomers via surfactant-free emulsion polymerization (Cheng et al., 2014b). The created amphiphilic polymer chains (by graft copolymerization of the hydrophobic polyvinyl monomer chains and starch molecules) acted as surfactant. The latex was formulated to be used in paper coating applications and it was shown that the surface strength of coated paper was improved. In a similar work, the same authors used seeded surfactant-free emulsion polymerization to synthesize the enzymatically hydrolyzed starch-graft-P(BA-co-MMA) in the presence of cerium sulfate ($Ce(SO_4)_2$) and KPS as complex initiators. Fig. 17 shows the mechanism of the graft copolymer synthesis. The authors showed that the folding resistance and picking resistance, as well as the migration problems, were improved compared to the use of a simple blending of latex and oxidized starch (Cheng et al., 2018).

Another approach to graft starch on polymer chains for paper coating applications was presented by Mou et al. in which cationic acetylated starch (CAS) was grafted on a St and BA copolymer by surfactant-free emulsion polymerization (CAS-g-P(St-co-BA)). A ferrous sulfate-hydrogen peroxide ($FeSO_4$-H_2O_2) redox system was used as initiator. It was shown that the grafting efficiency was dependent on the H_2O_2 and monomer concentration, obtaining a maximum grafting percentage of 56%. Film forming capacity of starch was greatly improved. The papers sized with the starch grafted copolymer, showed less water absorption, higher ring crush index and brushing strength (Mou et al., 2012).

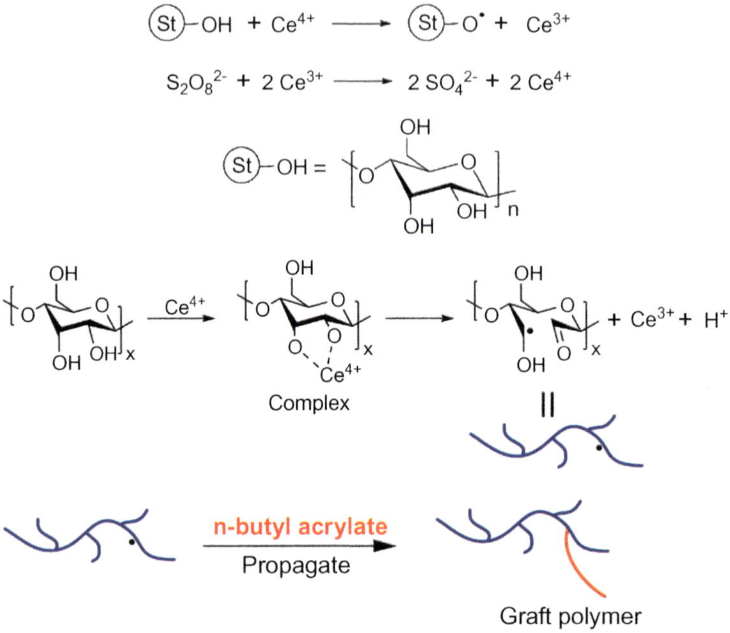

Fig. 17 Mechanism of the initiation and synthesis of the graft polymer using ES. *Reprinted with permission from Cheng S, Zhao Y, Wu Y: Surfactant-free hybrid latexes from enzymatically hydrolyzed starch and poly(butyl acrylate-methyl methacrylate) for paper coating,* Prog Org Coat *118:40–47, 2018. doi: 10.1016/j.porgcoat.2018.01.015.*

Another example of using renewable resources in surfactant-free emulsion polymerization to stabilize the latex is the use of casein, which is a natural protein from bovine milk. Casein film suffers from problems such as poor water resistance and brittleness. To overcome this problem, casein/synthetic polymer hybrids have gained attention in which, ideally, casein is grafted to the polymer chains. Pei-Li et al. introduced for the first time the synthetic method of amphiphilic core-shell latex particles in which hydrophobic vinylic polymer was grafted to a hydrophilic polymer containing amino groups. During the polymerization, generated free radicals on the amine nitrogen initiate the graft copolymerization with vinylic monomers (Li et al., 2002a). Different amine-substituted biopolymers such as casein, gelatin, albumin bovine serum (BSA), and chitosan were grafted to MMA chains (Li et al., 2002b). Fig. 18 shows the general schematic representation of the synthesis of amphiphilic core-shell particles.

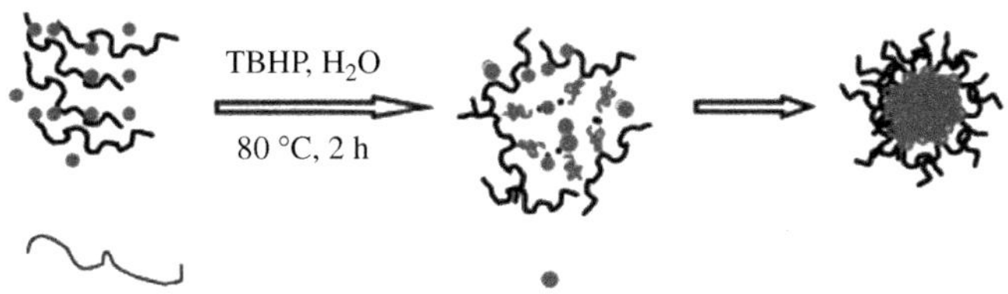

Fig. 18 Schematic representation of the synthesis of amphiphilic core-shell particles. *Reprinted with permission from Li P, Zhu J, Sunintaboon P, et al: New route to amphiphilic core—shell polymer nanospheres: graft copolymerization of methyl methacrylate from water-soluble polymer chains containing amino groups,* Langmuir *18:8641–8646, 2002b.*

Using the same methodology, caprolactam-butylacrylate co-modified casein (CA-CPL-BA) latex was synthesized via emulsifier-free emulsion polymerization by Ma et al. as film-former for leather finishing (Ma et al., 2013). Fig. 19 shows the schematic illustration of the formation of core-shell latex particles of CA-CPL-BA and the mechanism of the free radical incorporation of casein backbone and BA. It was shown that, the performance of the leather coated with the casein/synthetic polymer hybrid is comparable to the one finished with available commercial leather finishing agents.

In another work, core-shell casein-based silica nano-composite latex was synthesized via double-*in situ* emulsion polymerization in which acrylate chains were grafted to the casein matrix. In this system caprolactam modified casein particles are playing part in emulsifying the acrylate monomers and hence stabilizing the core-shell latex. Fig. 20 shows schematically the reaction mechanism. It was shown that the hybrid films showed higher water resistance and mechanical properties (Xu et al., 2013b).

Deeper studies on the microstructure and performance of these hybrid materials were done and it was found that their main drawback was the low grafting degree of the acrylic polymer to the casein (Picchio et al., 2014, 2015a,b). To address the low compatibility of casein with the (meth)acrylic polymer, Picchio et al. proposed the use of methacrylated casein which was synthesized by amine-glycidyl ether reaction (Picchio et al., 2016). Using this functionalized protein, the authors showed that the grafting degree increased up to 82%. This higher compatibility led to a significant improvement in some film properties such as the water resistance and the thermal and mechanical behavior of the film (Picchio et al., 2017).

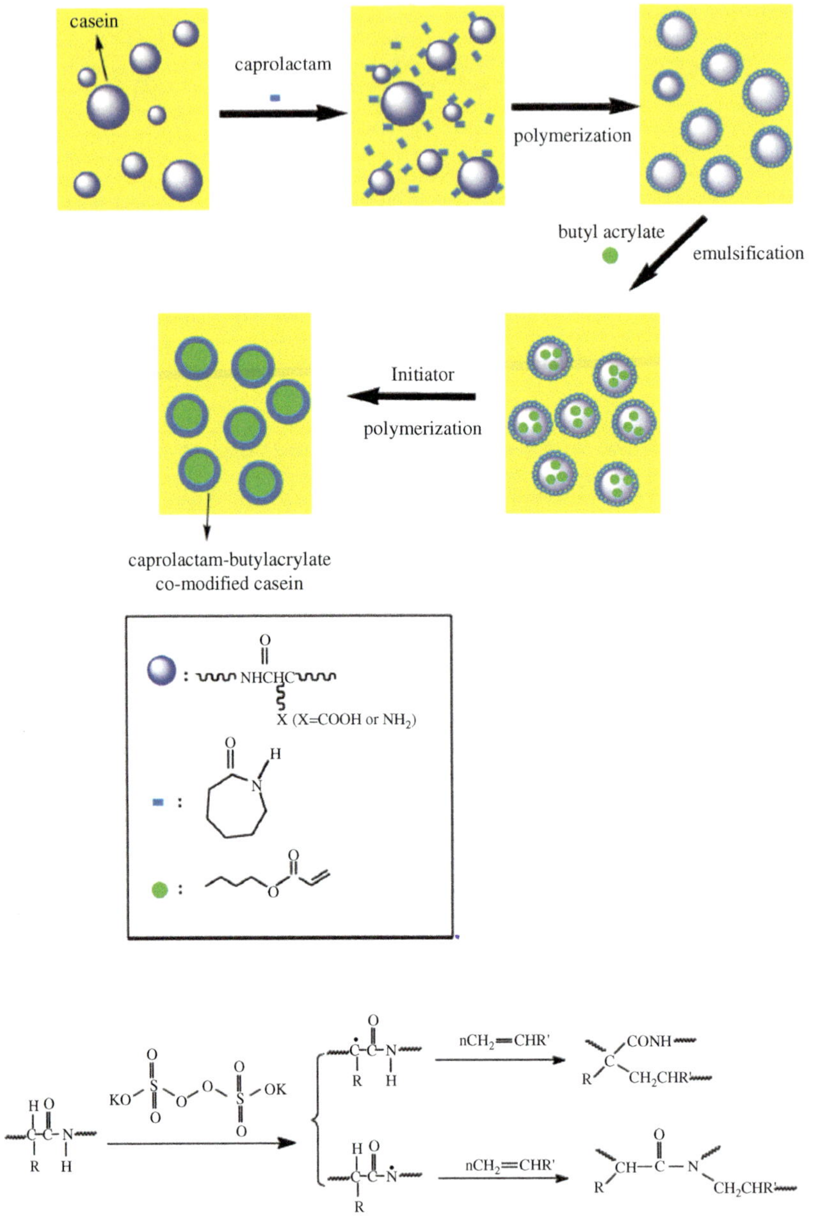

Fig. 19 Schematic illustration of the formation of core-shell latex particles of CA-CPL-BA and the mechanism of the free radical incorporation of casein backbone and butyl acrylate. *Reprinted with permission from Ma J, Xu Q, Zhou J, et al: Nano-scale core-shell structural casein based coating latex: synthesis, characterization and its biodegradability,* Prog Org Coat *76:1346–1355, 2013. doi: 10.1016/j.porgcoat.2013.04.006.*

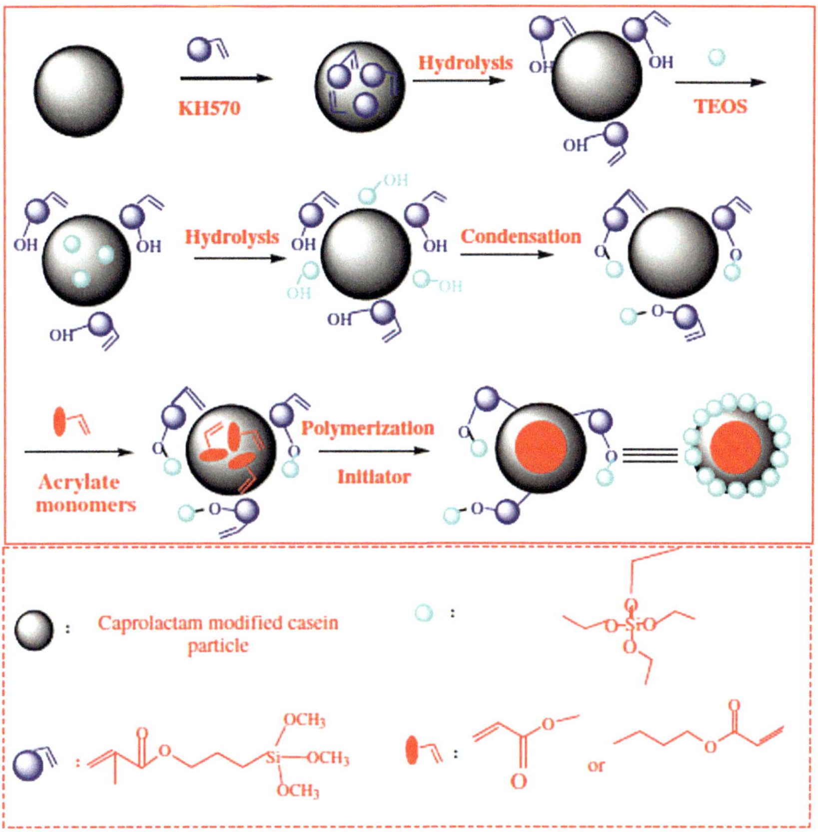

Fig. 20 Scheme of the formation of core-shell casein-based silica nano-composite latex. *Reprinted by permission from Xu Q, Ma J, Zhou J, et al: Bio-based core-shell casein-based silica nano-composite latex by double-in situ polymerization: synthesis, characterization and mechanism,* Chem Eng J *228:281–289, 2013b. doi: 10.1016/j.cej.2013.04.079.*

2.3 Crosslinkers

As mentioned above, crosslinkers are typical ingredients in emulsion polymerization reactions, especially in adhesive formulations. They are used to tune the microstructure of the final polymer together with CTAs (see Figure 21) and therefore tune the properties of the final material. Alternatives to reduce the use of petroleum based crosslinkers has also been investigated. In this line, crosslinkers derived from Casein (Qiang et al., 2014) or castor oil (Karak et al., 2009) have been used.

C22 acyloin (Fig. 21), derived from castor oil, was used as crosslinker in emulsion copolymerization process of acrylate and methacrylic acid (MAA). Its performance was similar to the one of diallylphtalate crosslinker (Van der

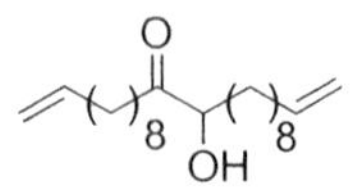

Fig. 21 Chemical structure of the C22 acyloin (Van der Steen et al., 2013).

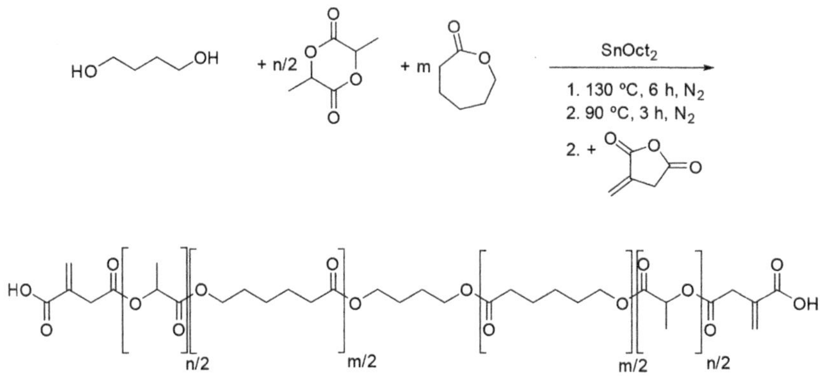

Fig. 22 Ring opening polymerization (ROP) of the symmetric crosslinker. *Reprinted with permission from Wenzel F, Agirre A, Aguirre M, et al: Incorporation of novel degradable oligoester crosslinkers into waterborne pressure sensitive adhesives: towards removable adhesives,* Green Chem *22:3272–3282, 2020. doi: 10.1039/D0GC00463D.*

Steen et al., 2013). As phthalates have endocrine disrupting properties, this renewable based crosslinker can be a good potential alternative for diallyl phthalate, although its efficiency in different emulsion polymerization systems has not been studied yet.

Wenzel et al. developed a PSA incorporating a degradable crosslinker synthesized using renewable monomers such as lactide and ε-caprolactone (Wenzel et al., 2020) as shown in Fig. 22). The final films presented good adhesive properties and the degradability was tested in an aqueous basic solution. It was demonstrated that the PSAs were degradable in alkali water in short times (30–45 min) and that the longer the length of the crosslinker used the fastest the degradation rate due to the larger number of ester groups in the crosslinkers. The authors claimed that this novel removable PSAs present adhesive performances comparable to commercial PSAs used in glass bottle labels. Notwithstanding, the novel PSAs would make the process of removing the labels more sustainable because of the lower temperatures and lower amounts of harsh solvents required to remove these PSAs from glass bottles.

2.4 Chain transfer agents (CTA)

D-limonene is a monoterpene which contains two unconjugated double bonds in its structure that could be used as CTA as it is shown in Fig. 23.

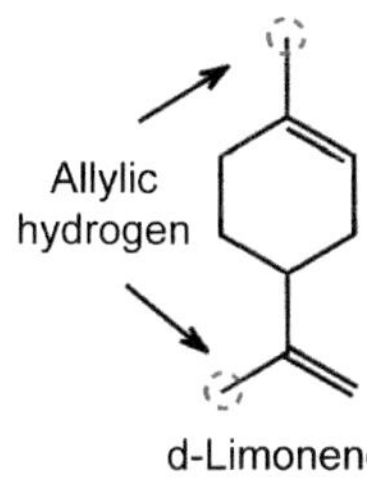

Fig. 23 D-limonene monoterpene structure. *Reprinted with permission from Ren S, Dubé M. A: Modification of latex microstructure and adhesive performance using d-Limonene as a chain transfer agent,* Int J Adhes Adhes *75:132–138, 2017. doi: 10.1016/j.ijadhadh.2017.03.003.*

Ren et al. took advantage of the two allylic hydrogens in the molecules to produce core-shell particles using BA, St and AA monomers in a two-step emulsion polymerization process (Ren and Dubé, 2017). Adjusting the ratio between the D-limonene and the DVB crosslinker, allowed turning the microstructure of the PSA. Even though, the D-limonene is not as effective as other CTAs, such as thiols, it could be a good candidate to modify the microstructure of the polymer latexes replacing fossil-fuel compounds.

2.5 Nanofillers

Nanofillers are incorporated into waterborne products to improve/maintain performance while reducing costs or to add additional properties to the coatings. Different renewable source nanofillers were used in emulsion polymerization including organic ones (such as cellulose nanocrystals and starch nanoparticles) and non-organic ones (such as montmorillonite clay). In this chapter, we will only focus on the renewable source based organic nanofillers which have been used in emulsion polymerization with applications on coating and adhesives.

CNCs can be used as sustainable reinforcing agents which have been used in many *in situ* emulsion and miniemulsion polymerizations in an unmodified form to enhance matrix reinforcement and improve the mechanical (Ben et al., 2011a,b; Ben Mabrouk et al., 2011, 2014; Dastjerdi et al., 2017; Limousin et al., 2019c), thermal (Ben Mabrouk et al., 2011), electrical properties (Atifi and Hamad, 2016) as well as the adhesive performance (Dastjerdi et al., 2018; Ouzas et al., 2018, 2019; Shamsabadi and Moghbeli, 2017). It is worth mentioning that in all these works additional emulsifiers/stabilizers were used to stabilize the polymer particles.

One of the first works in which CNCs were incorporated *in situ* during (mini)emulsion polymerization was the miniemulsion co-polymerization of styrene/ethylhexylacrylate in the presence of CNCs. Sodium dioctylsulfano succinate (DOSS) as emulsifier and hexadecane as co-stabilizer were used in the formulation. Using a reactive silane (c-metha-cryloxypropyl triethoxysilane), CNCs were anchored to the surface of poly(styrene-co-2-ethylhexyl acrylate) latex particles. A significant enhancement of storage modulus of the film with the addition of CNCs were achieved. The material showed a potential to be used as binder in waterborne coatings with desired hardness and blocking resistance (Ben Mabrouk et al., 2011). In another approach, miniemulsion polymerization of polybutylmethacrylate and CNCs were performed in the presence of dodecylpyridinium chloride (DPC) as a cationic surfactant. The electrostatic interaction between the positively charged droplets and negatively charged CNCs led to the anchoring of the nanofiller around the polymer particles. Fig. 24 shows the schematic representation of the synthetic route together with the Field Emission Scanning Electron microscopy (FE-SEM) image of the nanocomposite with 4% of CNC in the formulation. Good optical properties of the films prepared by casting confirmed a good dispersion of CNCs within the film (Ben et al., 2011b).

Dastjerdi and coworkers polymerized MMA and BA in the presence of a water dispersion containing CNCs (Dastjerdi et al., 2017). The CNCs content was changed from 0.5 to 2 wbm%. It was observed that small amounts of CNCs were enough to improve considerably the mechanical properties of the films. This was attributed to the crosslinking ability of the CNCs' surface hydroxyl groups that favor intermolecular interactions between the polymer and the surface of the CNCs.

CNCs have been used to improve the adhesive properties of the PSAs. It was shown that the PSAs that are synthesized by *in situ* seeded semibatch emulsion polymerization demonstrated substantially better performance in adhesive properties comparing to the PSAs which were prepared by blending. This was attributed to the better interaction of CNCs with polymer matrix and better wettability and film formation ability in the *in situ* emulsion polymerization of CNC-BA/MMA latex. Tack, peel and shear strength, increased almost 4, 6 and 20 times compared to the blank experiment without CNC, respectively (Dastjerdi et al., 2018). Different monomer systems were used *in situ* semibatch emulsion polymerization for synthesizing CNC nanocomposites as PSAs (Ouzas et al., 2018, 2019). Using up to 1 wt% of CNC in the formulation of IBA (isobutyl acrylate)/BA/MMA terpolymer, tack, peel and shear strength increased simultaneously. However, the viscosity increased significantly by adding CNC to the formulation (up to 3 times) (Ouzas et al., 2019). It was shown that the

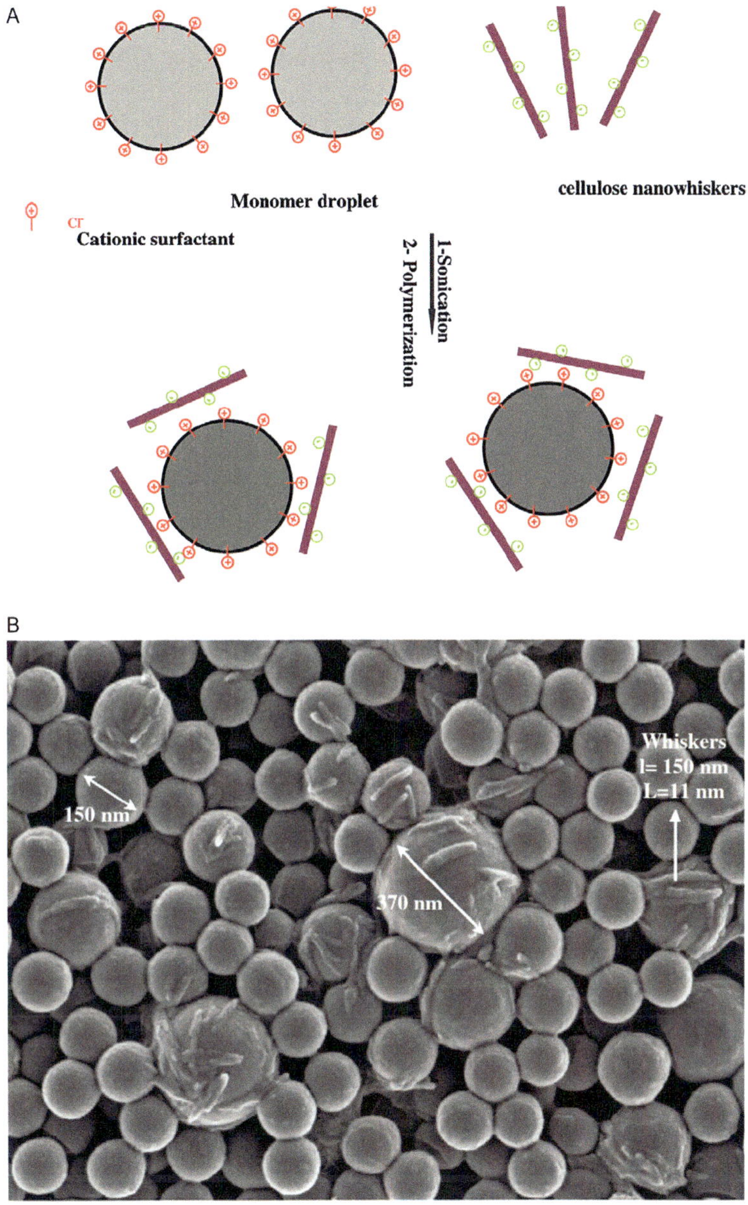

Fig. 24 (A) Schematic representation of the synthetic route of CNC/polymer nanocomposites, (B) FE-SEM image of nanocomposite dispersions with 4% CNC. *Reprinted with permission from Ben A, Rei M, Magnin A, et al: Synthesis and characterization of cellulose whiskers/polymer nanocomposite dispersion by mini-emulsion polymerization,* J Colloid Interface Sci *363:129–136, 2011b. doi: 10.1016/j.jcis.2011.07.050.*

hydrophobicity of the monomers in the copolymer system led to less compatibility and therefore a nonuniform distribution of the CNCs in the PSA films was obtained (Ouzas et al., 2018).

In situ miniemulsion was also assessed to incorporate CNCs to waterborne adhesives. PSAs were synthesized in the presence of the CNCs by miniemulsion copolymerization of BA/2EHA. It was shown that the adhesive performance of the PSAs was increased in terms of shear resistance (518%), peel strength (176%) without decreasing the probe tack (13%) by incorporation of up to 4 wt% CNC in the formulation (Shamsabadi and Moghbeli, 2017).

In spite of these works, incompatibility between cellulose and hydrophobic polymers and problems due to the processing, such as aggregation, makes the incorporation of CNC in the polymer latex challenging. Surface-modified CNC systems are studied to overcome this issue (Kedzior et al., 2018; Yu et al., 2019). Kedzior et al. tried to increase the compatibility with the polymer matrix and incorporate the CNCs into the core of the polymer particles. They achieved this by absorbing α-bromoisobutyryl bromide (Bib) on the surface of the CNCs to perform surface-initiated atom transfer radical polymerization (SI-ATRP) of both PMMA and PBA to disperse CNCs into the MMA monomer droplets (Kedzior et al., 2018). Using this methodology acrylate monomers, BA for instance, can be polymerized from the surface of the CNCs leading to polymer particles with CNCs in the core and PBA in the shell (Fig. 25). It was shown that the length of the grafted polymer is

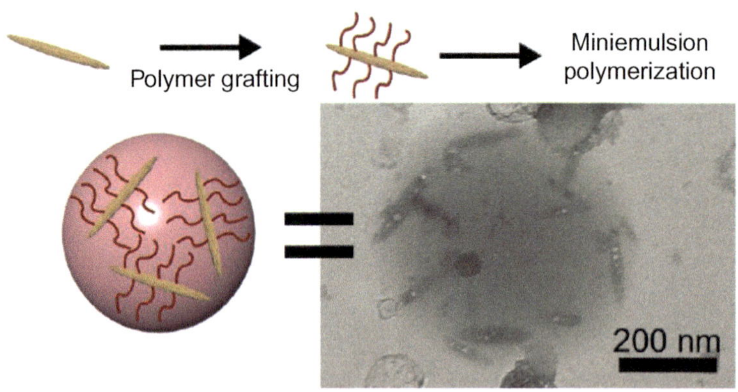

Fig. 25 Schematic of surface modification of CNC with grating low molar mass polymer on the surface and its incorporation into the core of latex particles. *Reprinted by permission from Kedzior SA, Kiriakou M, Niinivaara E, et al: Incorporating cellulose nanocrystals into the core of polymer latex particles via polymer grafting,* ACS Macro Lett *7:990–996, 2018. doi: 10.1021/acsmacrolett.8b00334.*

critical and too long grafted polymer resulted in CNC aggregation. Better dispersion of CNCs in the latex potentially improve the mechanical and adhesive properties of the films. This work as a proof of concept showed that by different modification of the surface of CNCs it is possible to increase the compatibility and control the location of the CNCs in the latex particles.

Functionalized CNC with 3-methacryloxypropyltrimethoxysilane coupling agent (γ-MPS) was used in emulsion polymerization of acrylic PSA (Yu et al., 2019). It was shown that incorporation of both CNC or functionalized CNC, increased at the same time the tack, shear and peel strength of the adhesives. However, functionalized CNC showed a higher improvement in adhesive performance comparing to the unmodified one due to the better compatibility and dispersion of functionalized CNC in the polymer latex (Fig. 26).

Starch nanoparticles (SNPs) provide barrier and mechanical properties to the final films due to their crystallinity (Sorrentino et al., 2007). In addition, the biobased content of the polymer can be increased considerably up to 50 wt% (Cummings et al., 2019b). Zhang et al. devoted their research to hydrophobically modify SNPs and in order to be able to crosslink and encapsulate them inside acrylic polymer particles (Zhang et al., 2018). The latexes presented core-shell morphology with the SNPs encapsulated inside a BA/MMA/AA shell. These latexes showed enhanced tack and peel strength. However, the latexes synthesized with unmodified or partially modified SNPs presented worse adhesive performance mainly due to the lack of core-shell morphology.

3. Conclusions and future perspectives

In an attempt to reduce the environmental impact of polymer production, biobased polymers are attracting researches, in both industry and academia. This chapter reviews the use of green and sustainable alternatives in the production of waterborne polymeric dispersions for coating and adhesive applications. Not only renewable monomers have been studied but also surfactants, crosslinkers, CTA and nanofillers. The use of renewable feedstock raw materials opens new possibilities to a broad range of novel products with promising properties. However, it should be mentioned that, in some cases the use of renewable raw materials does not necessarily mean that the overall process is more sustainable. Even though the number of works published in this field is increasing, the real sustainability of the new designed processes is not assessed in most of the works. Up to now, very few scientist

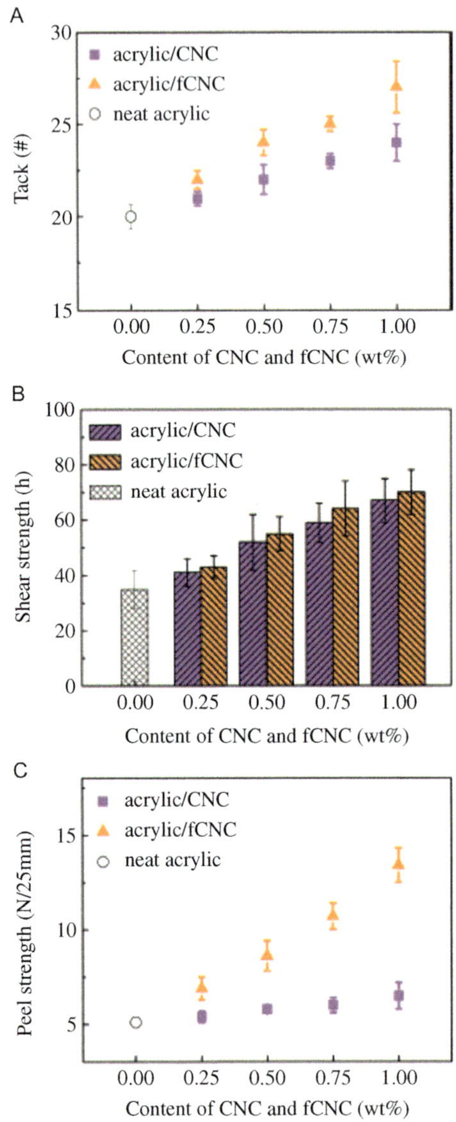

Fig. 26 Adhesive properties of the acrylic PSA. *Reprinted by permission from Yu Q, Yang W, Wang Q, et al: Functionalization of cellulose nanocrystals with γ-MPS and its effect on the adhesive behavior of acrylic pressure sensitive adhesives,* Carbohydr Polym *217:168–177, 2019. doi: 10.1016/j.carbpol.2019.04.049.*

(Blondiaux et al., 2019; Droesbeke and Du Prez, 2019) have evaluated they procedure using metrics such as the CHEM21 toolkit. Therefore, we strongly encourage researchers to start using quantitate metrics in order to objectively evaluate greenness of their new synthetic methods.

Finally, it must also be taken into account that the partial substitution of fossil fuels for biomass would require intensified farming of energy crops in competition with food production. These could lead to a dramatic rise in the cost of food threatening the health and sustenance of some countries, especially the developing ones (Mülhaupt, 2013). Therefore, when a new synthesis route is designed this aspect should also be taken into account.

References

Agarwal S, Greiner A: On the way to clean and safe electrospinning-green electrospinning: emulsion and suspension electrospinning, *Polym Adv Technol* 22:372–378, 2011. https://doi.org/10.1002/pat.1883.

Anastas PT, Warner JC: *Green chemistry: theory and practice,* New York, 1998, Oxford University.

Anastas PT, Zimmerman JB: Design through the 12 principles of green engineering, *Environ Sci Technol* 37:94–101, 2003.

Asua JM: *Polymeric dispersions: principles and applications,* 1997, Springer.

Asua JM: Emulsion polymerization: from fundamental mechanisms to process developments, *J Polym Sci A Polym Chem* 42:1025–1041, 2004.

Asua JM: *Polymer reaction engineering,* 2007, Blackwell Publishing.

Atifi S, Hamad WY: Emulsion-polymerized flexible semi-conducting CNCs–PANI–DBSA nanocomposite films, *RCS Adv* 6:65494–65503, 2016. https://doi.org/10.1039/C6RA13610A.

Badía A, Movellan J, Barandiaran MJ, Leiza JR: High biobased content latexes for development of sustainable pressure sensitive adhesives, *Ind Eng Chem Res* 57:14509–14516, 2018. https://doi.org/10.1021/acs.iecr.8b03354.

Badía A, Santos JI, Agirre A, Barandiaran MJ, Leiza JR: UV-tunable biobased pressure-sensitive adhesives containing piperonyl methacrylate, *ACS Sustain Chem Eng* 7:19122–19130, 2019. https://doi.org/10.1021/acssuschemeng.9b05067.

Badía A, Kastelijn M, Scheerder J, Leiza JR: Development of biobased waterborne coatings containing Ecomer®: an alkyl polyglucoside maleic acid ester monomer, *Prog Org Coat* 147:105708, 2020.

Barquero A, Agirre A, Barandiaran MJ, Leiza JR: Monitoring the evolution of the microstructure of vinyl silane monomer containing poly(vinyl acetate) based copolymer latexes during storage, *Eur Polym J* 121: 2019, 109299. https://doi.org/10.1016/j.eurpolymj.2019.109299.

Barquero A, Agirre A, Barandiaran MJ, Leiza JR: Emulsion copolymerization of vinyl acetate and vinyl silanes: kinetics and development of microstructure, *Macromol React Eng* 14: 2020, 1900043. https://doi.org/10.1002/mren.201900043.

BelHaaj S, Ben Mabrouk A, Thielemans W, Bouf S: A one-step miniemulsion polymerization route towards the synthesis of nanocrystal reinforced acrylic nanocomposites, *Soft Matter* 9:1975–1984, 2013. https://doi.org/10.1039/c2sm27190g.

Ben Mabrouk A, Kaddami H, Magnin A, Belgacem MN, Dufresne A, Boufi S: Preparation of nanocomposite dispersions based on cellulose whiskers and acrylic copolymer by miniemulsion polymerization: effect of the silane content, *Polym Eng Sci* 51:62–70, 2011. https://doi.org/10.1002/pen.

Ben Mabrouk A, Salon MCB, Magnin A, Belgacem MN, Boufi S: Cellulose-based nanocomposites prepared via mini-emulsion polymerization: understanding the chemistry of the nanocellulose/matrix interface, *Colloids Surf A Physicochem Eng Asp* 448:1–8, 2014. https://doi.org/10.1016/j.colsurfa.2014.01.077.

Ben Mabrouk A, Dufresne A, Boufia S: Cellulose nanocrystal as ecofriendly stabilizer for emulsion polymerization and its application for waterborne adhesive Ayman, *Carbohydr Polym* 229: 2019, 115504. https://doi.org/10.1016/j.neubiorev.2019.07.019.

Ben A, Magnin A, Naceur M, Boufi S: Melt rheology of nanocomposites based on acrylic copolymer and cellulose whiskers, *Compos Sci Technol* 71:818–827, 2011a. https://doi.org/10.1016/j.compscitech.2011.01.012.

Ben A, Rei M, Magnin A, Naceur M, Boufi S: Synthesis and characterization of cellulose whiskers/polymer nanocomposite dispersion by mini-emulsion polymerization, *J Colloid Interface Sci* 363:129–136, 2011b. https://doi.org/10.1016/j.jcis.2011.07.050.

Benedek I, Heymans LJ: *Pressure-sensitive adhesives technology*, New York, 1997, Marcel Dekker Inc.

Blondiaux E, Bomon J, Smoleń M, et al: Bio-based aromatic amines from lignin-derived monomers, *ACS Sustain Chem Eng* 7:6906–6916, 2019. https://doi.org/10.1021/acssuschemeng.8b06467.

Bullermann J, Spohnholz R, Friebel S, Salthammer T: Synthesis and characterization of polyurethane ionomers with trimellitic anhydride and dimethylol propionic acid for waterborne self-emulsifying dispersions, *J Polym Sci A Polym Chem* 52:680–690, 2014. https://doi.org/10.1002/pola.27049.

Bunker SP, Wool RP: Synthesis and characterization of monomers and polymers for adhesives from methyl oleate, *J Polym Sci A Polym Chem* 40:451–458, 2002. https://doi.org/10.1002/pola.10130.

Bunker S, Staller C, Willenbacher N, Wool R: Miniemulsion polymerization of acrylated methyl oleate for pressure sensitive adhesives, *Int J Adhes Adhes* 23:29–38, 2003. https://doi.org/10.1016/S0143-7496(02)00079-9.

Caillol S: Lifecycle assessment and green chemistry: a look at innovative tools for sustainable development. In *Environmental Impact on Polymers*, Hoboken, NJ, USA, 2014, John Wiley & Sons, Inc., pp 65–89. https://doi.org/10.1002/9781118827116.ch5

Chauvet J, Asua JM, Leiza JR: Independent control of sol molar mass and gel content in acrylate polymer/latexes, *Polymer (Guildf)* 46:9555–9561, 2005. https://doi.org/10.1016/j.polymer.2005.08.061.

Chen L: Application of green commercial surfactant in preparing purely acrylic latex via semi-continuous seeded emulsion polymerization, *J Surfactant Deterg* 16:197–202, 2013. https://doi.org/10.1007/s11743-012-1373-9.

Cheng S, Zhao Y: Synthesis of eccentric core-shell particles by surfactant-free emulsion polymerization with enzymatically hydrolyzed starch as the stabilizer, *Colloid Polym Sci* 295:2233–2241, 2017. https://doi.org/10.1007/s00396-017-4200-z.

Cheng S, Xu J, Wu Y: Preparation and characterization of oxidized starch-graftpoly(styrene-butyl acrylate) latex via emulsion polymerization, *J Polym Eng* 34:611–616, 2014a.

Cheng S, Zhang Y, Wu Y: Preparation and characterization of enzymatically degraded starch-g-poly (styrene-co-butyl acrylate) latex for paper coating, *Polym-Plast Technol Eng* 53:1811–1816, 2014b. https://doi.org/10.1080/03602559.2014.935397.

Cheng S, Zhao W, Wu Y: Optimization of synthesis and characterization of oxidized starch-graft-poly (styrene-butyl acrylate) latex for paper coating, *Starch/Stärke* 67:493–501, 2015. https://doi.org/10.1002/star.201400265.

Cheng S, Zhao Y, Wu Y: Surfactant-free hybrid latexes from enzymatically hydrolyzed starch and poly(butyl acrylate-methyl methacrylate) for paper coating, *Prog Org Coat* 118:40–47, 2018. https://doi.org/10.1016/j.porgcoat.2018.01.015.

Creton C: Pressure-sensitive adhesives: an introductory course, *MRS Bull* 28:434–439, 2003.

Cummings S, Zhang Y, Smeets N, Cunningham M, Dubé MA: On the use of starch in emulsion polymerizations, *Processes* 7:140–180, 2019a. https://doi.org/10.3390/pr7030140.

Cummings S, Trevino E, Zhang Y, Cunningham M, Dubé MA: Incorporation of modified regenerated starch nanoparticles in emulsion polymer latexes, *Starch/Starke* 71:1–13, 2019b. https://doi.org/10.1002/star.201800192.

Daniloska V, Carretero P, Tomovska R, Paulis M, Asua JM: High-performance adhesives resulting from spontaneous formation of nanogels within miniemulsion particles, *ACS Appl Mater Interfaces* 6:3559–3569, 2014. https://doi.org/10.1021/am405752k.

Dastjerdi Z, Cranston ED, Dubé MA: Synthesis of poly(n-butyl acrylate/methyl methacrylate)/CNC latex nanocomposites via in situ emulsion polymerization, *Macromol React Eng* 11:1–8, 2017. https://doi.org/10.1002/mren.201700013.

Dastjerdi Z, Cranston ED, Dubé MA: Pressure sensitive adhesive property modification using cellulose nanocrystals, *Int J Adhes Adhes* 81:36–42, 2018. https://doi.org/10.1016/j.ijadhadh.2017.11.009.

De San Luis A, Aguirreurreta Z, Pardo LM, et al: PS/PMMA-CdSe/ZnS quantum dots hybrid nanofibers for VOCs sensors, *Isr J Chem* 58:1347–1355, 2018. https://doi.org/10.1002/ijch.201800038.

Delatte D, Kaya E, Kolibal LG, Mendon SK, Rawlins JW, Thames SF: Synthesis and characterization of a soybean oil-based macromonomer, *J Appl Polym Sci* 131:1–10, 2014. https://doi.org/10.1002/app.40249.

Demchuk Z, Shevchuk O, Tarnavchyk I, et al: Free-radical copolymerization behavior of plant-oil-based vinyl monomers and their feasibility in latex synthesis, *ACS Omega* 1:1374–1382, 2016. https://doi.org/10.1021/acsomega.6b00308.

Demchuk Z, Li WSJ, Eshete H, Caillol S, Voronov A: Synergistic effects of cardanol- and high oleic soybean oil vinyl monomers in miniemulsion polymers, *ACS Sustain Chem Eng* 7:9613–9621, 2019. https://doi.org/10.1021/acssuschemeng.9b01137.

Desport JS, Moreno M, Barandiaran MJ: Fructose-based acrylic copolymers by emulsion polymerization, *Polymers (Basel)* 10:1–11, 2018. https://doi.org/10.3390/polym10050488.

Droesbeke MA, Du Prez FE: Sustainable synthesis of renewable terpenoid-based (meth)acrylates using the CHEM21 green metrics toolkit, *ACS Sustain Chem Eng* 7:11633–11639, 2019. https://doi.org/10.1021/acssuschemeng.9b01578.

Du Y, Zang YH, Du J: Effects of starch on latex migration and on paper coating properties, *Ind Eng Chem Res* 50:9781–9786, 2011. https://doi.org/10.1021/ie200807w.

Dubé MA, Salehpour S: Applying the principles of green chemistry to polymer production technology, *Macromol React Eng* 8:7–28, 2014. https://doi.org/10.1002/mren.201300103.

Errezma M, Ben Mabrouk A, Magnin A, Dufresne A, Bouf S: Surfactant-free emulsion Pickering polymerization stabilized by aldehyde-functionalized cellulose nanocrystals, *Carbohydr Polym* 202:621–630, 2018.

Fernandes AM, Moreno M, Adboudzadeh MA, Gracia R, Barandiaran MJ, Mecerreyes D: Odorless polymer latexes based on renewable protic ionic liquids for pressure-sensitive adhesives, *Green Mater* 2:24–30, 2014. https://doi.org/10.1680/gmat.13.00010.

González I, Asua JM, Leiza JR: Crosslinking in acetoacetoxy functional waterborne crosslinkable latexes, *Macromol Symp* 243:53–62, 2006a. https://doi.org/10.1002/masy.200651106.

González I, Arzamendi G, Asua JM, Leiza JR: Unexpected crosslinking during acetoacetoxy group protection on waterborne crosslinkable latexes, *Macromol Mater Eng* 291:1185–1193, 2006b. https://doi.org/10.1002/mame.200600235.

Grand View Research: *Emulsion polymers market size, share & trends analysis report by product, by application and segment forecasts, 2019–2025,* 2019.

Guinée J: Handbook on life cycle assessment—operational guide to the ISO standards, *Int J Life Cycle Assess* 6:255, 2001. https://doi.org/10.1007/BF02978784.

Gutsche C: Pressure-sensitive adhesives and applications. István Benedek, second edition, 2004 Marcel Dekker, Inc., New York, ISBN~0-8247-5059-4, 747 pages, price $ 185, *Colloid Polym Sci* 283:465, 2005. https://doi.org/10.1007/s00396-004-1201-5.

Haaj SB, Magnin A, Bouf S: Starch nanoparticles produced via ultrasonication as a sustainable stabilizer in Pickering emulsion polymerization, *RSC Adv* 4:42638–42646, 2014a. https://doi.org/10.1039/C4RA06194B.

Haaj SB, Thielemans W, Magnin A, Boufi S: Starch nanocrystal stabilized Pickering emulsion polymerization for nanocomposites with improved performance, *ACS Appl Mater Interfaces* 6:8263–8273, 2014b. https://doi.org/10.1021/am501077e.

Habibi Y, Lucia LA, Rojas OJ: Cellulose nanocrystals: chemistry, self-assembly, and applications, *Chem Rev* 110:3479–3500, 2010. https://doi.org/10.1021/cr900339w.

Imam SH, Bilbao-Sainz C, Sen Chiou B, Glenn GM, Orts WJ: Biobased adhesives, gums, emulsions, and binders: current trends and future prospects, *J Adhes Sci Technol* 27:1972–1997, 2013. https://doi.org/10.1080/01694243.2012.696892.

IUPAC: Monomer, 1996. https://goldbook.iupac.org/terms/view/M04019.

IUPAC: Macromonomer, 1997. https://goldbook.iupac.org/terms/view/M03668.

Jensen AT, Couto de Oliveira AC, Gonçalves SB, Gambetta R, Machado F: Evaluation of the emulsion copolymerization of vinyl pivalate and methacrylated methyl oleate, *J Appl Polym Sci* 133:1–11, 2016. https://doi.org/10.1002/app.44212.

Jiménez Saelices C, Save M, Capron I: Synthesis of latex stabilized by unmodified cellulose nanocrystals: the effect of monomers on particle size, *Polym Chem* 10:727–737, 2019. https://doi.org/10.1039/c8py01575a.

Johns A, Edwards K, Inglesby S, Quirino R: Emulsion polymerization of tung oil-based latexes with asolectin as a biorenewable surfactant, *Coatings* 6:56, 2016. https://doi.org/10.3390/coatings6040056.

Jovanović R, Dubé MA: Emulsion-based pressure-sensitive adhesives: a review, *J Macromol Sci Polym Rev* 44:1–51, 2004. https://doi.org/10.1081/MC-120027933.

Karak N, Rana S, Cho JW: Synthesis and charcterization of castor-oil-modified hyperbranched polyurethanes, *J Appl Polym Sci* 112:736–743, 2009. https://doi.org/10.1002/app.

Kaya E, Mendon SK, Delatte D, Rawlins JW, Thames SF: Emulsion copolymerization of vegetable oil macromonomers possessing both acrylic and allylic functionalities, *Macromol Symp* 324:95–106, 2013. https://doi.org/10.1002/masy.201200072.

Keddie AF, Joseph R: *Fundamentals of latex film formation,* Dordrecht, Netherlands, 2010, Springer Laboratory. https://doi.org/10.1007/978-90-481-2845-7.

Kedzior SA, Kiriakou M, Niinivaara E, et al: Incorporating cellulose nanocrystals into the core of polymer latex particles via polymer grafting, *ACS Macro Lett* 7:990–996, 2018. https://doi.org/10.1021/acsmacrolett.8b00334.

Kingsley K, Shevchuk O, Demchuk Z, Voronov S, Voronov A: The features of emulsion copolymerization for plant oil-based vinyl monomers and styrene, *Ind Crop Prod* 109:274–280, 2017. https://doi.org/10.1016/j.indcrop.2017.08.043.

Kobayashi S: Green polymer chemistry: recent developments, *Adv Polym Sci* 262:141–166, 2013. https://doi.org/10.1007/12_2013_236.

Ladmiral V, Jeannin R, Fernandes Lizarazu K, et al: Aromatic biobased polymer latex from cardanol, *Eur Polym J* 93:785–794, 2017. https://doi.org/10.1016/j.eurpolymj.2017.04.003.

Lamm ME, Li P, Hankinson S, Zhu T, Tang C: Plant oil-derived copolymers with remarkable post-polymerization induced mechanical enhancement for high performance coating applications, *Polymer* 174:170–177, 2019. https://doi.org/10.1016/j.polymer.2019.04.072.

Lazaridis N, Alexopoulos AH, Kiparissides C: *Semi-batch emulsion copolymerization of vinyl acetate and butyl acrylate using oligomeric nonionic surfactants,* 2001, Wiley Online Library.

P.P. Li, J.M. Zhu, F. Harris, Amphiphilic core-shell latexes, US Patent 2002/0143081 A1, 2002a.

Li P, Zhu J, Sunintaboon P, Harris FW: New route to amphiphilic core—shell polymer nanospheres: graft copolymerization of methyl methacrylate from water-soluble polymer chains containing amino groups, *Langmuir* 18:8641–8646, 2002b.

Li WSJ, Negrell C, Ladmiral V, et al: Cardanol-based polymer latex by radical aqueous miniemulsion polymerization, *Polym Chem* 9:2468–2477, 2018. https://doi.org/10.1039/c8py00167g.

Lim JC, Lee MC, Lim T, Kim BJ: Synthesis of sorbital based nonionic surfactants and characterization of interfacial and adhesive properties for waterborne pressure sensitive adhesives, *Colloids Surf A Physicochem Eng Asp* 446:80–89, 2014. https://doi.org/10.1016/j.colsurfa.2014.01.018.

Limousin E, Ballard N, Asua JM: Soft core–hard shell latex particles for mechanically strong VOC-free polymer films, *J Appl Polym Sci* 136:, 2019https://doi.org/10.1002/app.47608.

Limousin E, Ballard N, Asua JM: The influence of particle morphology on the structure and mechanical properties of films cast from hybrid latexes, *Prog Org Coat* 129:69–76, 2019b. https://doi.org/10.1016/j.porgcoat.2019.01.015.

Limousin E, Ballard N, Asua JM: Synthesis of cellulose nanocrystal armored latex particles for mechanically strong nanocomposite films, *Polym Chem* 10:1823–1831, 2019c. https://doi.org/10.1039/c8py01785a.

Lopez A, Degrandi-Contraires E, Canetta E, Creton C, Keddie JL, Asua JM: Waterborne polyurethane-acrylic hybrid nanoparticles by miniemulsion polymerization: applications in pressure-sensitive adhesives, *Langmuir* 27:3878–3888, 2011. https://doi.org/10.1021/la104830u.

Lopez A, Reyes Y, Degrandi-Contraires E, Canetta E, Creton C, Asua JM: Waterborne hybrid polymer particles: tuning of the adhesive performance by controlling the hybrid microstructure, *Eur Polym J* 49:1541–1552, 2013. https://doi.org/10.1016/j.eurpolymj.2013.01.030.

Lu Y, Larock RC: New hybrid latexes from a soybean oil-based waterborne polyurethane and acrylics via emulsion polymerization, *Biomacromolecules* 8:3108–3114, 2007. https://doi.org/10.1021/bm700522z.

Lu Y, Larock RC: Soybean-oil-based waterborne polyurethane dispersions: effects of polyol functionality and hard segment content on properties, *Biomacromolecules* 9:3332–3340, 2008. https://doi.org/10.1021/bm801030g.

Lu Y, Larock RC: Synthesis and properties of grafted latices from a soybean oil-based waterborne polyurethane and acrylics, *J Appl Polym Sci* 119:3305–3314, 2011. https://doi.org/10.1002/app.

Lu Y, Xia Y, Larock RC: Surfactant-free core-shell hybrid latexes from soybean oil-based waterborne polyurethanes and poly(styrene-butyl acrylate), *Prog Org Coat* 71:336–342, 2011. https://doi.org/10.1016/j.porgcoat.2011.03.027.

Ma J, Xu Q, Zhou J, Gao D, Zhang J, Chen L: Nano-scale core-shell structural casein based coating latex: synthesis, characterization and its biodegradability, *Prog Org Coat* 76:1346–1355, 2013. https://doi.org/10.1016/j.porgcoat.2013.04.006.

Magalhães S, Alves L, Medronho B, et al: Brief overview on bio-based adhesives and sealants, *Polymers (Basel)* 11:1–20, 2019. https://doi.org/10.3390/polym11101685.

Mariz IdFA, Leiza JR, De la Cal JC: Competitive particle growth: a tool to control the particle size distribution for the synthesis of high solids content low viscosity latexes, *Chem Eng J* 168:938–946, 2011. https://doi.org/10.1016/j.cej.2011.02.023.

McElroy CR, Constantinou A, Jones LC, Summerton L, Clark JH: Towards a holistic approach to metrics for the 21st century pharmaceutical industry, *Green Chem* 17:3111–3121, 2015. https://doi.org/10.1039/c5gc00340g.

McKinsey: https://www.mckinsey.com/industries/chemicals/our-insights/how-plastics-waste-recycling-could-transform-the-chemical-industry, 2019.

Mecerreyes D: Polymeric ionic liquids: broadening the properties and applications of polyelectrolytes, *Prog Polym Sci* 36:1629–1648, 2011. https://doi.org/10.1016/j.progpolymsci.2011.05.007.

Min TI, Klein A, El-Aasser MS, Vanderhoff JW: Morphology and grafting in polybutylacrylate-polystyrene core-shell emulsion polymerization, *J Polym Sci, Polym Chem Ed* 21:2845–2861, 1983. https://doi.org/10.1002/pol.1983.170211001.

Molina-Gutiérrez S, Ladmiral V, Bongiovanni R, Caillol S, Lacroix-Desmazes P: Radical polymerization of biobased monomers in aqueous dispersed media, *Green Chem* 21:36–53, 2019. https://doi.org/10.1039/c8gc02277a.

Moreno M, Goikoetxea M, Barandiaran MJ: Biobased-waterborne homopolymers from oleic acid derivatives, *J Polym Sci A Polym Chem* 50:4628–4637, 2012. https://doi.org/10.1002/pola.26287.

Moreno M, Miranda JI, Goikoetxea M, Barandiaran MJ: Sustainable polymer latexes based on linoleic acid for coatings applications, *Prog Org Coat* 77:1709–1714, 2014a. https://doi.org/10.1016/j.porgcoat.2014.05.016.

Moreno M, Goikoetxea M, Barandiaran MJ: Surfactant-free miniemulsion polymerization of a bio-based oleic acid derivative monomer, *Macromol React Eng* 8:434–441, 2014b. https://doi.org/10.1002/mren.201300162.

Moreno M, Goikoetxea M, De La Cal JC, Barandiaran MJ: From fatty acid and lactone bio-based monomers toward fully renewable polymer latexes, *J Polym Sci A Polym Chem* 52:3543–3549, 2014c. https://doi.org/10.1002/pola.27422.

Moreno M, Lampard C, Williams N, et al: Progress in organic coatings eco-paints from bio-based fatty acid derivative latexes, *Prog Org Coat* 81:101–106, 2015. https://doi.org/10.1016/j.porgcoat.2015.01.001.

Mou J, Li X, Wang H, Fei G, Liu Q: Preparation, characterization, and water resistance of cationic acetylated starch-g-poly(styrene-butyl acrylate) surfactant-free emulsion, *Starch/Stärke* 64:826–834, 2012. https://doi.org/10.1002/star.201200041.

Mubofu EB: Castor oil as a potential renewable resource for the production of functional materials, *Sustain Chem Process* 4:11–23, 2016. https://doi.org/10.1186/s40508-016-0055-8.

Mülhaupt R: Green polymer chemistry and bio-based plastics: dreams and reality, *Macromol Chem Phys* 214:159–174, 2013. https://doi.org/10.1002/macp.201200439.

Murawski A, Quirino RL: Vegetable oils as a chemical platform. In Thakur VK, editor: *Polymer gels, Gels horizons from Science to smart materials,* 2018, Springer Nature Singapore, pp 125–152.

Ouzas A, Niinivaara E, Cranston ED, Dubé MA: In situ semibatch emulsion polymerization of 2-ethyl hexyl acrylate/n-butyl acrylate/methyl methacrylate/cellulose nanocrystal nanocomposites for adhesive applications, *Macromol React Eng* 12:, 2018, 1700068. https://doi.org/10.1002/mren.201700068.

Ouzas A, Niinivaara E, Cranston ED, Dub MA: Synthesis of poly (isobutyl acrylate/n-butyl acrylate/methyl methacrylate)/CNC nanocomposites for adhesive applications via in situ semi-batch emulsion polymerization, *Polym Compos* 40:1365–1377, 2019. https://doi.org/10.1002/pc.24869.

Pei X, Tan Y, Xu K, Liu C, Lu C, Wang P: Pickering polymerization of styrene stabilized by starch-based nanospheres, *Polym Chem* 7:3325–3333, 2016. https://doi.org/10.1039/c6py00341a.

Petri DFS: Xanthan gum: a versatile biopolymer for biomedical and technological applications, *J Appl Polym Sci* 132:42035–42048, 2015. https://doi.org/10.1002/app.42035.

Picchio ML, Minari RJ, Gonzalez VDG, et al: Waterborne acrylic-casein nanoparticles. Nucleation and grafting, *Macromol Symp* 344:76–85, 2014. https://doi.org/10.1002/masy.201400013.

Picchio ML, Passeggi MCG, Barandiaran MJ, Gugliotta LM, Minari RJ: Waterborne acrylic-casein latexes as eco-friendly binders for coatings, *Prog Org Coat* 88:8–16, 2015a. https://doi.org/10.1016/j.porgcoat.2015.06.012.

Picchio ML, Minari RJ, Gonzalez VDG, Barandiaran MJ, Gugliotta LM: New strategy to improve acrylic/casein compatibilization in waterborne hybrid nanoparticles, *J Appl Polym Sci* 132:1–9, 2015b. https://doi.org/10.1002/app.42421.

Picchio ML, Passeggi MCG, Barandiaran MJ, Gugliotta LM, Minari RJ: Acrylic/casein latexes with controlled degree of grafting and improved coating performance, *Prog Org Coat* 101:587–596, 2016. https://doi.org/10.1016/j.porgcoat.2016.10.002.

Picchio ML, Minari RJ, Gugliotta LM: Enhancing the coating properties of acrylic/casein latexes with high protein content, *J Coat Technol Res* 14:543–553, 2017. https://doi.org/10.1007/s11998-016-9877-2.

Pokeržnik N, Krajnc M: Synthesis of a glucose-based surfmer and its copolymerization with n-butyl acrylate for emulsion pressure sensitive adhesives, *Eur Polym J* 68:558–572, 2015. https://doi.org/10.1016/j.eurpolymj.2015.03.038.

Prokopov NI, Gritskova IA: Characteristic features of heterophase polymerisation of styrene with simultaneous formation of surfactants at the interface, *Russ Chem Rev* 70:791–800, 2001. https://doi.org/10.1070/rc2001v070n09abeh000669.

Pu G, Dubay MR, Zhang J, Severtson SJ, Houtman CJ: Polyacrylates with high biomass contents for pressure-sensitive adhesives prepared via mini-emulsion polymerization, *Ind Eng Chem Res* 51:12145–12149, 2012. https://doi.org/10.1021/ie301492v.

Qiang XH, Xue Q, Zhang H, et al: Preparation and characterization of acrylic resin/protein composite crosslinked films, *J Coat Technol Res* 11:923–931, 2014. https://doi.org/10.1007/s11998-014-9599-2.

Quintero C, Mendon SK, Smith OW, Thames SF: Miniemulsion polymerization of vegetable oil macromonomers, *Prog Org Coat* 57:195–201, 2006. https://doi.org/10.1016/j.porgcoat.2006.08.011.

Ren S, Dubé MA: Modification of latex microstructure and adhesive performance using d-Limonene as a chain transfer agent, *Int J Adhes Adhes* 75:132–138, 2017. https://doi.org/10.1016/j.ijadhadh.2017.03.003.

Roberge S, Dubé MA: Emulsion-based pressure sensitive adhesives from conjugated linoleic acid/styrene/butyl acrylate terpolymers, *Int J Adhes Adhes* 70:17–25, 2016. https://doi.org/10.1016/j.ijadhadh.2016.05.003.

Scott PJ, Meenakshisundaram V, Hegde M, et al: 3D printing latex: a route to complex geometries of high molecular weight polymers, *ACS Appl Mater Interfaces* 12:10918–10928, 2020. https://doi.org/10.1021/acsami.9b19986.

Shamsabadi MK, Moghbeli MR: Cellulose nanocrystals (CNCs) reinforced acrylic pressure-sensitive adhesives (PSAs) prepared via miniemulsion polymerization, *Int J Adhes Adhes* 78:155–166, 2017. https://doi.org/10.1016/j.ijadhadh.2017.06.024.

Smeets NMB, Imbrogno S, Bloembergen S: Carbohydrate functionalized hybrid latex particles, *Carbohydr Polym* 173:233–252, 2017. https://doi.org/10.1016/j.carbpol.2017.05.075.

Sorrentino A, Gorrasi G, Vittoria V: Potential perspectives of bio-nanocomposites for food packaging applications, *Trends Food Sci Technol* 18:84–95, 2007. https://doi.org/10.1016/j.tifs.2006.09.004.

Stubbs JM, Sundberg DC: Core-shell and other multiphase latex particles-confirming their morphologies and relating those to synthesis variables, *J Coat Technol Res* 5:169–180, 2008. https://doi.org/10.1007/s11998-007-9060-x.

Tabone MD, Cregg JJ, Beckman EJ, Landis AE: Sustainability metrics: life cycle assessment and green design in polymers, *Environ Sci Technol* 44:8264–8269, 2010. https://doi.org/10.1021/es101640n.

Tellica MI: https://www.marketintellica.com/report/MI27877-global-pressure-sensitive-adhesive-sales-market, 2019.

Urban D, Takamura K: *Polymer dispersions and their industrial applications,* 2002a, Wileyhttps://doi.org/10.1002/3527600582.

Urban D, Takamura K: *Polymer dispersions an their industrial applications,* 2002b, Wiley-VCH Weinheim.

Van der Steen M, Bretz I, Kabasci S, Stevens CV: Synthesis of biobased multivalent crosslinkers from a castor oil-derived C22-acyloin, *Ind Crop Prod* 46:238–245, 2013. https://doi.org/10.1016/j.indcrop.2013.01.017.

Van Herk A: *Chemistry and technology of emulsion polymerisation,* 2013, Wiley.

Wenzel F, Agirre A, Aguirre M, Leiza JR: Incorporation of novel degradable oligoester crosslinkers into waterborne pressure sensitive adhesives: towards removable adhesives, *Green Chem* 22:3272–3282, 2020. https://doi.org/10.1039/D0GC00463D.

Wisser T, Riedel A, Gscheidmeier M, Maginot J: Process for the preparation of camphene by isomerisation of alpha-pinene. EP. Patent 0539990, 1997.

Xu J, Hu H: Preparation and characterization of styrene acrylate emulsion surface sizing agent modified with rosin, *J Appl Polym Sci* 123:611–616, 2012. https://doi.org/10.1002/app.

Xu J, Long L, Hu H: Preparation of starch-based styrene acrylate emulsion used as surface-treatment agent for decorative base paper, *J Polym Eng* 33:323–330, 2013a. https://doi.org/10.1515/polyeng-2012-0168.

Xu Q, Ma J, Zhou J, Wang Y, Zhang J: Bio-based core-shell casein-based silica nano-composite latex by double-in situ polymerization: synthesis, characterization and mechanism, *Chem Eng J* 228:281–289, 2013b. https://doi.org/10.1016/j.cej.2013.04.079.

Yu Q, Yang W, Wang Q, Dong W, Du M, Ma P: Functionalization of cellulose nanocrystals with γ-MPS and its effect on the adhesive behavior of acrylic pressure sensitive adhesives, *Carbohydr Polym* 217:168–177, 2019. https://doi.org/10.1016/j.carbpol.2019.04.049.

Yuan J, Mecerreyes D, Antonietti M: Poly(ionic liquid)s: an update, *Prog Polym Sci* 38:1009–1036, 2013. https://doi.org/10.1016/j.progpolymsci.2013.04.002.

Zeeb B, Thongkaew C, Weiss J: Theoretical and practical considerations in electrostatic depositioning of charged polymers, *J Appl Polym Sci* 131:1–11, 2014. https://doi.org/10.1002/app.40099.

Zhang Y, Dube MA: *Green emulsion polymerization technology,* 2017, Springer International Publishing. https://doi.org/10.1007/12.

Zhang L, Cao Y, Wang L, Shao L, Bai Y: Polyacrylate emulsion containing IBOMA for removable pressure sensitive adhesives, *J Appl Polym Sci* 133:1–7, 2016. https://doi.org/10.1002/app.42886.

Zhang Y, Cunningham MF, Smeets NMB, Dubé MA: Starch nanoparticle incorporation in latex-based adhesives, *Eur Polym J* 106:128–138, 2018. https://doi.org/10.1016/j.eurpolymj.2018.07.014.

CHAPTER SIX

Toward a digital polymer reaction engineering

Stefano Lazzari*, Andree Lischewski, Yury Orlov, Peter Deglmann, Andreas Daiss, Eduard Schreiner, and Hugo Vale*

Digitalization in Materials Science R&D, BASF SE, Ludwigshafen am Rhein, Germany

*Corresponding authors: e-mail address: stefano.lazzari@basf.com; hugo.vale@basf.com

Contents

Abstract

What is digitalization, and why do we need it? What does digitalization mean for research and development in polymer reaction engineering (PRE)? In this chapter, we address these questions, starting from a global perspective, briefly analyzing the current situation. We then illustrate our views on how the virtual representation of a polymerization process, i.e., a digital twin, would enable us to widen the spectrum of research and development in PRE. The main tasks of the digital twin are, in our opinion, to approach two yet unsolved problems: linking reaction conditions to final material properties and producing materials in an optimal way. Our strategy to tackle these issues is to combine physical and data science models. We analyze how the predictive ability of deterministic approaches, such as kinetic, thermodynamic and fluid mechanics models is improved by parameters and mechanistic understanding provided by quantum chemistry and molecular dynamics. While we constantly aim for a deeper physico-chemical understanding to improve these physical models, we recognize the power of statistics and machine learning, as they enable us to approach highly complex

Advances in Chemical Engineering, Volume 56
ISSN 0065-2377
https://doi.org/10.1016/bs.ache.2020.07.001

problems, thanks to their large flexibility. When analyzing the current limitations of each technique, we believe that concrete progress can only be achieved if we systematically centralize our knowledge. Therefore, we conclude our perspective on digitalization in PRE by advocating for initiatives that foster an open source sharing of published experimental data, parameters, models, and algorithms.

1. What is digitalization and why do we need it?

Today, the world is changing more rapidly than ever before, driven by demographic change and new digital technologies. We are facing diverse social and environmental challenges due to limited natural resources, climate change and the increasing demands of a growing global population (BASF, 2019b).

The world inhabitants were 7.7 billion people as of 2019, and are expected to become 10 billion by 2060 (Roser et al., 2013; United Nations, Department of Economic and Social Affairs, Population Division, 2019). In addition, the share of population aged 60 and over, will double from 2020 to 2050 worldwide (BASF, 2019b). Climate change threatens our ability to ensure global food security, eradicate poverty and achieve sustainable development. Greenhouse gas (GHG) emissions from human activity and livestock are a significant driver of climate change, trapping heat in the earth's atmosphere and triggering global warming (IPCC, 2018, 2019).

Chemistry is key to solving many of these challenges and several concrete initiatives are already underway, worldwide and across sectors. Guided by the Sustainability Development Goals, identified by the United Nations to steer the world's efforts toward the most relevant issues, countries, NGOs, and companies are stepping up to contribute to the solution (United Nations, 2015). BASF is planning to achieve CO_2-neutral growth until 2030 by keeping total GHG emissions and energy purchases at the 2018 level, while growing production volumes (BASF, 2019a). Different approaches to permanently sequester CO_2 from the atmosphere are now available,[a,b] NGOs and companies are investing in reforestation,[c,d] and the car industry is heavily investing in electric mobility (Mosquet et al., 2020). Moreover, new processes are

[a] https://www.climeworks.com/.
[b] https://globalthermostat.com/.
[c] https://www.treedom.net/en/.
[d] https://onetreeplanted.org/.

being developed to recycle CO_2 to produce syngas,[e] and more than 40 chemical companies have founded the non-profit "Alliance to end plastic waste" and are tackling the problem of oceanic plastic waste.[f]

What are the challenges for the chemical industry? While the pressure to innovate is higher than ever, given the current world situation, the competition is increasing, feedstock-related difficulties arise, and uncertain trade and tax policies have developed (Countryman et al., 2019; Islam, 2018). In this increasingly complex environment, many consulting companies point to digitalization as one of the key levers to improve materials and processes to the required environmental and sustainable standards, while keeping economic feasibility under control. Accenture has estimated that digitalization in R&D could bring profits in the order of \$500 billion to the chemical industry (Countryman et al., 2019), McKinsey estimates \$100 billion for pharma and healthcare (Chilukuri et al., 2017), while Roland Berger mentions benefits of more than \$1 trillion for the European manufacturing industry, if a well rolled-out digitalization infrastructure is created (Bloching et al., 2015).

Such large potential, in the order of hundreds of billions of dollars, is expected to unveil from a whole range of directions, which can be clustered in four branches (Bloching et al., 2015):

(i) *digital data*, that enables to describe and predict processes and optimize decisions;

(ii) *automation*, fostered by combining machine learning with traditional technologies, with the aim to reach autonomous, "self-driving" plants, increasing production speed, cutting costs, save resources, and reducing error rates;

(iii) *connectivity* of production, marketing and customer demand, that synchronizes the supply chain and shortens production cycles;

(iv) *digital access*, providing new services to customers, giving them full operational transparency.

The benefits resulting from these segments have been extensively deepened (Islam, 2018; Klei et al., 2017; Ulbrich and Eschenröder, 2018) and are beyond the scope of our analysis.

Here, we want to take a more specific perspective, discussing digitalization in research and development within the polymer reaction engineering (PRE) community. PRE research is a well-established field, with significant

[e] https://www.basf.com/global/en/media/news-releases/2019/12/p-19-420.html.

[f] https://endplasticwaste.org/.

scientific contributions dating back to the 1930s. Tremendous advances followed in the subsequent decades, that contributed in shaping the polymer industry (Asua, 2007; Dotson et al., 1995; Flory, 1953).

We believe that digitalization will expand the spectrum of R&D in PRE providing new research directions and questions, affecting all the above-mentioned digitalization areas (digital data, automation, connectivity, digital access). This will happen through the establishment of one crucial concept, that elegantly connects research and digitalization, and along whose line we will orient our discussion: the digital twin (Fig. 1).

Digital twins are a virtual, dynamic representation of a system, enabling its understanding, control, and optimization. The ideal digital twin of a polymerization process should describe its main features, mostly related to the reaction environment, as polymeric materials are products-by-process. Besides realistically reflecting the geometry of process units (Fig. 1), digital twins should:

(i) *predict the polymerization outcome* (e.g., product yield, molecular weight distribution, branching/crosslinking degree, particle size distribution);
(ii) *link polymer properties to application/*material properties;
(iii) *control and optimize the reaction* in real-time for one or more variables (specific product properties, productivity, safety);
(iv) *monitor the quality of inputs and output flows*, to adjust the process if a change in raw materials or product quality occurs;
(v) *enable predictive maintenance*, by monitoring the state of the reactor, impeller speed, fouling/corrosion presence;
(vi) *coordinate with other digital twins* in a plant, to automatically run processes, based on customer demands, market needs, logistics, and current state of reactor pools available.

Fig. 1 The digital twin of a powerplant.

In what follows, we will discuss the development of a digital twin for PRE, along these six directions, from an R&D perspective. We will analyze which combination of techniques is best suited to develop the features of the digital twin (Section 2), and summarize which efforts are currently being made toward its development (Section 3). We will then provide an outlook of which future steps could be undertaken, and where digitalization will lead PRE (Sections 4 and 5).

2. Digital twins for polymer reaction engineering

We can reorganize the features of the digital twin around two big unsolved problems:

(1) link recipe and reaction environments to polymer and material properties, features (i), (ii);

(2) produce a material, based on clients demands, while controlling, optimizing and maintaining a reactor pool, features (iii)–(vi).

Defining these two issues as the holy grails of PRE would hardly be an exaggeration. Our community is tackling them since at least the 1960s (Chappelear and Simon, 1969; Ray, 1972). Despite the significant body of literature produced so far, and the physicochemical understanding gained, many fundamental and technical facets of these two issues remain open. The opportunity to advance in the solution of these two long-standing problems is provided by digitalization, that enables the combination of complementary modeling approaches.

We will divide the approaches into two broad categories: physical modeling and data science (Table 1). In the former, we place modeling techniques such as kinetic and thermodynamics models, computational fluid dynamics, and molecular modeling (e.g., quantum chemistry and molecular dynamics). In general, any modeling approach that is based on physicochemical reasoning falls in this category. Conversely, approaches that purely rely on data, their collection, organization and correlation, fall into the data science category. Here, we include statistics and machine learning techniques. All approaches are sketched in Table 1.

Note that the definitions chosen here (Table 1), clarify which facets of these otherwise broad disciplines are being considered. The terms approach and technique will be used interchangeably throughout the text.

A deeper discussion of the different approaches, along with their advantages, limitations and interdependencies, is presented in Section 3. First, we will analyze which of these techniques is most suitable to approach the two, long-standing problems.

Table 1 Description of physical modeling and data science techniques, framed in the PRE context.

	Symbol used	Technique	Description	Information provided in PRE context	Exemplary references
Physical modeling		Kinetic and thermodynamic (TD) models	Mass and energy balances (kinetics), equation of states (TD)	Kinetics: conversions, reaction rates and heat, process simulation, polymer architecture, and particle size distribution TD: solubilities, pressures, reaction enthalpies	Asua (2007), Dotson et al. (1995), Flory (1953), Gmehling et al. (2019), Kontogeorgis et al. (2020), Kontogeorgis and Folas (2010)
		Computational Fluid Dynamics (CFD)	Navier-Stokes equation, mass and energy balances	Fluid flows, reactor design, heat exchange and mixing parameters	Buchelli et al. (2005), Elgebrandt et al. (2006), Kolhapure et al. (2005), Kolhapure and Fox (1999), Pohn et al. (2013), Wells and Ray (2005)
	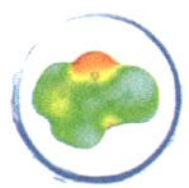	Molecular modeling (Quantum Chemistry and Molecular Dynamics)	Schrödinger equation (Quantum Chemistry), Newton's equation of motion, coupled with suitable energy potentials (Molecular Dynamics)	Quantum chemistry: kinetic rates of reactions (e.g., propagation, backbiting, etc.) in different media Molecular dynamics: solubilities and aggregation behavior of emulsifiers	Coote (2009), Dossi et al. (2011), Mavroudakis et al. (2015)
Data science		Statistical methods and Machine Learning (ML)	Statistical methods: linear/non-linear regressions, clustering, factor analysis, multivariate analysis. Machine learning algorithms: based on statistical models, learn from examples or experience	Identifying correlations between variables, when physical understanding is insufficient, e.g., to link sensor information and need to perform reactor maintenance, or polymer with application properties	Bishop (2006), Chaloner and Verdinelli (1995), Chen et al. (2020), Fisher (1935), Oakland (2007), Hastie et al. (2009), Traini et al. (2019), Wu et al. (2013)

2.1 Linking recipe to polymer and material properties

Understanding which reaction conditions lead to the desired material and its end-properties, implies solving three sub-problems: (i) predicting the polymer architecture, based on the reaction conditions, (ii) relating the polymer architecture to the polymer properties, and (iii) linking the polymer properties to the application properties. Note that with the term "polymer architecture" we refer to the combination of polymer chain composition, molecular weight and particle size distributions, and polymer microstructure. "Polymer properties" includes characteristics such as morphology, solubility, and suspension stability. Application properties reflect end-properties of polymeric materials and formulations in their application realm (e.g., mechanical resistance, swelling, stickiness, wettability).

Depending on the available physicochemical knowledge and resources, different approaches can be employed to address the three sub-problems.

We believe that the first two sub-problems (i.e., describing polymer architecture and properties) are best approached by physical modeling techniques. Provided enough understanding is available, precise questions can be answered, employing a combination of models based on kinetics, thermodynamics, computational fluid dynamics, and molecular modeling (i.e., molecular dynamics and quantum chemistry).

The availability of physicochemical understanding linking the properties of the polymer to the macroscopic end-properties of the material is, except for a few simple cases, much scarcer. Thus, to approach the third sub-problem, a data science approach, relying on statistics and machine learning, provides the most advantages (Fig. 2).

The rationale behind this division is that the maturity of physical modeling enables us to describe the outcome of complex polymerizations (i.e., polymer architecture and properties) by identifying a limited number of parameters. The same cannot be said with regards to linking the polymer properties to application/product properties. Only data science approaches, with their larger number of parameters and flexibility, can bridge the two domains, bypassing our current lack of physicochemical understanding (Chen et al., 2020). The additional parameters compensate our limited knowledge about the connection of microscopic to macroscopic properties.

As an example, consider an emulsion system with monomers fed from the gas phase. What do the physical modeling and the data science realms have to offer?

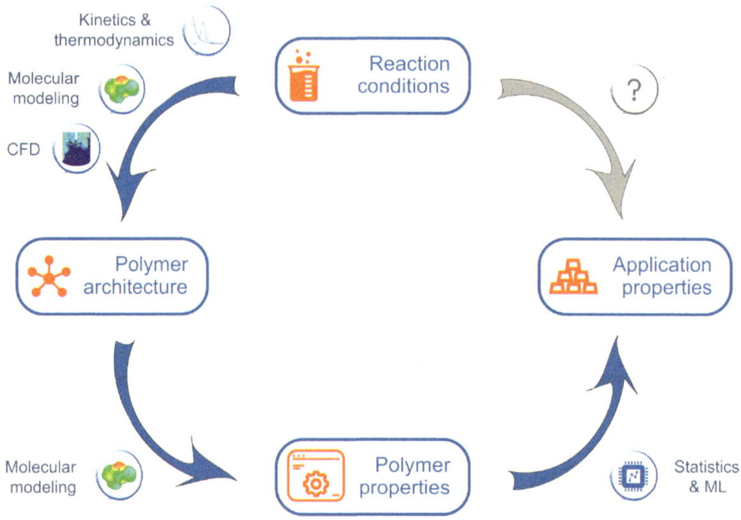

Fig. 2 Linking reaction conditions to application properties, leveraging physical modeling and data science.

Kinetic models provide us with conversion, particle size and chain length distributions, as well as crosslinking/branching degrees and other features of the molecular weight distribution (Asua, 2007; Dotson et al., 1995). Thermodynamic models ensure the correct prediction of the monomer partition among the various phases, which plays a key role in the kinetics and is crucial for process safety, as it determines the system pressure. Computational fluid dynamics (CFD) models enable us to design the reactor properly, providing the viscosity and heat-exchange properties of the system. When using new molecules in such a process, we can rely on quantum chemistry to estimate the elementary rate coefficients in a medium of a given polarity (Deglmann et al., 2018). Molecular dynamics can provide the tendency of a novel surfactant to form micelles, diffusion constants, interfacial loading by surfactants or the morphology of a polymer particle surface. While heterogeneous polymerization is far from being a closed topic (Vale and McKenna, 2005), it is fair to say that the ensemble of physical modeling approaches available today gives us a detailed perspective on the polymerization outcome, especially in terms of polymer microstructure.

As soon as we wish to link polymer properties to application or end-material properties, such as the performance of a paint or an adhesive, data science approaches become the natural choice. Statistical methods and machine learning approaches provide us with powerful correlations, based on large data pools, that relate properties of the polymer to the end-application ones (Chen et al., 2020). While these approaches remain empirical in nature, it is worth to recall how much so-called effective or engineering approaches have enabled a technological progress.

Overall, we select a technique from the physical modeling pool whenever we understand the problem well enough, and there is a reasonable number of parameters to be estimated through a limited number of experiments or with the help of another modeling technique.

Whenever our lack of physicochemical knowledge is high, and the number of parameters and interplaying factors rises exponentially, we tend to prefer a data science approach.

It is worth stressing that despite this conceptual distinction, the combination of the different techniques is crucial to solve our problem of linking the reaction conditions to the application properties. Moreover, it is desirable that, as our understanding grows with time, physically motivated models will be able to link the microscopic properties of the polymer to the macroscopic properties of the material.

2.2 Optimization and control

Another challenge that has sparked research and technological advancement is the optimization and control of polymerization processes. From a broader perspective, the aim is to satisfy customer's needs, i.e., producing a material that fulfills certain specifications, while running the process in an optimal way. Having solved the problem described in Section 2.1, we know which recipe and reactor environment satisfy certain product properties.

However, there are still some issues we need to consider, such as the selection of the best available reactor, its maintenance, and its optimized and controlled operation.

Fig. 3 shows how a combination of physical modeling and data science techniques enable the solution of the problem.

Let us take the previous example of a multi-phase emulsion polymerization process. Imagine a customer request aligning with this process, and a new production campaign start. Once the request arrives, a scheduling algorithm optimally addresses the need. Predictive maintenance, guided by a machine learning approach, will support the decision, by indicating the most suitable reactor in the plant to perform the polymerization. The on-line control of the process, in terms of efficiency, safety and product quality, will be steered by thermodynamic and kinetic models, that enable

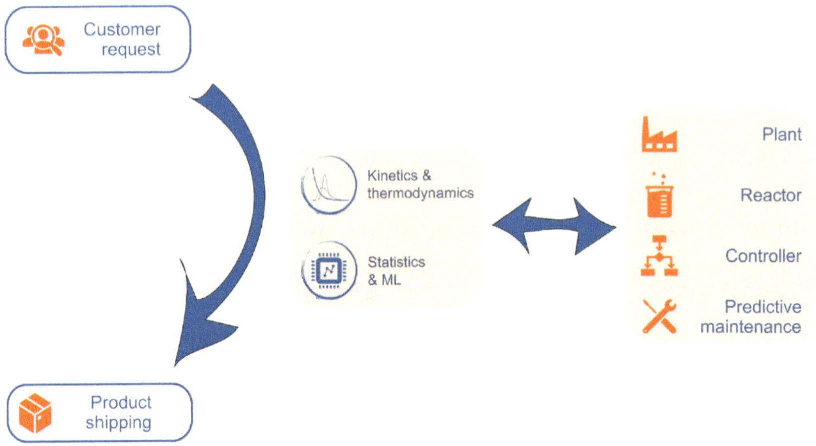

Fig. 3 To produce and ship the products required by customers, a combination of data science and physical modeling enables the optimized control and maintenance of polymerization reactions in a powerplant.

the prediction of the polymerization outcome, based on the trajectory of the reactor states monitored/estimated by hard- and soft-sensors and analytics (Gerlinger et al., 2019). This allows to keep reagents purity and product quality at the desired level. If the production continues in time, the reactor could undergo fouling, resulting in poorer heat exchange and potentially off-spec products. Machine learning algorithms predict when maintenance is due, and when it is meaningful to stop a production line and switch to a different one. The development of such algorithms relies on a large and organized sensor data pool.

The decision to entrust a task to a given technique reflects our current physicochemical understanding and the complexity of a given process. Selecting machine learning approaches to address the fouling issue and the predictive maintenance is a natural choice, when we consider the complexity behind the fouling process, our limited knowledge about them, and the general difficulties of predicting when a machine will cease to be trustworthy (Golightly, 2019; Traini et al., 2019). Currently, only an algorithm, able to learn from experience, may be able to approach such a problem, given the hundreds (if not thousands) of interplaying variables.

Once these approaches become more robust and integrate the diverse information each of them delivers, self-guided plants will become closer to reality.

3. Toward digital twins

We have established that the development of a digital twin is based on physical modeling and data science approaches. In this section, we describe how each approach contributes to the two long-standing PRE questions, i.e., how to link reaction conditions to material properties, and how to optimally produce a desired material, while controlling, optimizing, and maintaining a reactor pool. We then suggest how to realize the full potential of the approaches in developing digital twins. Note, the subsequent analysis and cited literature reflect the authors' experience and is not meant to be exhaustive. The historic development and the state of the art of the techniques, out of the scope of this perspective, are well documented in the literature, and some references are provided throughout the text.

3.1 Kinetic models

Kinetic models relate reaction conditions to the polymer chain length distribution and its architecture. Crosslinks, branching points, cycles, chain length, and chain composition affect the polymer microstructure, and nearly all final application properties (e.g., biodegradability, mechanical resistance, adhesion, swelling).

Literature examples of models that show how recipes affect polymer architecture are extensive. Monomer concentrations and feeding strategies have been related to crosslinking degrees and intramolecular cycles, that affect gel points, swelling and mechanical resistance of hydrogels (Diederich et al., 2017; Dotson et al., 1995; Tobita and Hamielec, 1990). Branching degrees are related to backbiting and beta-scission, that may be controlled by temperature and monomer selection (Ballard and Asua, 2018; Ballard et al., 2018; Barth et al., 2010; Wang and Hutchinson, 2011). Degradability and biodegradability have been shown to be affected by several effects, including pH, molecular weight, polymer crystallinity and species diffusivity (Casalini et al., 2014; Laycock et al., 2017; Lazzari et al., 2014).

An industrial example of kinetic models providing control over the polymer properties is shown in Fig. 4. First a kinetic model was compared with experimental data (Fig. 4, left panel) to estimate the missing parameter values. In the process, two reaction domains were identified: the first, where the polymer was built-up and the molecular weight kept at bay by backbiting and scission, and the second one, where transfer to polymer dominated and sharply increased the molecular weight. Thanks to this knowledge,

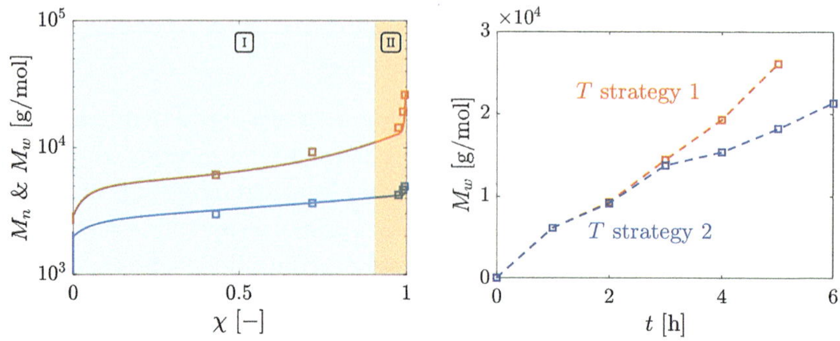

Fig. 4 Left panel: Number average (M_n) and weight average (M_w) molecular weights against conversion. Model (continuous line) and experimental data (squares) in good agreement, showing the existence of two regimes. Right panel: experimental data only (squares plus dashed lines to guide the eye) of two temperature profiles. The desired weight average molecular weight is obtained by a step-decrease in the temperature.

we selected a temperature profile that decreased the occurrence of transfer to polymer, and obtained the desired molecular weight and viscosity (Fig. 4, right panel).

Kinetic models provide a crucial building block also in the frame of reactor optimization (see Fig. 3), as they describe the current state of the reactor in terms of reaction rates and heat production, and allow to predict the next state, enabling controllers to be activated if deviations from the desired product specifications are predicted, or if a variation in reagent purity is detected.

Kinetic models are defined by three essential building blocks. The first is the reaction network, that clarifies which kinetic steps (e.g., initiation, propagation, termination) need to be considered in a given polymerization. Identifying side-reactions (e.g., backbiting, scission, transfer to polymer) is also crucial: despite the slow rates at which they proceed, they determine the polymer microstructure and have a strong impact on the final material properties.

Once we know which reactions occur, we need to know their kinetic rate coefficients (the second building block), that determine the species concentrations and the heat produced in the reactor. In the case of a bimolecular reaction among the species i and j, with concentrations C_i, C_j:

$$r_x = k_x(T, \underline{y})C_iC_j$$

where $k_x(T, \underline{y})$ is the apparent kinetic rate constant of reaction step x, (e.g., propagation, chain transfer, crosslinking, termination). $k_x(T, \underline{y})$ depends on the temperature T, and other properties, $\underline{y}$, such as the conversion of the system, the electrolyte concentrations, or the polarity of the system (Fischer et al., 2018; Lacík et al., 2004).

The third and last building block is the ensemble of physical parameters, such as monomer solubilities, emulsifier parameters (e.g., critical micellar concentration, hydrophilic-to-lipophilic balance (HLB) value) or reaction enthalpies.

The development of the polymerization reaction network (i.e., the first building block) is based on the nature of the polymerization (e.g., free-radical, ionic, step-growth). To narrow down the number of reactions steps to be included, we can leverage experimental results, quantum chemistry (QC) calculations or machine learning approaches.

For instance, to include or exclude the transfer to polymer in a homopolymerization, we can monitor the degree of long-chain branching via NMR. QC approaches may reduce the experimental efforts by simulating different reaction pathways and identifying the most probable ones. This is particularly useful when dealing with complex reaction networks (e.g., catalytic cycles), but also in apparently more obvious cases, such as the homopolymerization of *N*-vinyl-pyrrolidone, where it turned out that chain transfer to monomer initiates a sequence of reactive steps that lead to fouling (Deglmann et al., 2019).

Machine Learning (ML) approaches are not yet extensively used in polymer reaction engineering, but their potential in identifying reaction outcomes has been proven in small molecules (Coley et al., 2017, 2018). Gathering a proper amount of experimental data for polymerization reactions, as the one available in databases such as Reaxys,[g] is of undisputable relevance when ML approaches are to be adopted (Coley et al., 2017, 2018; Yang et al., 2019).

The second building block of kinetic models, i.e., kinetic rate constants, give access to rates of reaction and heat production. The identification of kinetic rate constants relies on experimental approaches, quantum chemistry, or kinetic models.

The most established experimental techniques to identify kinetic rate constants are pulsed-laser polymerization with size-exclusion chromatography (PLP-SEC), single-pulse PLP electron paramagnetic resonance (SP-PLP-EPR), and so-called diffusion clocks (Drawe and Buback, 2016; Drawe et al., 2015; Kattner et al., 2017; Lacík et al., 2003; Wittenberg et al., 2013). Other experimental methods (i.e., high-pressure liquid

[g] https://www.elsevier.com/solutions/reaxys.

chromatography (HPLC), gas chromatography (GC), nuclear magnetic resonance (NMR)) have also proved valuable in providing information about rate coefficients (Fischer et al., 2017).

An alternative to performing experiments is relying on quantum chemistry (QC) (Mavroudakis et al., 2015). QC provides approximate kinetic parameters of great value, as some reaction rates (e.g., backbiting or beta-scission) are not directly observable. QC has also helped relating reactivity to solvent polarity, a feature that is usually only observable from an experimental perspective (Deglmann et al., 2018).

In many cases, the rate coefficients have been obtained by comparing experiments with the kinetic model itself (Lazzari et al., 2014; Wang and Hutchinson, 2011; Yu et al., 2011). However, it is very desirable to minimize the number of parameters to be fitted this way, in the interest of obtaining a robust kinetic model, and to minimize the number of experiments to be performed.

The third and final building block of kinetic models are physical parameters, (e.g., monomer solubilities, reaction enthalpies, aggregation parameters of surfactants, etc.). These may be obtained by experimental techniques, thermodynamic models, molecular dynamics or machine learning approaches.

Classical solubility experiments (Prausnitz, 1999) have been automated in recent years, leveraging microfluidics (Abolhasani et al., 2015). This change improved speed and reliability of the measurements, and minimized the overall measurement cost. Even though microfluidics is more employed in the "small molecules" field, positive examples in the PRE realm are found, suggesting that a further deployment of microfluidics is potentially useful (Li et al., 2008).

The experimental information can be rationalized in thermodynamic models, that provide solubility expressions scaling with reaction parameters (e.g., temperature, system composition) (Flory, 1953; Gross and Sadowski, 2001; Schaffer et al., 2003).

An alternative to estimate solubilities is given by machine learning approaches, that have proven their capability in predicting solubility of small molecules based on solubility databases (Yang et al., 2019). Refining these models and extending their use in the polymer field is very desirable, given the great number of experimental measurements that could be spared.

To obtain information about emulsifier aggregation parameters, molecular dynamics (MD) offers an advantageous approach with respect to a trial-and-error formulation approach. MD helps reducing the experimental efforts to identify suitable combinations of amphiphilic molecules to improve stability, or at least narrow down the number of molecules to be screened experimentally (Weiss et al., 2016).

As new materials are constantly sought, new functional monomers are synthesized, and new monomer combinations are screened. To accelerate product or process development, kinetic models are needed.

However, the experimental identification of the reaction rate networks, kinetic rate constants, and physical parameters, would require a very significant time and resource investment. For instance, a set of more than 30 parameters was necessary to simulate the system in Fig. 4. About 10 parameters related to initiation and cross-propagation (Skeist, 1946) were determined independently. All others had to be obtained from other sources, such as quantum chemistry simulations, or producing large enough experimental data sets to fit the kinetic model against.

Systems consisting of charged monomers, or those proceeding by ionic polymerizations, where the rates depend on the environment polarity, present the additional difficulty of lack of fundamental physicochemical understanding.

While many modeling alternatives exist, they do not yet work in optimal coordination. Experimental data, models, and parameters are shared in open-access or standard publications, but they are not yet centralized, and limit the realization of the full potential of kinetic models. The solution to this conundrum might be enabled by data sharing (see Section 4).

3.2 Thermodynamic models

Thermodynamic modeling plays an important role in bridging reaction conditions to the polymer architecture (see Fig. 2), and in model-based process optimization (see Fig. 3). The success of simulation studies is critically dependent on how accurately the thermodynamic properties of pure components and mixtures are estimated for the investigated system.

Thermodynamic properties in industrial systems are difficult to evaluate, due to the lack of experimental data covering the broad range of complex reactive environments encountered. This difficulty underlines the importance of thermodynamic models which reliably predict properties of interest with a limited amount of available information.

For example, during process optimization of thermoplastic foams, several questions have been raised regarding the safe design of the reactor and selection of operation conditions to get maximum blowing strength, using CO_2

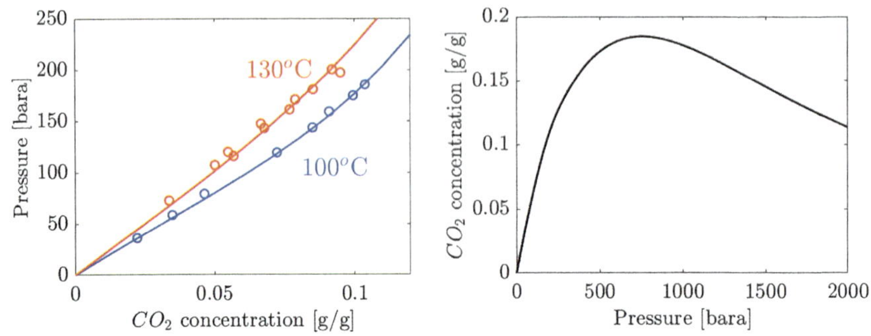

Fig. 5 Left panel: PC-SAFT model parameterization based on experimental CO_2 gas solubility data. Right panel: model-based extrapolation over the broad pressure range.

as a blowing agent. To approach this problem, the following tasks have been addressed by thermodynamic modeling:

- Calculate reactor pressure as a function of CO_2 concentration and temperature (*safety relevant issue,* Fig. 5, *left panel*);
- For a given temperature, find the pressure leading to the optimal CO_2-solubility in polystyrene foams (*product relevant issue,* Fig. 5, *right panel*).

In the first step, the Perturbed-Chain Statistical Associating Equation of State (PC-SAFT EoS) (Gross and Sadowski, 2001) was parameterized for the above described system. All pure component model parameters for the polymer and CO_2 were taken from literature (Enders et al., 2005; Peters et al., 2013). The remaining unknown binary interaction parameter has been adjusted using the available experimental data (Enders et al., 2005). In the second step, using this model it was possible to extrapolate the gas solubility in the polymer at the desired process conditions. The predictions have shown that high pressures lead to the decrease in the CO_2-solubility and at which pressures the gas solubility exhibits a maximum.

The setup of the working thermodynamic model consists of three building blocks:

- The central element in model development is the correct assessment of the dominant physicochemical contributions of the studied system, that control enthalpic and entropic terms in the Gibbs free energy. Moreover, the model calibration strategy and the availability of accurate experimental data, necessary to obtain physically meaningful model parameters, are as important as the model itself.
- Rigorous multiphase computations require efficient flash algorithms, allowing robust conversion by minimizing the Gibbs free energy of

the investigated system. This is especially relevant for highly nonideal systems, e.g., polydisperse polymers to consider their molecular weight distributions or multicomponent reactive mixtures.
- Finally, the thermodynamic models should be accessible to the practicing product developers and engineers. Nowadays different application programming interfaces (API) are available, enabling data transfer between the simulation tools.

Most thermodynamic models developed for engineering applications fall into one of three categories: (i) thermodynamic property correlation equations, (ii) models for extrapolation and (iii) models for screening purposes. Selecting the representative model and its parameters, depends on various factors: application task, required predictive accuracy, capabilities and limitations of the status-quo thermodynamic approaches, availability of the experimental data. A comprehensive overview on pairing models with applications is given in the textbooks by Kontogeorgis and Folas (2010) and Gmehling et al. (2019).

Applied thermodynamics has a long and successful history of technological innovations (Chen and Mathias, 2002). Several review articles originating from academia and industry discuss the achieved advances, unsolved problems and industrial requirements for thermodynamic models and physical property data (Abildskov and Kontogeorgis, 2004; Dohrn and Pfohl, 2002; Hendriks et al., 2010; Prausnitz, 1999; Prausnitz, 2001; Zeck, 1991). In the remaining of this section, we will focus on one of the many unsolved issues.

Recently, Kontogeorgis et al. (2020) published an excellent review article, summarizing key limitations and open questions regarding the use of molecular-based equations of state, derived from the Statistical Associating Fluid Theory (SAFT) (Chapman et al., 1989, 1990). One of the central topics in that manuscript is the complexity of the pure component parameterization. One can only confirm the fundamental relevance of this problem area. So far, it is unclear how to choose the right set of pure component parameters for associating compounds to extrapolate reliably the thermodynamic properties of the mixtures. Very often, this bottleneck leads to an insufficient prediction accuracy for new process/product design and time-consuming model recalibration trials. This is an evident but crucial problem, which hinders the realization of the full application potential of the SAFT-approach in industry.

In most SAFT-model variants, the associating term accounting for the hydrogen bonding remains unchanged. Which information, obtained from dedicated thermodynamic measurements, model predictions (e.g., COSMO-RS, force field molecular simulations) and statistical approaches (e.g., design of experiments), is more useful in identifying SAFT-associating parameter values (energy and volume)? Which SAFT-association scheme is the most realistic one, among the available choices? Answering these questions is an essential prerequisite, if we want to model vapor-liquid and liquid-liquid phase equilibria for complex multicomponent mixtures with adequate accuracy.

3.3 Computational fluid dynamics models

Computational fluid dynamics (CFD) is a powerful, yet computationally costly tool for the analysis of chemical reactors. If and how it is used must therefore be decided on a case-by-case basis: If chemical reactions are fast compared to mass and heat transfer processes, mixing has an influence on the result of these reactions, and CFD analysis is worth pursuing, and represents a crucial step in identifying the proper reaction environment that results in the desired polymer architecture (see Fig. 2).

For many polymerization processes, changes of the chemical composition due to chemical reactions are relatively slow compared to mixing. Hence, ideal reactors provide an accurate enough description, and a detailed analysis of mixing processes with CFD is unnecessary. Examples are many radical polymerization reactions, operated in batch and semi-batch mode, that last hours, whereas typical mixing time scales in stirred tanks are of the order of a few minutes or less.

Anionic or cationic polymerization reactions, however, are extremely fast and therefore strongly influenced by mixing and a conceptual separation of mixing and chemical processes in not possible anymore. Other examples where mixing and polymerization reactions cannot be separated, are continuous processes, where fouling may occur or where high temperatures, pressures or monomer concentrations are encountered (e.g., ethylene polymerization in high pressure tubular or stirred tank reactors), or where heat transfer limitations occur (Buchelli et al., 2005; Kolhapure and Fox, 1999; Kolhapure et al., 2005).

Depending on the specific chemistry and reactor set-up, CFD can support polymer reaction engineering for design and scale-up purposes in three different ways:

(1) For very fast reaction systems, a direct coupling between chemical kinetics and CFD can be done to get spatially resolved concentration and temperature distributions, from which engineering quantities such as conversion or selectivity can be calculated (Kolhapure and Fox, 1999).

(2) In cases where distinct zones can be distinguished in a reactor, and where there is no back-coupling of the chemical reactions on the fluid flow, CFD can be used to parameterize compartment models (determination of compartment volumes and mass exchange fluxes) (Elgebrandt et al., 2006; Pohn et al., 2013; Wells and Ray, 2005).

(3) Finally, for all cases where a complete separation between chemical reactions and fluid mechanics is possible, CFD can be used to characterize reactors in terms of heat transfer, power input or residence time behavior which can then be used in polymer reaction engineering models.

The quality of CFD results in the frame of polymer reaction engineering strongly depends on the physical phenomena that are involved. For single phase systems, which do not exhibit viscoelastic behavior, engineering precision can be achieved for many calculated quantities. In the case of multiphase systems, results are often qualitative or semi-quantitative, as the CFD simulation output strongly depends on specific system features (e.g., type and number of co-existing phases, volume fractions of the different phases, occurring physical phenomena, such as coalescence and breakage of disperse phase entities).

In the following, three cases are discussed, exemplifying the previously described situations where CFD supports polymer reaction engineering.

Fig. 6 shows a CFD analysis of a nonisothermal, single phase, step-growth polymerization process in a static mixer reactor by approach (1) in order to compare the fouling risk for different reactor scales. In this case, a detailed kinetic model has been included in the CFD simulation to determine the level of gel formation in the reactor and the related fouling. The quality of the predictions is solely limited by the quality of the kinetic model and material property data used.

An example for the second approach is given in Fig. 7. For the description of the reaction kinetics of a radical polymerization under harsh reaction conditions, CFD calculations are coupled with a simplified reaction kinetics scheme. The purpose is to understand the difference in the performance of different reactor designs and not to quantitatively predict the microstructure of the resulting polymer. As can be seen in Fig. 7, the considered reactor designs lead to very different radical concentration distributions. With the computed radical concentration profiles, both reactors can be split into compartments, and the impact of mixing can be studied by means of detailed polymer reaction engineering models.

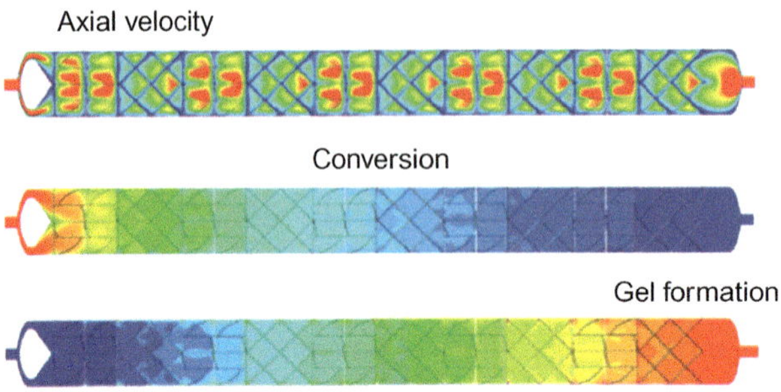

Fig. 6 Analysis of gel formation in a continuous static mixer reactor which is used for a step-growth polymerization process. Top to bottom: Contours of axial velocity component, end group conversion and gel formation in a cross section through the middle of the reactor. Flow from left to right. Blue = Low values of quantity. Red = High values of quantity.

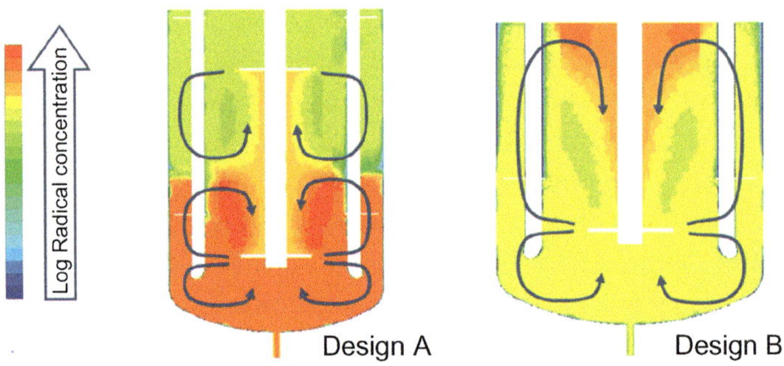

Fig. 7 Analysis of reaction conditions for a free radical polymerization under harsh reaction conditions in different continuous stirring tank reactors. Center and right: Distribution of radical concentration for the two different reactor set-ups. The qualitative differences are triggered by the different flow patterns and the flow regime in the reactors.

Finally, Fig. 8 shows an example of a bubble column reactor in which a tracer component shall be quickly mixed into the reaction mixture before the start of the reaction. Mixing is induced by a gas stream that induces a vortex in the column. The tracer component is fed with the gas stream. Two different reactor designs are compared with respect to their mixing times. As can be seen, mixing proceeds significantly faster for design B than for design A. Here, the CFD calculations are not linked to any kinetics model but solely mixing times are computed as characteristic engineering quantities.

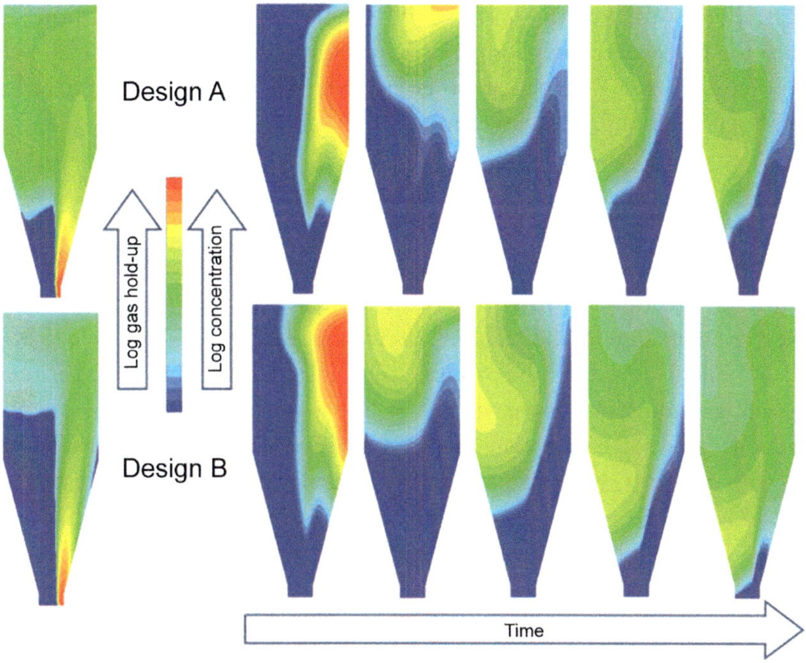

Fig. 8 Analysis of a gas induced tracer component mixing process in a bubble column. Comparison of two different column designs. Top: Design A. Bottom: Design B. Left: Logarithm of gas hold-up. Right: Evolution of logarithm of tracer component concentration with time.

To further promote the utilization of CFD in polymer process engineering, it is important to improve and automate simulation workflows, to make model setup less time-consuming, such that CFD is included more as a standard tool in development projects as a digital alternative to experiments. This requires, among many other things, that input data is available in a digital form (e.g., CAD representation of geometries) and that model and material property databases are built-up and seamlessly connected to CFD simulation tools.

It is also desirable to further improve and extend CFD models in fields such as multiphase flows where CFD calculations are still, as stated, often only qualitative or semi-quantitative. However, increasing model complexity is not always key to improved results. It is the opinion of the authors that, at least in an industrial environment, a successful application of CFD methods relies on good and justified simplifications of the real process to come to useful results, both timely and cost efficiently.

It is believed, that in future, CFD simulations will be used more and more for the characterization of process equipment to build digital twins of polymerization reactors or other polymer specific devices. Therefore, the above described automated simulation workflows will be a prerequisite.

3.4 Quantum chemistry models

Quantum chemistry (QC) models, providing information on rate coefficients and reaction pathways, are crucial in establishing the link between reactive environment and polymer architecture (see Fig. 2).

QC calculations offer the promise, by being ab initio, that no experimental data is required to answer questions like if or to what extent a chemical reaction takes place (reaction thermodynamics) or via which mechanism, and thus how fast, it will be (reaction kinetics). This is done by computing activation energies and entropies (required for the Eyring-equation which relates these to rate coefficients) via approximately solving the molecular Schrödinger equation.

QC unveils its great potential when augmenting a basic kinetic scheme (e.g., initiation, propagation, termination) by rare, yet very relevant, side reactions. Consider a radical polymerization leading to a typical linear vinylic functional polymer. In each polymer chain, 100 to 10,000 monomers are incorporated. For such systems, chemical analytics cannot resolve the full molecular structure anymore: it will yield information on main repeating units, maybe even on side reactions that can occur many times upon growing a single chain (e.g., short chain branching). However, kinetic information on other side reactions, even more relevant for product properties (e.g., transfer reactions), is only accessible from indirect evidence, i.e., molecular weight distributions.

Thus, it is not surprising that the use of quantum chemical predictions (in particular of rate coefficients) in polymer reaction engineering has been discussed by various authors, both in research papers (Coote, 2009; Dossi et al., 2011) and in monographs (Soroush, 2018).

In contrast to other applications of quantum chemistry, where calculation methods and numerical settings leading to mediocre accuracy suffice, the prediction of chemical kinetics requires the use of the most accurate computational protocols which are still computationally affordable. This is due to the exponential dependence of kinetics on both activation energies and entropies. Thus, even employing a high computational accuracy, the predicted absolute rates may be far off for a direct use in a process model. Nevertheless, kinetic predictions can be very useful for determining relative rates, i.e., when there is a known rate constant for one reaction, and the ratio of this rate to that of a reaction where kinetics are unknown is often predicted with good accuracy. An example for the latter case is a comparative study of aliphatic tertiary amines for polyurethane catalysis (Muuronen et al., 2019).

Also excluding some reactions (e.g., because of too high activation barriers) and providing reasonable starting values for the rates of other reactive steps, represent valuable information that QC provide as input to kinetic models.

Besides information on the kinetics of new reactions, also known kinetic pathways can be studied in different reaction media by QC, leveraging refined implicit solvation models. We can rationalize the variation of the rate coefficients at changing environment in a thermodynamic way, quantifying the activity coefficients of the reactants and the transition state (Deglmann et al., 2017, 2018). This approach and the impact of the solvent environment is exemplified in Fig. 9. Here we show to which extent the

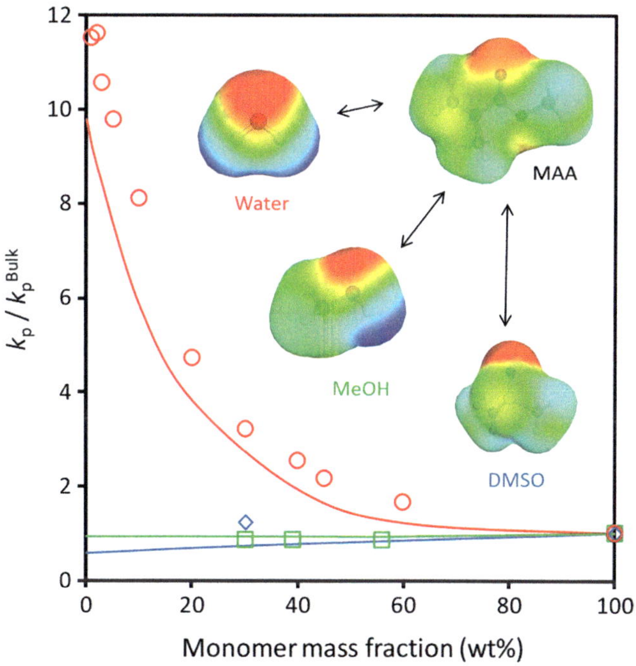

Fig. 9 Relative propagation rate of methacrylic acid as a function of solvent type and monomer mass fraction. *Reproduced with permission from Deglmann, P., Hungenberg, K.-D., Vale, H. M.: Dependence of propagation rate coefficients in radical polymerization on solution properties: a quantitative thermodynamic interpretation,* Macromol React Eng 12*(4):1800010.*

propagation rate coefficient of methacrylic acid scales with respect to its bulk value, as a function of different solvents and monomer mass fractions.

For the quantum chemical study of molecular reactivity, in principle both dynamic and static approaches are possible. Although it appears most intuitive to study reactive events by ab-initio molecular dynamics (MD) using a simulation box containing all required reactants plus solvent molecules, this does currently not represent a realistic option for screening purposes because of the huge computational resources it requires (in spite of the fact that typically steered MD flavors are used that enforce the desired reaction to take place many orders of magnitude faster than it would need if the system is allowed to evolve at a given temperature without a bias).

Thus, for studies addressing the abovementioned questions, typically static approaches that consider only stationary points on potential energy hypersurfaces (i.e., relaxed zero-kelvin structures of reactants and transition states) are applied. This is done by considering molecular models which do not look too dissimilar from how a chemist would write down a reaction (e.g., a propagating radical is represented as an oligomer with very few repeating units and low molecular species are modeled as they are).

In typical computational protocols to predict chemical kinetics, structural relaxation takes place at a computationally cheaper level, followed by more accurate single point energy calculations at a higher level. The lower level is almost always density functional theory (DFT), whereas the higher level can either be DFT or post-Hartree-Fock methods with large basis sets. To perform statistical thermodynamics and compute enthalpies and entropies at temperatures other than 0 K, also molecular vibrational spectra at the structural relaxation level have to be computed. Another ingredient, potentially more important than the energy calculation method, is the solvation model, that accounts for the reaction occurring in a condensed phase in bulk or solution, rather than between two isolated species colliding in the gas phase. Depending on the reaction of interest, it is also necessary to consider conformational partition functions of all species or tunneling. The numerous contributions to be included in such computational protocols are summarized in the literature (Deglmann et al., 2009).

The application of quantum chemical methods in technical applications would strongly benefit from further advances in the solvation treatment of chemical reactions. The number of available implicit solvation models is still rather small. There are continuum methods like PCM or COSMO, which consider a solute in an environment with a specific dielectric constant and refinements on these, like COSMO-RS or the SM-family of models, where also interactions like hydrogen bonding are accounted for more explicitly. Furthermore, methods that better capture the reaction dynamics would be desirable. In QC, kinetics is approached as reaction thermodynamics, treating reactants and products as reactants and transition states; the often complex path from reactants to a transition state that can also have an influence on rates (reflected by the transmission factor in Eyring theory) is so far mostly overlooked.

3.5 Molecular dynamics models

Molecular dynamics (MD), in its widest sense, ranges from atomistic MD to dissipative particle dynamics (DPD) and provides access to a variety of properties and phenomena.

To elucidate the contribution of MD in linking the reaction environment to material properties (see Fig. 2), it is instructive to consider the design of a polymer tailored to a specific use (Weiss et al., 2016), shifting our perspective. So far, we have considered situations where we knew which monomers we wanted to use, and we predicted the polymer architecture (through kinetics and QC), and the impact of reactor choices (through thermodynamics and CFD). But we can look at the problem from another angle, i.e., having a specific polymer property in mind, we can ask ourselves: which monomers and which polymer architecture would best satisfy our needs? To answer this question, we need to relate the polymer architecture and composition to the morphology, solubility, and account for the compatibility with other compounds. While the compatibility question usually is more efficiently addressed using thermodynamic methods guiding the choice of monomers, the morphology and compatibility issues are best investigated in a particle simulation. As not only bulk polymer, but also complex formulations are potential target products, we may need to consider the impact of further processing steps, e.g., extrusion, or the preparation of a full formulation. Molecular simulations can provide insight into the behavior of a complex formulation including possible dilution or mechanical stresses. With the help of MD, we are now in possession of the polymer design able to provide the envisioned polymer, i.e., monomer candidates, relative composition and polymer architecture. To synthesize the desired polymer is the "direct" problem that we have been discussing in the other sections, and that is typically solved by a kinetic model, aided by thermodynamics, QC and CFD. MD may help also in this context, by providing specific parameters, e.g., diffusion constants, that are relevant in kinetic models.

Molecular dynamics simulations can be used in generating input parameters for optimization and control. A series of papers (Chen et al., 2019a,b) shows the use of an energy-minimization multi-scale (EMMS) approach, for the liquid-liquid flow in rotor-stator (RS) mixing devices, to predict droplet size distribution using CFD and a population balance model. Within the proposed framework, a single-film model was used to describe the transport of surfactants. This surfactant transport model was used to calculate a coalescence efficiency corrector required by the population balance model. Coarse-grained MD simulations were used to calculate the necessary input parameters, namely, the diffusion constant of surfactant molecules in water, the adsorption and desorption constants of surfactants to the oil/water interface as well as the maximal adsorption amount of the surfactant at the interface.

As the two examples indicate, molecular simulations are most useful if the question addressed is as precise as possible. There is no need to reproduce reality in all its complexity within the simulation, because the answer, if obtainable, will be as complex as the reality. In the practical application it means that one should strive to construct a simplified model still containing all the necessary physics/chemistry. To do that, it is necessary to know the composition of the real system as precisely as possible. Additionally, there must be enough understanding of the underlying physicochemical problem to ask a specific question and to design an "in-silico experiment" to address that question. Moreover, as soon as it comes to performing the simulation, interaction potentials of some sort are necessary. Those could be atomistic force fields (Cornell et al., 1995; Kaminski et al., 2001; MacKerell et al., 1998; Oostenbrink et al., 2004), coarse-grained force fields (Marrink et al., 2007; Shinoda et al., 2008) or soft interaction potentials (Español and Warren, 1995; Groot and Warren, 1997; Hoogerbrugge and Koelman, 1992; Weiss et al., 2016). Finally, in an industrial environment, an automatized and industrially robust framework is necessary for building three-dimensional particle systems, submitting simulations to computing facilities, analyzing the simulation output and bookkeeping of all the simulations (In 't Veld and Rutledge, 2003; Weiss et al., 2016). The two mentioned examples show that the choice of a particular model depends on the question. While DPD was used in the tailored polymer design for an agricultural formulation, the MARTINI coarse-grained force field was used to calculate the input parameters for coalescence efficiency corrector. Not only the choice of the interaction model, but also the very shape of the simulation cell may depend on the question: from small elongated orthorhombic cells to study phenomena at a flat interface, to very large cubic cells to study general self-organization. Usually, sets of simulations are performed to study the influence of some parameter on the targeted observable. This is where automated workflows become invaluable.

Although molecular simulations are used since decades, there is constant room for improvement. The major bottlenecks are: the chemical scope and precision of the interaction model, statistical sampling of the ensemble through the simulation, and the sampling of the parameter space relevant for the scientific question. The last point gets already some attention by using appropriate design of experiments techniques (Reizman and Jensen, 2016). Also, the problem of sampling during the simulation is being addressed for decades already. Apart from various enhanced sampling techniques (Darve and Pohorille, 2001; Grubmuller, 1995; Laio and Parrinello, 2002; Marinari and Parisi, 1992; Miao and McCammon, 2017) more recent methods involving different flavors of machine learning (Noe et al., 2019; Wang et al., 2019; Zhu et al., 2002) promise an optimistic future. Machine learning can also contribute to the development of alternative and potentially more precise interaction potentials (Behler and Parrinello, 2007; Schutt et al., 2018; Smith et al., 2017). It remains to be seen how generalizable these models will be. These efforts complement the long ongoing quest for extended chemistry (Huang and Roux, 2013; Mayne et al., 2013; Vanommeslaeghe et al., 2010; Wang et al., 2004) and accuracy of force fields (Lamoureux and Roux, 2003; Wang et al., 2013; Wu et al., 2012) as well as soft interaction potentials (Kacar et al., 2013; Liyana-Arachchi et al., 2015). Finally, it should be noted that results obtained using different methods could vary substantially (Lindorff-Larsen et al., 2012; Martin, 2006). In a larger scope, it means that the full potential of particle dynamics models can only be achieved if we devise and then adhere to best practice protocols.

3.6 Statistics & machine learning

Machine learning models identify and quantify statistical relations between measured and/or simulated input and output data. More specifically, a predictive (supervised) machine learning model takes as input a specified set of input features (e.g., recipe components and/or process conditions) and uses a fitting algorithm in order to estimate a functional relation between inputs and target properties (e.g., chemical or mechanical performance metrics for the product under consideration):

$$f : \text{features} \rightarrow \text{target properties}$$

One key difference with respect to first-principles, or physical, modeling is that the functional form of *f* is not derived from physical laws but obtained by, e.g., fitting a specified class of functions to the data as well as possible. Thus, machine learning is no ab initio method but constrained by the quality and quantity of the data at hand.

The functional relation *f* obtained this way can subsequently be used to predict target properties for feature combinations that are not present in the historical data. This allows for in silico screening or exploration of new hypothetical feature combinations via interpolation and partially also extrapolation. Moreover, a parameterized machine learning model can be used as input for a numerical optimization scheme which aims at identifying optimal feature combinations in order to achieve certain desirable target properties.

Due to these abilities, machine learning models are essential tools providing the link between intermediate computed physicochemical properties and application/performance properties of interest (see Fig. 2) (Chen et al., 2020). In many cases, a rigorous physical model is not available in order to quantify this relation.

In addition, most machine learning modeling approaches also come with built-in feature importance metrics. These are ranking schemes that list the significance of each feature w.r.t predicting the target property from a statistical point of view. In this way, some insight into the "black box" machine learned function can be obtained.

There is no commonly accepted definition regarding the distinction between "classical" statistics (such as analysis of variance and linear regressions) and machine learning techniques (such as neural networks or other nonlinear modeling techniques). Typically, elements from both disciplines are required in the framework outlined in this chapter.

The remainder of this section will have a focus on relevant machine learning concepts and workflows related to materials science research and product development. However, it should be noted that many of these concepts also apply to predictive maintenance for production plants for instance as discussed in Section 2. In this context, historical data are used to derive functional relations between plant operating modes (encoded in production sensor time series for instance) and product quality measures (Traini et al., 2019). Ideally, such models are able to forecast problematic operating states or off-spec product quality. The results are often made available to end users via easy-to-interpret control charts (Oakland, 2007).

It is important to stress that machine learning relies heavily on the data used to train the model. In general, the input data for a machine learning model, based on physicochemical properties will come in the following format:

- It is a data table of input X where each row X_i denotes a data point, i.e., an individual recipe or experiment. The columns, indexed with j, denote recorded features of this particular experiment.
- Often, additional features are derived from raw input data using physical models. Using such derived inputs leads to higher accuracy and a larger applicability domain of the machine learning model to be developed (Chen et al., 2020).
- The relevant application properties or targets of this experiment are recorded in a separate table Y.

The accuracy and applicability domain of a machine learning model depend crucially on each of the following criteria that the data should meet:

- *Completeness:* It is paramount that all parameters potentially affecting performance are identified and recorded in X. These contain recipe parameters, process conditions as well as derived physicochemical properties. For instance, if a target property of interest is known to depend on a certain process condition during synthesis (pressure, temperature etc.) then these conditions should be recorded in the input data set X. Ideally, the input data contain all information such that the recorded experiments are reproducible in the sense that repeating experiment i with all features as recorded in X will produce the application properties as listed in Y plus some minor statistical noise term which is beyond control.
- *Data-generating mechanism:* In many real-world scenarios, the input data for a machine learning model is historical data. By this we mean that they record the inputs and outputs of past lab experiments. Variations of individual input parameters were often done following expert knowledge and a trial-and-error scheme. However, it is advisable and important that the data for a machine learning model are generated according to the *statistical design of experiment (DoE)*. This summarizes a variety of algorithms proposing recipe configurations that are reasonably spread in the allowed input space and allow for an optimal determination of the effects that the identified input features have w.r.t performance variation. The techniques range from classical design schemes (Fisher, 1935) to more iterative techniques based on Bayesian statistics (Chaloner and Verdinelli, 1995). Ideally, such a DoE is not done on raw recipe data but in the intermediate space of computed physicochemical descriptors.

It is important to stress that to work most effectively, a machine learning model needs data that covers the input domain of interest as well as possible (ideally DoE generated). Frequently, the question arises of "how many data points are needed?" for a ML model to work. The answer to this question depends crucially on the mechanism under which data was generated. 100 carefully chosen recipes with a design of experiment can have more explanatory power that 1000 historical experiments clustered along a very narrow region in the relevant chemical space.

In order to come from raw data to a machine learning model that can make predictions for new hypothetical feature combinations, the following general workflow applies:

- *Data exploration and feature engineering:* First, the data should be explored using standard plotting techniques and elementary statistics. In this way, obvious outliers or otherwise problematic values or patterns in the data can be identified and discussed. Also, the data can be enriched in this step by computing new features from existing ones.
- *Unsupervised feature preprocessing:* (Hastie et al., 2009) Typically, the input data originating from chemical recipes and derived physicochemical descriptors show a rich internal structure. By this we mean that input features are statistically correlated (due to chemical mixing rules, for instance) and high-dimensional (many features are present initially). Such correlations can be identified using pairwise correlation plots and dimensionality reduction techniques such as Principal Component Analysis (PCA). Redundant input features can be collapsed.
- *Model building:* The actual model building phase quantifies the statistical relation between input and outputs: An initial class of functions, f_γ, depending on a set of free parameters γ, is chosen, and the parameters are specified by fitting the function to the data, e.g., by solving

$$\min \sum_i \left\| f_\gamma(x_i) - y_i \right\|^2$$

w.r.t γ. f_γ is chosen such that an optimal balance between complexity, performance (measured on hold-out test data) and accuracy is achieved. When dealing with physicochemical descriptors, there are often many input features compared to training data ("large p-small n problem"). This calls for techniques such as regularized linear regression or multivariate regression splines (Hastie et al., 2009). In most cases, there is not enough data for deep learning neural network techniques (danger of overfitting).

- *Iteration cycles:* Model results are discussed with process experts and fine-tuned. Also, an applicability domain on which the model predictions can be trusted is defined from a statistical point of view. Ideally, predictions made by a machine learning model for new hypothetical recipes are evaluated by real-world experiments. In this way the model quality can be assessed, and the model can be refined.

This workflow is applied to a range of materials science research problems that fall under the name "descriptor modeling" (as described above) (Wu et al., 2013). In most real-world applications in chemical research, there are several bottlenecks when it comes to using the full power of ML models.

As already emphasized, the potential of machine learning approaches is limited by quality and quantity of the input data. Often, in chemical research, data is not recorded automatically, leading to errors when transferring it. Also, not all relevant input data w.r.t reproducibility of the recorded experiments is always quantified and listed. Furthermore, statistical design of experiments for generating good input data is not always systematically applied.

Moreover, ML models could unfold more potential by viewing them as modules of an automated learning loop which describes an autonomous research machine, or AI system (Fig. 10).

Such a research machine is a physical system that perceives its environment, comprehends it, and based on these observations takes a rational (i.e., based on mathematical optimization) decision. After this initial loop, the AI system re-assesses the situation, and reiterates, if necessary, the process. Machine learning models are essential building blocks of the "comprehend" and "decide" modules in which new information in form of data is processed via a statistical model. In such a way, autonomous iterations could be set up, that tune given input parameters w.r.t specified goals without requiring heavy human interaction.

Based on this discussion, two aspects can be identified that would leverage the potential of machine learning models in the context of process research engineering:

On one hand, appropriate data-generating mechanisms (such as statistical DoE) have to be used more consistently and automated data workflows need to be established that would allow for a smoother integration of machine learning models into physical environments. On the other hand, we need modeling architectures providing a better interplay between

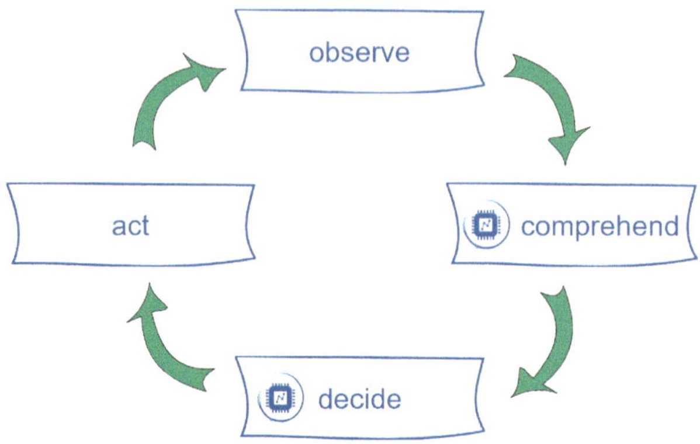

Fig. 10 Role of machine learning within an autonomous research machine, or AI system.

scientific modeling techniques and statistical black-box modeling, such that these two disciplines balance each other's strengths and weaknesses in an optimal way.

4. Outlook

So far, we have seen how advances in physical models and data science have been shaping our polymer reaction engineering knowledge, resulting in countless improved products and processes. It turned out that the synergic combination of the different techniques is central to this progress.

Yet the questions that guided our discussion seem to remain elusive. How can we get closer to "self-driving" plants and to a better material design, fulfilling future needs and requirements?

When analyzing the steps forward that should be taken to foster the advances within each approach, a commonality emerges: shared, organized, information.

Since the first academic journals in the second half of the 17th century, and the first encyclopedia, about a century later, the scientific community has come a long way in gathering of knowledge and results.

Thousands of scientific journals exist, and many online platforms push the sharing of experimental data and code.

Notable examples from the Computer Science community are the open-source GitHub repository for sharing code,[h] and Kaggle, an online community of Data Scientists to share, among other things, data sets.[i]

[h] https://github.com/.

[i] https://www.kaggle.com/.

Similar approaches exist in the chemistry realm (Warr, 2014), such as the Elsevier platform Reaxys, where chemistry information and data from published literature can be retrieved.[g] Another noteworthy example, is CASREACT, provided by the American Chemical Society, gathering single and multistep reactions, providing for instance information concerning reaction conditions, yields and catalysts.[j]

The PRE community has the chance, in the authors' opinion, to join these efforts, or initiate similar projects. Recently, the introduction of a representation system for macromolecules opened up the possibility to share and disseminate polymer-related research results in a systematic way (Lin et al., 2019). Building on this, the IUPAC has launched an initiative to construct a machine-accessible kinetic database for radical polymerizations.[k].

Taking inspiration from these initiatives (Lin et al., 2019),[k] the PRE community could pursue the development of a web-based, open-source, *polymerization toolbox*, gathering an experimental and a modeling-centered workflow. The former would include published experimental results as well as kinetic, physical and thermodynamic parameters. The latter a repository for existing models and algorithms relevant in the PRE field, ideally coded in a modular structure facilitating their interaction.

The benefits from the creation of such polymerization toolbox are at least twofold. On one hand, it fosters the interaction of the PRE community with scientists from very diverse fields (e.g., data science, statistics, chemometrics). On the other hand, centralizing knowledge and speeding up the retrieval of information, is a crucial step to keep the pace with the ever-evolving research questions, reflecting the new challenges our society faces.

The polymerization toolbox could increase the capabilities in performing:

- *data analytics and machine learning,* on the large amount of experimental results available;
- *model validation,* comparing against different experimental data sets produced by different research groups;
- *quantification and validation of reaction pathways,* through comprehensive parameter and experimental data comparison;
- *model deployment/refinement,* by sharing all the employed models and algorithms.

[j] https://www.cas.org/support/documentation/reactions.

[k] https://iupac.org/projects/project-details/?project_nr=2019-045-1-400.

Such a complex architecture could be developed in different stages, progressively involving different sub-communities, and accommodating models, parameters, and experimental data, in a modular way.

The PRE field has reached a certain degree of maturity, gathering about a century worth of research and knowledge. Sharing and centralizing our tools and results would certainly help to further push our understanding, and potentially translate in a new wave of technological progress.

5. Conclusions

In our rapidly growing, complex, global world, the polymer industry finds in digitalization a powerful lever, offering both technological advances and economic opportunities.

In our opinion, developing digital twins for polymerization processes is the way forward, expanding the spectrum of polymer reaction engineering, gaining deeper scientific knowledge, while creating innovative products.

Physical models and data science approaches are the building blocks of digital twins, which can bring us closer to solving two long-standing open questions in polymer reaction engineering. Linking reaction environments to final material properties and producing materials in an optimal way are ambitious goals, that seem finally closer to be achieved.

The different techniques that constitute physical and data science approaches, have been working in an increasingly synergic way. We have witnessed how the predictive ability of deterministic approaches, such as kinetic, thermodynamic and fluid mechanics models, is improved by parameters and mechanistic understanding provided by quantum chemistry and molecular dynamics. While constantly aiming for a deeper physicochemical understanding, statistics and machine learning approaches enable us to solve highly complex problems, thanks to their large flexibility.

So how do we get to a fully digitalized polymer industry? One where customers' demands for a new product are swiftly met by a rapid response of the research and development departments, that coordinate with marketing, and production (Fig. 11)?

At the heart of this still long path, there are our physical and data science models, that we need to solidify and further develop in the coming years.

Progress can only be achieved if we systematically centralize our knowledge, joining or creating initiatives that foster an open-source sharing of published experimental data, parameters, models, and algorithms.

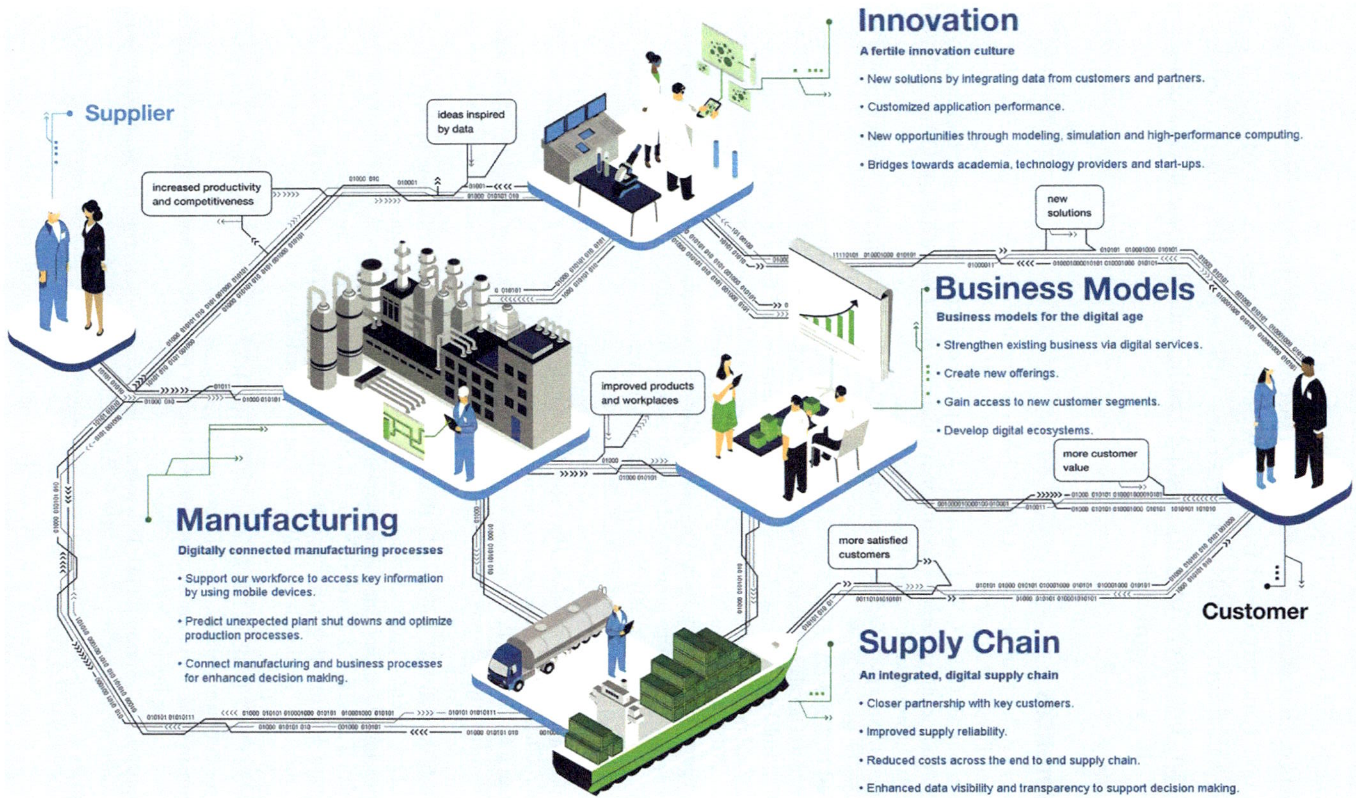

Fig. 11 An integrated, digitalized polymer industry.

Scientific and technological advances have always been achieved through collaboration. The realization of the first computer in 1941, for instance, leveraged a century of discoveries, sparked by the Analytic Engine, conceptualized in 1837. Whether the realization of digital twins and "self-driving" polymer processes will require less than a century is unclear, but surely only by sharing results and ideas we have a chance to fulfill this paradigm shift.

References

Abildskov J, Kontogeorgis GM: Chemical product design, *Chem Eng Res Des* 82(11):1505–1510, 2004.

Abolhasani M, Coley CW, Jensen KF: Multiphase oscillatory flow strategy for in situ measurement and screening of partition coefficients, *Anal Chem* 87(21):11130–11136, 2015.

Asua J: *Polymer reaction engineering*, 2007, Wiley.

Ballard N, Asua JM: Radical polymerization of acrylic monomers: an overview, *Prog Polym Sci* 79:40–60, 2018.

Ballard N, Veloso A, Asua JM: Mid-chain radical migration in the radical polymerization of n-butyl acrylate, *Polymers (Basel)* 10(7):765, 2018.

Barth J, Buback M, Hesse P, Sergeeva T: Termination and transfer kinetics of butyl acrylate radical polymerization studied via SP-PLP-EPR, *Macromolecules* 43(9):4023–4031, 2010.

BASF, 2019a Energy and climate protection. https://report.basf.com/2019/en/managements-report/responsibility-along-the-value-chain/environmental-protection-health-and-safety/energy-and-climate-protection.html (accessed 27.03.2020).

BASF, Strategy report, 2019b, http://report.basf.com/2019/en/managements-report/our-strategy.html (accessed 27.03.2020).

Behler J, Parrinello M: Generalized neural-network representation of high-dimensional potential-energy surfaces, *Phys Rev Lett* 98(14):146401, 2007.

Bishop CM: *Pattern recognition and machine learning*, 2006, Springer.

Bloching B, Leutiger P, Oltmanns T, et al.: *The digital transformation of industry*, 2015, Roland Berger. https://www.rolandberger.com/sv/Publications/The-digital-transformation-industry.html.

Buchelli A, Call ML, Brown AL, Bird A, Hearn S, Hannon J: Modeling fouling effects in LDPE tubular polymerization reactors. 3. computational fluid dynamics analysis of a reacting zone, *Ind Eng Chem Res* 44(5):1493–1501, 2005.

Casalini T, Rossi F, Lazzari S, Perale G, Masi M: Mathematical modeling of PLGA microparticles: from polymer degradation to drug release, *Mol Pharm* 11(11):4036–4048, 2014.

Chaloner K, Verdinelli I: Bayesian experimental design: a review, *Stat Sci* 10(3): 273–304, 1995.

Chapman WG, Gubbins KE, Jackson G, Radosz M: SAFT: equation-of-state solution model for associating fluids, *Fluid Phase Equilib* 52:31–38, 1989.

Chapman WG, Gubbins KE, Jackson G, Radosz M: New reference equation of state for associating liquids, *Ind Eng Chem Res* 29(8):1709–1721, 1990.

Chappelear DC, Simon RHM: Polymerization reaction engineering, In *Addition and condensation polymerization processes*, 1969, American Chemical Society, pp 1–24.

Chen C-C, Mathias PM: Applied thermodynamics for process modeling, *AIChE J* 48(2):194–200, 2002.

Chen C, Guan X, Ren Y, et al.: Mesoscale modeling of emulsification in rotor-stator devices, *Chem Eng Sci* 193:171–183, 2019a.

Chen C, Guan X, Ren Y, et al.: Mesoscale modeling of emulsification in rotor-stator devices, *Chem Eng Sci* 193:156–170, 2019b.

Chen G, Shen Z, Iyer A, et al.: Machine-learning-assisted de novo design of organic molecules and polymers: opportunities and challenges, *Polymers (Basel)* 12(1):163, 2020.

Chilukuri S, Fleming E, Westra A. 2017, Digital in R&D: the $100 billion opportunity. McKinsey & Company. https://www.mckinsey.com/industries/pharmaceuticals-and-medical-products/our-insights/digital-in-r-and-d-the-100-billion-opportunity.

Coley CW, Barzilay R, Jaakkola TS, Green WH, Jensen KF: Prediction of organic reaction outcomes using machine learning, *ACS Cent Sci* 3(5):434–443, 2017.

Coley CW, Green WH, Jensen KF: Machine learning in computer-aided synthesis planning, *Acc Chem Res* 51(5):1281–1289, 2018.

Coote ML: Quantum-chemical modeling of free-radical polymerization, *Macromol Theory Simul* 18(7–8):388–400, 2009.

Cornell WD, Cieplak P, Bayly CI, et al.: A second generation force field for the simulation of proteins, nucleic acids, and organic molecules, *J Am Chem Soc* 117(19):5179–5197, 1995.

Countryman T, Quilligan A, Narsalay R, Sen A: Transforming growth with digital innovation. 2019, Accenture. https://www.accenture.com/za-en/insights/industry-x-0/chemicals.

Darve E, Pohorille A: Calculating free energies using average force, *J Chem Phys* 115 (20):9169–9183, 2001.

Deglmann P, Müller I, Becker F, Schäfer A, Hungenberg K-D, Weiss H: Prediction of propagation rate coefficients in free radical solution polymerization based on accurate quantum chemical methods: vinylic and related monomers, including acrylates and acrylic acid, *Macromol React Eng* 3(9):496–515, 2009.

Deglmann P, Hungenberg K-D, Vale HM: Dependence of propagation rate coefficients in radical polymerization on solution properties, *Macromol React Eng* 11(1), 2017, 1600037.

Deglmann P, Hungenberg K-D, Vale HM: Dependence of propagation rate coefficients in radical polymerization on solution properties: a quantitative thermodynamic interpretation, *Macromol React Eng* 12(4), 2018, 1800010.

Deglmann P, Hellmund M, Hungenberg KD, Nieken U, Schwede C, Zander C: Side reactions in aqueous phase polymerization of N-Vinyl-pyrrolidone as possible source for fouling, *Macromol React Eng* 13(5), 2019, 1900021.

Diederich VEG, Villiger T, Storti G, Lattuada M: Modeling of the degradation of poly(ethylene glycol)-co-(lactic acid)-dimethacrylate hydrogels, *Macromolecules* 50(14): 5527–5538, 2017.

Dohrn R, Pfohl O: Thermophysical properties—industrial directions, *Fluid Phase Equilib* 194-197:15–29, 2002.

Dossi M, Storti G, Moscatelli D: Quantum chemistry: a powerful tool in polymer reaction engineering, *Macromol Symp* 302(1):16–25, 2011.

Dotson NA, Galvan R, Laurence RL, Tirrell M: *Polymerization process modeling*, 1995, Wiley.

Drawe P, Buback M: The PLP-SEC method: perspectives and limitations, *Macromol Theory Simul* 25(1):74–84, 2016.

Drawe P, Buback M, Lacík I: Radical polymerization of alkali acrylates in aqueous solution, *Macromol Chem Phys* 216(12):1333–1340, 2015.

Elgebrandt RC, Fletcher DF, Gomes VG, Romagnoli JA, Marquardt W, Pantelides C, editors: A framework for modeling particle size effects in emulsion polymerization systems using computational fluid dynamics linked to a detailed population balance model, In *Computer aided chemical engineering*, vol. 21, 2006, Elsevier, pp 551–556.

Enders S, Kahl H, Winkelmann J: Interfacial properties of polystyrene in contact with carbon dioxide, *Fluid Phase Equilib* 228-229:511–522, 2005.

Español P, Warren P: Statistical mechanics of dissipative particle dynamics, *EPL* 30(4): 191–196, 1995.

Fischer EJ, Storti G, Cuccato D: Aqueous free-radical polymerization of non-ionized and fully ionized methacrylic acid, *Processes* 5(4):23, 2017.

Fischer EJ, Cuccato D, Storti G, Morbidelli M: Effect of the charge interactions on the composition behavior of acrylamide/acrylic acid copolymerization in aqueous medium, *Eur Polym J* 98:302–312, 2018.

Fisher RA: *The design of experiments*, Oxford, England, 1935, Oliver & Boyd. p xi, 251.

Flory PJ: *Principles of polymer chemistry*, 1953, Cornell University Press.

Gerlinger W, Asua JM, Chaloupka T, et al.: Dynamic optimization and non-linear model predictive control to achieve targeted particle morphologies, *Chem Ing Tech* 91(3): 323–335, 2019.

Gmehling J, Kleiber M, Kolbe B, Rarey J: *Chemical thermodynamics for process simulation*, 2019, Wiley-VCH.

Golightly R: *Predictive maintenance: when does old become new again?*, 2019, AspenTech. https://www.aspentech.com/en/blog/blog/Predictive_Maintenance_When_Does_Old_Become_New_Again.

Groot RD, Warren PB: Dissipative particle dynamics: bridging the gap between atomistic and mesoscopic simulation, *J Chem Phys* 107(11):4423–4435, 1997.

Gross J, Sadowski G: Perturbed-chain SAFT: an equation of state based on a perturbation theory for chain molecules, *Ind Eng Chem Res* 40(4):1244–1260, 2001.

Grubmuller H: Predicting slow structural transitions in macromolecular systems: conformational flooding, *Phys Rev E Stat Phys Plasmas Fluids Relat Interdiscip Topics* 52(3): 2893–2906, 1995.

Hastie T, Tibshirani R, Friedman J: The elements of statistical learning: data mining, inference, and prediction, In *Springer series in statistics*, ed 2., New York, 2009, Springer.

Hendriks E, Kontogeorgis GM, Dohrn R, et al.: Industrial requirements for thermodynamics and transport properties, *Ind Eng Chem Res* 49(22):11131–11141, 2010.

Hoogerbrugge PJ, Koelman JMVA: Simulating microscopic hydrodynamic phenomena with dissipative particle dynamics, *EPL* 19(3):155–160, 1992.

Huang L, Roux B: Automated force field parameterization for non-polarizable and polarizable atomic models based on Ab initio target data, *J Chem Theory Comput* 9(8): 3543–3556, 2013.

In 't Veld PJ, Rutledge GC: Temperature-dependent elasticity of a semicrystalline interphase composed of freely rotating chains, *Macromolecules* 36(19):7358–7365, 2003.

IPCC: Summary for policymakers, In Masson-Delmotte V, Zhai P, Pörtner H-O, Roberts D, Skea J, Shukla PR, Pirani A, Moufouma-Okia W, Péan C, Pidcock R, Connors S, Matthews JBR, Chen Y, Zhou X, Gomis MI, Lonnoy E, Maycock T, Tignor M, Waterfield T, editors: *Global warming of 1.5°C. An IPCC special report on the impacts of global warming of 1.5°C above pre-industrial levels and related global greenhouse gas emission pathways, in the context of strengthening the global response to the threat of climate change, sustainable development, and efforts to eradicate poverty*, Geneva, Switzerland, 2018, World Meteorological Organization. 32 pp.

IPCC: Summary for policymakers. In Shukla PR, Skea J, Calvo Buendia E, Masson-Delmotte V, Pörtner H-O, Roberts DC, Zhai P, Slade R, Connors S, van Diemen R, Ferrat M, Haughey E, Luz S, Neogi S, Pathak M, Petzold J, Portugal Pereira J, Vyas P, Huntley E, Kissick K, Belkacemi M, Malley J, editors: *Climate change and land: an IPCC special report on climate change, desertification, land degradation, sustainable land management, food security, and greenhouse gas fluxes in terrestrial ecosystems*. 2019, In press.

Islam K: A tipping point for the chemicals industry. PWC. 2018. http://usblogs.pwc.com/industrialinsights/2018/09/25/a-tipping-point-for-the-chemicals-industry/.

Kacar G, Peters EAJF, de With G: A generalized method for parameterization of dissipative particle dynamics for variable bead volumes, *EPL* 102(4), 2013, 40009.

Kaminski GA, Friesner RA, Tirado-Rives J, Jorgensen WL: Evaluation and reparametrization of the OPLS-AA force field for proteins via comparison with accurate quantum chemical calculations on peptides, *J Phys Chem B* 105(28):6474–6487, 2001.

Kattner H, Drawe P, Buback M: Chain-length-dependent termination of sodium methacrylate polymerization in aqueous solution studied by SP-PLP-EPR, *Macromolecules* 50(4):1386–1393, 2017.

Klei A, Moder M, Stockdale O, Weihe U, Winkler G: *Digital in chemicals: from technology to impact*, McKinsey & Company. 2017. https://www.mckinsey.com/industries/chemicals/our-insights/digital-in-chemicals-from-technology-to-impact.

Kolhapure NH, Fox RO: CFD analysis of micromixing effects on polymerization in tubular low-density polyethylene reactors, *Chem Eng Sci* 54(15–16):3233–3242, 1999.

Kolhapure NH, Fox RO, Daiβ A, Mähling F-O: PDF simulations of ethylene decomposition in tubular LDPE reactors, *AIChE J* 51(2):585–606, 2005.

Kontogeorgis GM, Folas GK: *Thermodynamic models for industrial applications*, 2010, Wiley.

Kontogeorgis GM, Liang X, Arya A, Tsivintzelis I: Equations of state in three centuries. Are we closer to arriving to a single model for all applications? *Chem Eng Sci X* 7, 2020, 100060.

Lacík I, Beuermann S, Buback M: PLP – SEC study into free-radical propagation rate of nonionized acrylic acid in aqueous solution, *Macromolecules* 36(25):9355–9363, 2003.

Lacík I, Beuermann S, Buback M: PLP-SEC study into the free-radical propagation rate coefficients of partially and fully ionized acrylic acid in aqueous solution, *Macromol Chem Phys* 205(8):1080–1087, 2004.

Laio A, Parrinello M: Escaping free-energy minima, *Proc Natl Acad Sci USA* 99(20): 12562–12566, 2002.

Lamoureux G, Roux BT: Modeling induced polarization with classical Drude oscillators: theory and molecular dynamics simulation algorithm, *J Chem Phys* 119(6): 3025–3039, 2003.

Laycock B, Nikolić M, Colwell JM, et al.: Lifetime prediction of biodegradable polymers, *Prog Polym Sci* 71:144–189, 2017.

Lazzari S, Codari F, Storti G, Morbidelli M, Moscatelli D: Modeling the pH-dependent PLA oligomer degradation kinetics, *Polym Degrad Stab* 110:80–90, 2014.

Li W, Pham HH, Nie Z, MacDonald B, Guenther A, Kumacheva E: Multi-step microfluidic polymerization reactions conducted in droplets: the internal trigger approach, *J Am Chem Soc* 130(30):9935–9941, 2008.

Lin TS, Coley CW, Mochigase H, et al.: BigSMILES: a structurally-based line notation for describing macromolecules, *ACS Cent Sci* 5(9):1523–1531, 2019.

Lindorff-Larsen K, Maragakis P, Piana S, Eastwood MP, Dror RO, Shaw DE: Systematic validation of protein force fields against experimental data, *PLoS One* 7(2), 2012, e32131.

Liyana-Arachchi TP, Jamadagni SN, Eike D, Koenig PH, Siepmann JI: Liquid-liquid equilibria for soft-repulsive particles: improved equation of state and methodology for representing molecules of different sizes and chemistry in dissipative particle dynamics, *J Chem Phys* 142(4), 2015, 044902.

MacKerell AD, Bashford D, Bellott M, et al.: All-atom empirical potential for molecular modeling and dynamics studies of proteins, *J Phys Chem B* 102(18):3586–3616, 1998.

Marinari E, Parisi G: Simulated tempering: a new Monte Carlo scheme, *EPL* 19(6):451–458, 1992.

Marrink SJ, Risselada HJ, Yefimov S, Tieleman DP, de Vries AH: The MARTINI force field: coarse grained model for biomolecular simulations, *J Phys Chem B* 111(27): 7812–7824, 2007.

Martin MG: Comparison of the AMBER, CHARMM, COMPASS, GROMOS, OPLS, TraPPE and UFF force fields for prediction of vapor–liquid coexistence curves and liquid densities, *Fluid Phase Equilib* 248(1):50–55, 2006.

Mavroudakis E, Cuccato D, Moscatelli D: On the use of quantum chemistry for the determination of propagation, copolymerization, and secondary reaction kinetics in free radical polymerization, *Polymers* 7(9):1789–1819, 2015.

Mayne CG, Saam J, Schulten K, Tajkhorshid E, Gumbart JC: Rapid parameterization of small molecules using the Force Field Toolkit, *J Comput Chem* 34(32):2757–2770, 2013.

Miao Y, McCammon JA: Gaussian accelerated molecular dynamics: theory, implementation, and applications, *Annu Rep Comput Chem* 13:231–278, 2017.

Mosquet X, Arora A, Xie A, Renner M: *Who will drive electric cars to the tipping point? BCG*, 2020. https://www.bcg.com/publications/2020/drive-electric-cars-to-the-tipping-point.

Muuronen M, Deglmann P, Tomovic Z: Design principles for rational polyurethane catalyst development, *J Org Chem* 84(12):8202–8209, 2019.

Noe F, Olsson S, Kohler J, Wu H: Boltzmann generators: sampling equilibrium states of many-body systems with deep learning, *Science* 365(6457), 2019, eaaw1147.

Oakland JS: *Statistical process control*, 2007, Butterworth-Heinemann.

Oostenbrink C, Villa A, Mark AE, van Gunsteren WF: A biomolecular force field based on the free enthalpy of hydration and solvation: the GROMOS force-field parameter sets 53A5 and 53A6, *J Comput Chem* 25(13):1656–1676, 2004.

Peters FT, Herhut M, Sadowski G: Extension of the PC-SAFT based group contribution method for polymers to aromatic, oxygen- and silicon-based polymers, *Fluid Phase Equilib* 339:89–104, 2013.

Pohn J, Cunningham M, McKenna TFL: Scale-up of emulsion polymerization reactors part I—development of a model framework, *Macromol React Eng* 7(8):380–392, 2013.

Prausnitz JM: Thermodynamics and the other chemical engineering sciences: old models for new chemical products and processes, *Fluid Phase Equilib* 158-160:95–111, 1999.

Prausnitz JM: Chemical engineering and the postmodern world, *Chem Eng Res Des* 79(7):777–788, 2001.

Ray WH: On the mathematical modeling of polymerization reactors, *J Macromol Sci C Polym Rev J* 8(1):1–56, 1972.

Reizman BJ, Jensen KF: Feedback in flow for accelerated reaction development, *Acc Chem Res* 49(9):1786–1796, 2016.

Roser M, Ritchie H, Ortiz-Ospina E: World population growth. 2013, Published online at OurWorldInData.org. Retrieved from: https://ourworldindata.org/world-population-growth [online resource].

Schaffer MA, McAuley KB, Cunningham MF, Marchildon EK: Prediction of water solubility in nylon melts based on flory-huggins theory, *Polym Eng Sci* 43(3):639–646, 2003.

Schutt KT, Sauceda HE, Kindermans PJ, Tkatchenko A, Muller KR: SchNet—a deep learning architecture for molecules and materials, *J Chem Phys* 148(24):241722, 2018.

Shinoda W, DeVane R, Klein ML: Coarse-grained molecular modeling of non-ionic surfactant self-assembly, *Soft Matter* 4(12):2454–2462, 2008.

Skeist I: Copolymerization: the composition distribution curve, *J Am Chem Soc* 68(9): 1781–1784, 1946.

Smith JS, Isayev O, Roitberg AE: ANI-1: an extensible neural network potential with DFT accuracy at force field computational cost, *Chem Sci* 8(4):3192–3203, 2017.

Soroush M, editor: *Computational quantum chemistry*, 2018, Elsevier.

Tobita H, Hamielec AE: Crosslinking kinetics in polyacrylamide networks, *Polymer* 31(8): 1546–1552, 1990.

Traini E, Bruno G, D'Antonio G, Lombardi F: Machine learning framework for predictive maintenance in milling, *IFAC-PapersOnLine* 52(13):177–182, 2019.

Ulbrich M, Eschenröder E: Digital disruption in the lab: the case for R&D digitalization in chemicals. 2018, Accenture. https://www.accenture.com/us-en/blogs/chemicals-and-natural-resources-blog/digital-disruption-in-the-lab-the-case-for-rd-digitalization-in-chemicals.

United Nations, 2015. Transforming our world: the 2030 agenda for sustainable development. https://sustainabledevelopment.un.org/post2015/transformingourworld/publication (accessed 27.03.2020).

United Nations, Department of Economic and Social Affairs, Population Division (2019). World population prospects 2019: highlights. ST/ESA/SER.A/423.

Vale HM, McKenna TF: Modeling particle size distribution in emulsion polymerization reactors, *Prog Polym Sci* 30(10):1019–1048, 2005.

Vanommeslaeghe K, Hatcher E, Acharya C, et al.: CHARMM general force field: a force field for drug-like molecules compatible with the CHARMM all-atom additive biological force fields, *J Comput Chem* 31(4):671–690, 2010.

Wang W, Hutchinson RA: A comprehensive kinetic model for high-temperature free radical production of styrene/methacrylate/acrylate resins, *AIChE J* 57(1):227–238, 2011.

Wang J, Wolf RM, Caldwell JW, Kollman PA, Case DA: Development and testing of a general amber force field, *J Comput Chem* 25(9):1157–1174, 2004.

Wang LP, Chen J, Van Voorhis T: Systematic parametrization of polarizable force fields from quantum chemistry data, *J Chem Theory Comput* 9(1):452–460, 2013.

Wang J, Olsson S, Wehmeyer C, et al.: Machine learning of coarse-grained molecular dynamics force fields, *ACS Cent Sci* 5(5):755–767, 2019.

Warr WA: A short review of chemical reaction database systems, computer-aided synthesis design, reaction prediction and synthetic feasibility, *Mol Inform* 33(6–7):469–476, 2014.

Weiss H, Deglmann P, In 't Veld PJ, Cetinkaya M, Schreiner E: Multiscale materials modeling in an industrial environment, *Annu Rev Chem Biomol Eng* 7:65–86, 2016.

Wells GJ, Ray WH: Methodology for modeling detailed imperfect mixing effects in complex reactors, *AIChE J* 51(5):1508–1520, 2005.

Wittenberg NFG, Buback M, Hutchinson RA: Kinetics and modeling of methacrylic acid radical polymerization in aqueous solution, *Macromol React Eng* 7(6):267–276, 2013.

Wu JC, Chattree G, Ren P: Automation of AMOEBA polarizable force field parameterization for small molecules, *Theor Chem Acc* 131(3):1138, 2012.

Wu K, Natarajan B, Morkowchuk L, Krein M, Breneman CM: From drug discovery QSAR to predictive materials QSPR, In *Informatics for materials science and engineering*, 2013, Butterworth-Heinemann, pp 385–422.

Yang K, Swanson K, Jin W, et al.: Analyzing learned molecular representations for property prediction, *J Chem Inf Model* 59(8):3370–3388, 2019.

Yu Y, Storti G, Morbidelli M: Kinetics of ring-opening polymerization ofl,l-Lactide, *Ind Eng Chem Res* 50(13):7927–7940, 2011.

Zeck S: Thermodynamics in process development in the chemical industry—importance, benefits, current state and future development, *Fluid Phase Equilib* 70(2–3):125–140, 1991.

Zhu Z, Tuckerman ME, Samuelson SO, Martyna GJ: Using novel variable transformations to enhance conformational sampling in molecular dynamics, *Phys Rev Lett* 88(10): 100201, 2002.

CHAPTER SEVEN

From batch to continuous free-radical polymerization: Recent advances and hurdles along the industrial transfer

Juri Ilare and Mattia Sponchioni*
Department of Chemistry, Materials and Chemical Engineering, Politecnico di Milano, Milano, Italy
*Corresponding author: e-mail address: mattia.sponchioni@polimi.it

Contents

Abstract

The care of safety and environment is nowadays an essential aspect in the conduction of every kind of factory. A decisive contribution to these aspects is given by the conversion from batch to continuous processes. In polymer manufacturing, this transition allows to better manage important safety and pollution concerns, such as energy transport phenomena and process emissions. This, in turn, facilitates the fulfilling of the more and more stringent requirements imposed by local and international regulatory agencies. In addition, through this process intensification, it is also possible to optimize both investment and operative costs, thus overcoming the current profit margin reduction in the field of polymerizations. The aim of this chapter is to provide an insight into the technical and process difficulties that are currently hampering the transition from batch to continuous free-radical polymerization (FRP). The discussion is driven by considering key points in polymer manufacturing, including safety, environmental concerns, product quality and cost management. For a seek of clarity, we drove the discussion by considering the two main modalities in which the FRP can be conducted, namely homogenous and heterogeneous reactions. Extensive reference to literature examples is made to highlight the state of art in the field and the recent innovations toward

Advances in Chemical Engineering, Volume 56
ISSN 0065-2377
https://doi.org/10.1016/bs.ache.2020.07.005

continuous processes. Finally, successful examples of such auspicable transition achieved on an industrial scale are presented, trying to highlight the features that can be generalized to reach the goal.

1. General features of continuous polymerization processes

Polymers are materials of fundamental impact in the market since the second half of the 20th century. Indeed, polymers are the main components in a wide variety of products, such as packaging materials (Al-Tayyar et al., 2020), electronics (Weng et al., 2020), paintings (Liu et al., 2013a), surface coatings (Manfredini et al., 2019; Wang et al., 2000), textiles, adhesives (Shen et al., 2014), wastewater treatment chemicals (Diani and Gall, 2006), health and home care products (Capasso et al., 2020; Kaur et al., 2018). More recently, they are also attracting significant attention in the biomedical field (Albuquerque and Malafaia, 2018; Ferrari et al., 2018; Kaur et al., 2018; Mekonnen et al., 2013; Perale and Hilborn, 2017; Yong et al., 2018). Their low cost and great versatility in terms of chemical and physical properties (e.g., composition, structure, molecular weight, particle size distribution and mechanical behavior (Albuquerque and Malafaia, 2018; Jiang et al., 2018; Palmiero et al., 2016; Plessis et al., 2001a; Wang et al., 2017a)) enabled their use in replacement of many inorganic and organic materials. The widespread adoption of polymer products in the market is corroborated by the data collected by PlasticsEurope, leader pan-European association of plastic manufacturers. According to these data, the worldwide plastic production has reached 360 million tons in 2018 (PlasticsEurope, 2020).

With particular reference to polymers produced through free-radical polymerization (FRP), most of the current production is realized in discontinuous or semi-continuous processes (Ghadipasha et al., 2016). In general, discontinuous or batch reactors are closed systems where the reactants are initially charged, left to react and then the desired product is recovered after a specific time, mainly determined by the reaction kinetics. In this configuration, the mixture is usually stirred by mechanical agitation in order to ensure homogeneous conditions inside the reaction volume. In most cases, the heat exchange system for the temperature control in the reactor is external (e.g., based on heating/cooling jackets) to prevent any contact with possibly corrosive and hazardous liquids as well as any impact on the process behavior. However, other approaches to heat removal are also applied. It is common, for example, to operate with a solvent at its boiling temperature, so that the temperature is constant and its vapor can be continuously condensed in an external unit and recycled back to the vessel after the removal of the latent heat of evaporation (Tobita and Hamielec, 2015; Wessels, 1977). The key feature of a discontinuous process is its flexibility. This justifies the adoption of batch reactors in industrial campaigns for the production of limited amounts of different products.

In semi-batch reactors, a stream of reactants or of reaction mixture is continuously fed to or removed from the vessel, according to the specific needs of the production process. Also in this case, mechanical stirring is usually applied along with an external heat exchange system. Most polymerization processes nowadays employ this strategy for safety reasons, mainly connected to the limited reactant holdup inside the reactor, and to better tailor the product properties (Plessis et al., 2000, 2001a,b).

A common limitation to batch and semi-batch configurations is their intrinsic poor productivity, due to the several dead times that affect the process at many levels. The main time losses occurring along the overall discontinuous process period can be subdivided in four main steps: (i) charging or feeding, (ii) reaction, (iii) discharging, and sometimes (iv) cleaning (Asua, 2016; Durand and Engell, 2016). These unavoidable stages represent interruption of the process continuity and, therefore, additional slack times, resulting in a general loss of productivity. This in turn determines high operating costs. In addition, since higher reactor volumes or the simultaneous running of more than one reactor are required to satisfy the raising market demand, discontinuous operations are also affected by high capital expenditures (Asua, 2016; Buchholz, 2010; Jas and Kirschning, 2003).

To face the increasing market demand of polymers as well as this high costs associated to discontinuous or semi-continuous reactors, we are currently witnessing the seek of a transition to continuous processes among many polymer manufacturers.

In continuous reactors, streams of reactants and products are continuously fed and withdrawn, respectively. The practical implementations of continuous reactors for FRP are based on stirred tank reactors (STRs) or tubular reactors (TRs). The main feature of a STR is the uniform composition and temperature in the reaction volume, both granted by intense mechanical stirring. On the other hand, in TRs mechanical agitation is quite uncommon even if active (mechanical stirring) and static mixing (e.g., static mixers) are frequently used in the case of polymer systems. This is mainly due to the high viscosity or multiphase nature of the reacting mixture, which requires agitation to ensure homogeneous conditions and avoid phase separation in the reactor (Carvalho et al., 2010; Chevrel et al., 2016a). A continuous conduction is typically favored when a high productivity of a specific polymer grade is required (Asua, 2016; Kohlmann et al., 2016) and this is one of the essential motivations explaining why this operating mode is becoming more and more popular in industry. The reduction of many dead times characterizing the discontinuous polymer production enables the cut of the operating costs of a continuous process. In addition, since the same productivity as that of a discontinuous process can be achieved with a smaller continuous reactor, also the capital expenditures can be decreased. Following these considerations, since many polymers exhibit low to medium value, continuous production processes are helpful in contrasting the present profit margin reduction, which is indeed a major issue in many polymer manufacturing companies.

In addition to the said process intensification, continuous processes alleviate important safety and environmental concerns, thus enabling to face the more and more stringent regulations from local and international organizations that are expected in the polymer industry in the near future (Rossetti and Compagnoni, 2016). In this sense, continuous processes offer many advantages compared to their discontinuous counterparts, mostly because of the typically lower reaction volumes treated. For example, many monomers commonly used to synthesize everyday polymeric products are flammable and/or harmful for human health. The use of these very dangerous compounds in chemical industries is currently leading to an escalation of the prevention practices for their transport, storage, use, and disposal. For instance, aiming at understanding better the potential disastrous consequences, Liu et al. (2015) designed thermal hazard analyses to evaluate the safety issues related to the use of organic peroxides and azo compounds, which are commonly adopted as thermal initiators in radical polymerizations. Their potential accumulation within batch reactors, where larger amounts of residual monomers are stored, result in the generation of higher explosion and emission risks that may involve surrounding areas (Rossetti and Compagnoni, 2016). Thus, the authors studied in detail the concept of self-accelerating decomposition temperature (SADT), which nowadays is a parameter regulated by United Nations (UN) to make the chemical transport and storage steps safer. The results confirmed that the proposed methodology is helpful in elucidating how to deal with most of the safety concerns (i.e., safe handling, storage, transportation, and disposal procedures) characterizing organic peroxides and azo compounds.

Another critical safety issue, particularly recurrent in FRP, is the runaway phenomenon. This arises from the exothermicity of FRP (Chern, 2006; Copelli et al., 2013). In fact, if an effective dissipation of the produced thermal energy cannot be achieved, an incontrollable increase in the reaction temperature is experienced. This is even more problematic in polymer manufacturing, where the raise in viscosity at high monomer conversion reduces the heat dissipation from the reaction medium (Williams, 2007). The abovementioned temperature growth augments the propagation constant k_p, with a consequent further generation of heat in a loop that culminates with the complete loss of temperature control, the increase in the pressure and, in the end, the reactor explosion. For this reason, the heat exchange performance is a crucial aspect and the heat transport within the reaction mixture and through the reactor vessel must be carefully evaluated. As long as batch reactors have lower productivity compared to continuous settings, they sometimes require bigger reaction volumes to satisfy the customer demands. Consequently, the magnitude of the runaway risk (i.e., the consequences of a catastrophic event) is higher. In addition, as long as the heat exchangers are engineered to avoid the runaway phenomenon, the reaction temperature is kept well below dangerous values, thus reducing the reaction rate and increasing the time required to reach high conversions. Since heat exchange systems are expensive, this fact also means higher operational and investment efforts.

For the proper design of the heat exchange system, also the structural properties of the reactor should be taken into consideration. As an example, if the heat is removed by means of an external jacket, the effective reactor external surface has to be considered. In fact, a semi-continuous reactor is usually only partially filled, due to the feeding of the reaction mixture or the removal of the product. This leads to a reduced surface area involved in the heat exchange. However, even if the vessel is not completely filled and, in the case of an external jacket, the exchange surface is not totally used, the feeding or the product withdrawal typical of semi-batch reactors permits to contrast sudden temperature variations. In order to investigate the connection between the feeding strategy and the product quality, Zeaiter et al. (2002) proposed sensitivity analyses of styrene emulsion polymerization in a semi-batch configuration. Their results pointed out the important influence of temperature on the product molecular weight distribution (MWD), which is one of the main polymer properties. Instead, the efficiency of semi-batch processes in the heat management in very exothermic polymerization reactions was analyzed by Abel et al. (2000), who modeled temperature variations in the case of cooling system failure proving that it is possible to guarantee a safe process also in critical situations. First of all, the authors implemented and experimentally validated a dynamic process model. In order to achieve a thermally safe condition, even if also in the eventuality of the cooling system failure an active exchange is still present, they considered an adiabatic environment as conservative limit. Thus, it was possible to run the semi-batch system in critical situation, with time reduction up to 35%, exploiting the relationship between inlet flow and temperature profiles.

On the other hand, the intrinsic configuration of continuous reactors guarantees a higher exchange surface area per unit volume (S/V). As an example, while semi-batch reactors are only partially filled at the beginning, continuous STRs, in the case of cooling/heating jacket systems, typically employ the whole reactor volume, so that the entire external surface is involved in the heat exchange. This means a higher efficiency in the thermal exchange, with the additional possibility of conducting the polymerization reactions at higher temperatures and therefore with higher reaction rates. Higher S/V ratios and, in turn, more efficient heat management are easily achieved in TRs by adopting long tubes with small diameters (Hamer and Ray, 1986; Hamert et al., 1986). Furthermore, this configuration minimizes dead volumes and possible gas phase accumulation, thus limiting possible explosion issues (Abad et al., 1995; Petty et al., 2002; Wang et al., 2017b).

Last but not least, the achievement of stable and reproducible product quality is an essential aspect in order to make the transition of industrial polymerization processes from batch to continuous industrially feasible and appealing. For instance, a recurrent issue is that monomers charged in the reactor contain variable amounts of inhibitors or impurities, which negatively affect the reaction rate. In the worst case, the reaction starts only when the inhibitors are fully depleted, with an unpredictable delay of the process time. Along with the intrinsic irreproducibility typical of discontinuous processes, this latter aspect causes the well-known batch-to-batch variability of the end-product (Harmain and Chan, 2018). Therefore, products discharged from batch vessels must be inspected in terms of chemical and physical features and often are addressed to further manufacturing with the aim of reaching the market specifications, at cost of relevant operational expenditures (Asua, 2016; Kohlmann et al., 2016). In order to highlight the importance of controlling and understanding the polymer quality and to be able to make every batch campaign reproducible, Richards and Congalidis (2006) have presented the implementation of a hierarchical approach to the control system design. Furthermore, batch operations are often disregarded because of intrinsic limits in the control over the main characteristics of the final product. An example is the production of copolymers with a pre-determined composition that cannot be obtained due to the well-known compositional drift characterizing radical processes. If monomers with different reactivity are statistically copolymerized in a batch reactor, polymer chains with a wide variability in the composition are produced, following the Mayo-Lewis law (Kamesh et al., 2016; Sponchioni et al., 2019; Szkudlarek et al., 2017). To overcome this extensive inhomogeneity, the industrially adopted solution generally consists in the gradual feeding of the more reactive monomer, with a spontaneous transition toward the more flexible semi-batch. For instance, Vega et al. (1997) studied the emulsion copolymerization of acrylonitrile and butadiene. First, they modeled an industrial batch reactor in order to evaluate the kinetic and physicochemical parameters necessary for predicting conversion, MWD, number-average particle diameter, and copolymer composition. Successively, this model was applied to a semi-batch configuration in order to improve the process productivity and overcome the typical compositional drift recurrent in batch configurations. In fact, some applications adopt a semi-batch reactor in which the initial reactant mixture is composed only by the less reactive monomer and the more reactive one is added by means of an inlet stream in order to maintain constant the final polymer composition. Li et al. (2005) proposed a semi-batch experimental setup of a high temperature butyl acrylate and butyl methacrylate free-radical solution copolymerization. They were able to demonstrate that, under starved-feed conditions, the products remained homogeneous with a composition similar to that of the monomer mixture and, moreover, these approaches guaranteed safe conditions in case of runaway. The drawback in this policy may be related to the excessively long reaction time which is a fundamental parameter in all industrial polymerizations (Kozub and MacGregor, 1992; Richards and Congalidis, 2006). The same partialized feeding of the monomers can be carried out using continuous reactors, which in turn alleviate the composition control-productivity trade-off that is typical of semi-continuous processes.

In the case of heterogeneous polymerizations the proper product quality is also assessed in terms of particle size distribution. In this sense, the semi-batch mode allows to tune this parameter, which is particularly important in many application fields such as biomedical, cosmetic, and coatings. As shown by Sponchioni et al. (2017), the semi-batch operating mode guarantees a better control of the nanoparticle size compared to other systems. In addition, Zeaiter et al. (2002) reported the possibility of controlling the particle size by varying the monomer feed rate.

The mentioned advantages in terms of improved productivity and easier management of safety and environmental issues justify the current interest in the transferring of polymer production toward continuous processes. This is actually not trivial, since to ensure the same product quality as for batch productions, the three main aspects of the synthesis process, i.e., thermodynamics, kinetics, and fluid dynamics require to be carefully considered (Rossetti and Compagnoni, 2016). Indeed, all of them are affected by the specific process configuration and the understanding of this influence (i.e., reactor typology and operating conditions) is crucial to ensure the desired final material performances (Asua, 2016; Kohlmann et al., 2016; Meimaroglou et al., 2016; Plessis et al., 2001a; Richards and Congalidis, 2006). Due to the uncertainties related to this transition, model-based approaches are often pursued to save experimental effort.

In this chapter, we will discuss about the current technologies developed in the field of polymer production via FRP to reach this auspicable transition from batch to continuous processes. In particular, we will distinguish between homogenous processes, characterized by a lower conceptual and technical complexity and where the fluid dynamics plays a less relevant role, and heterogeneous processes, where on the other hand, the fluid dynamics strongly affects the product quality and the necessity to handle multiphase systems increases the technical complexity. After the presentation of these concepts, the microreactor technology, one of the currently most interesting strategies for continuous polymerizations, is briefly reviewed. Finally, some of the most relevant practical examples taken from industry and research activity, which are adopted to bring this transition into reality, are presented.

2. Homogeneous polymerization

2.1 Bulk polymerization

Bulk processes use the monomer itself as reaction medium. This procedure is particularly suitable to obtain pure, solid polymers. Unfortunately, the medium viscosity strongly raises up during the reaction, with the incurrence of major limitations in heat and mass transports. These difficulties deeply affect the kinetic behavior and, consequently, the final product features. Therefore, the heat removal efficiency must be carefully evaluated, taking always into account its lowering toward industrial scale due to the decreasing surface per unit volume ratio (Vladimir et al., 2000). Dealing with energy fluxes, hurdles can be solved using some alternative cooling utilities, e.g., internal cooling coils, external heat exchanger, reflux cooling. Internal cooling coils are not handy to be applied in bulk processes because they could interfere with the agitation, fundamental requirement in this reaction. External heat exchangers must be designed so to remove high amount of heat, but the pumping system implies important energy effort due to the high viscosity of the polymerizing mixture. Reflux cooling subtracts heat by means of solvent or monomer evaporation. Thus, the generated vapor has to be mixed back with the reactive mass. Problems arise when high conversions do not allow to overcome the mass and heat transport limitations because of the pronounced viscosity.

Regarding the mass transfer limitations, the achievement of continuous operations is demanding. Mechanical stirring is extremely difficult and it is possible to conduct the process only using high-shear mechanical equipment. TRs with internal static and dynamic mixers, e.g., Kneader type, which are basically extruders as shown in Fig. 1, are an interesting technology able to guarantee this essential structural characteristic and represent the most recurrent examples of continuous reactors for bulk polymerizations.

The most common configuration uses twin-shafts useful to maintain the reaction environment clean exploiting frictions inside the reaction environment. The avoidance of fouling is particularly important given its impact on the residence time distribution (RTD) that, in turn, can dramatically affect the product quality. Therefore, it is crucial to carefully design the shaft in order to avoid depositions. On the other hand, it is necessary to preserve

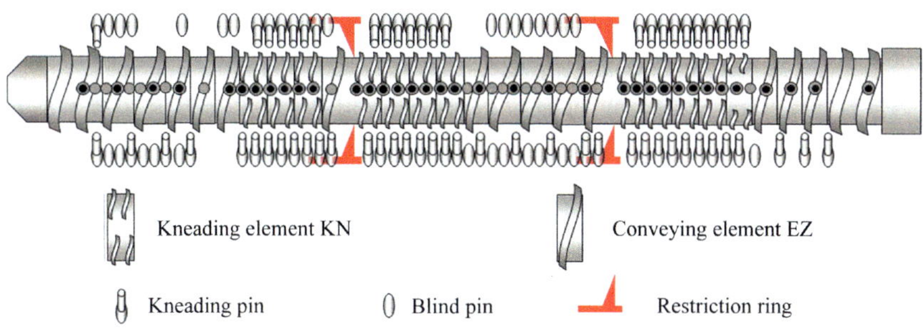

Fig. 1 Example of Kneader reactor screw configuration. *Figure reproduced from Salinier, A, et al: Influence of specific mechanical energy during compounding in a co-kneader on electrical and rheological properties of multiscale composite materials,* Compos Commun *17:154–162, 2020 with permission from Elsevier.*

the mechanical simplicity of the structures in order to safeguard the internal element robustness. For instance, Fleury (2006) polymerized MMA and co-polymerized acrylate/acrylic monomers in Kneader reactors with single shaft and back-mixing. The experimental results showed productivity two times higher than classic PFRs and, thanks to axial and radial homogeneous conditions, produced co-polymer structures without compositional drift. Recently, Monchatre et al. (2015) used a co-Kneader reactor to analyze the time-path followed by the compounds involved within the structure. This TR typology is equipped with axially moved rotating screws. The authors operated sensitivity analyses on some parameters (i.e., feeding, screw dynamic, screw design) that could induce variation in RTD. Thus, by means of modeling, they virtually reproduced the analyzed system. Their main focus was the RTD evaluation and chose series of CSTRs to reproduce the fluid dynamic conditions within the tube. Interesting results elucidated RTD broadening due to back-mixing induced by the high number of mixing elements. Furthermore, the independency of the distribution from the internal equipment configuration was proven. Finally, they confirmed good agreement between the experimental data and the model.

Another interesting continuous process strategy, which could help to overcome the fluid dynamics difficulties in bulk polymerizations, is based on the use of STRs in the initial moments of the reaction to provide a good mixing, followed by a TR with high-shear mechanical equipment to face the viscosity increase. In this way, when the conversion and hence the viscosity are still low, a good mixing can be achieved. Then, at higher polymer fractions, the reactive mixture is provided with the required higher mixing energy from the high-shear devices. Gharaghani et al. (2012) implemented a model aimed at describing the bulk polymerization of styrene in this kind of configuration. In particular, they designed a three-stages tubular reactor in the second section. Some typical important parameters in polymer field were simulated (i.e., conversion, polydispersity, number-average molecular weight). By means of the cited model, they proposed two strategies to control the TR. Both of them relied on the temperature control of the external jacket and at the outlet of each tubular part, respectively. The second strategy generated better responses, concerning polymer features, in case of dynamic system conduction.

2.2 Solution polymerization

In solution polymerization, the mass and heat transfer limitations recurring in bulk processes can be mitigated by using a solvent. The solvent has to dissolve homogeneously both the monomer and the polymer, reducing the system viscosity and favoring the removal of the heat generated during the reaction (Tobita and Hamielec, 2015). The transition toward continuous processes for this polymerization is not particularly challenging, but it is worth analyzing more in detail important process parameters so to highlight drawbacks and advantages in different reactor configurations.

The RTD has great influence on the polymer microstructure, which is a fundamental feature since it affects the macroscopic properties of the material. Such distribution expresses the probability of a specific element of fluid to leave the system at a certain time. Typically, the RTD of stirred batch and tubular reactors are quite narrow, while that of continuous stirred tanks are much broader, thus affecting important process features, such as conversion and yield, in very different ways. Consequently, a complete understanding of the phenomena involved in RTD tuning is of fundamental importance (Oechsler et al., 2018).

In the case of homogeneous FRPs, when the polymer chains do not restart their growth after their formation (i.e., irreversible terminations are operative), the residence time within the reaction environment is much larger than the time requested for their growth. Accordingly, the MWD of a polymer produced inside a stirred batch or a tubular reactor is broader than that of a polymer leaving a reactor approaching an ideal CSTR. This means that the constant concentration effect is more important than that of the RTD. However, in the frequent instance of high viscosity polymerizing mixtures, the micromixing inside a CSTR is compromised and a broadening in MWD similar to that found in stirred batch and TFR configurations is experienced. In fact, Roudsari et al. (2013) published a CFD study of MMA solution polymerization in CSTR. They focused on the importance of micromixing in polymerizations. More in detail, they demonstrated that the impeller has a great impact on the homogeneity within the tank, which in turn influences some essential parameters such as conversion and product structure. Furthermore, substantial increasing values of conversion were observed with higher residence time.

Instead, if dead chains are capable to polymerize again (as in the case of chain transfer to polymer reactions), the reactor RTD has a great impact on the final MWD and the best choice is the TFR (Meira and Oliva, 2011). In fact, the polymer hold up in stirred tanks can induce secondary reactions which generate chains with different features.

Kohlmann et al. (2016) proposed a smart strategy to design a continuous process for the conduction of solution polymerizations, highlighting the essential requirements and hurdles briefly illustrated in Scheme 1.

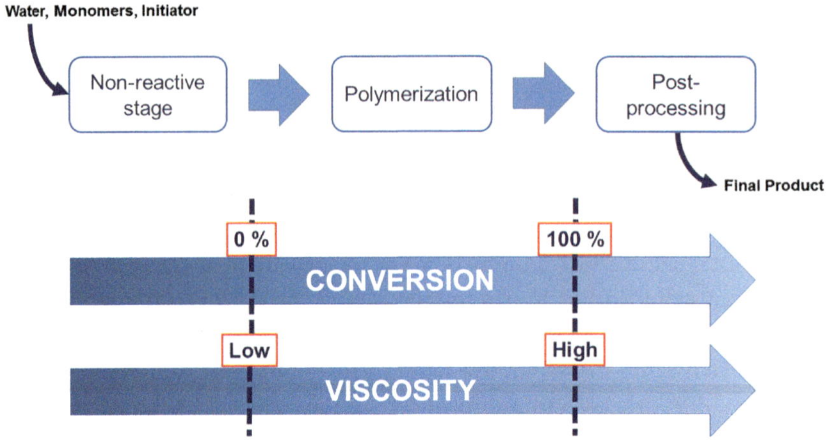

Scheme 1 Key points in continuous solution polymerizations as proposed by Kohlmann et al. (2016).

Particularly, it is necessary to pay attention to mass and heat transfer limitations, strictly related to the mixing, in order to operate toward high conversions and convenient MWDs. Thus, the features of the final product (e.g., polymer concentration, pH, ionic strength) can be tuned through post-processing operations in order to reach the specifications required by the market.

The use of TRs in homogeneous FRPs is currently quite limited due to the typically high viscosity that in turn introduces the risk of clogging. On the other hand, the anticipated higher flexibility, shorter slack times during start up, shutdown and grade transition, and better heat management make these systems appealing for these polymerizations. Therefore, a huge research effort has been spent for developing strategies to integrate TRs in solution FRPs. Melo et al. (2003) formulated an interesting model of solution polymerization in a loop reactor (Fig. 2).

This tubular configuration is actually divided in two reactors. A portion of the reaction mixture at the outlet of the first section is recycled to the second one, in order to obtain a modulation of the RTD that approaches that of a CSTR reactor. The steep conversion profile and the good micromixing provided by the recycle pump make this system industrially appealing. The authors also investigated the impact of the recycling pump on the process dynamic and stability. The results showed that the pump worsens the heat exchange during the reaction as well as, together with other external mechanical structures (e.g., tube structures, fittings, valves), the process

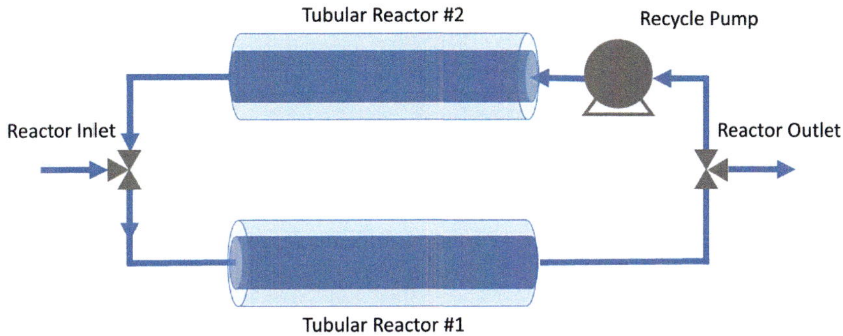

Fig. 2 Loop reactor schematization.

stability. Also Goerke et al. (2016) proposed an interesting feasibility study on the transition from semi-batch to PFR polymerizations. They investigated the product quality reproducibility in a generic solution copolymerization with a kinetic behavior similar to the acrylic acid one. Difficulties in realizing, using TRs, the same composition obtained by means of controlled feeding in semi-continuous configuration are well known. Therefore, using a continuous discretized side injection model, they elucidated hurdles in sequence length distribution control. Finally, they defined an area of transferability on the base of some parameters (i.e., number-average molecular weight distribution, polydispersity). Instead, in homopolymerizations, when the key feature is only the chain length distribution, a broader window for the transition to continuous processes was delineated. Branched polymers are also mainly produced using the semi-batch mode as long as this particular polymer structure is obtained via chain transfer to polymer reactions that are favored in the high polymer concentration mixture of the semi-continuous reactors (Cheng, 2003; Cheng et al., 2005). In these configurations, it is necessary to pay attention when high conversion values are achieved. If the viscosity increases, also fluid pumping requires a higher energy effort, other than for fouling issues and pressure drops.

3. Heterogeneous polymerization

In heterogeneous processes, the monomer and polymer are not soluble in the reaction medium (generally water). Typically, a dispersed monomer phase is initially formed by stirring and the final polymer is produced as dispersion of particles. Heterogeneous polymerization can be in suspension or emulsion depending upon the type of the selected initiator, either oil- or water-soluble, respectively. In suspension polymerization, the initial monomer droplets become polymer particles and each particle is a micro-bulk reactor. On the other hand, the particles are nucleated as a new phase in emulsion polymerization, while the initial monomer droplets are no other than reservoirs of reactant. While particle sizes from 10^4 nm up to more than 1 mm are obtained in suspension, colloidal particles are typically produced by emulsion polymerization, with size ranging from tens to few hundreds nanometers. In both cases, the dispersed system guarantees a good control over the heat transfer as well as over the system viscosity. In fact, the dispersant is effective in the dissipation of the heat produced during the polymerization. However, the need of suitable additives (surfactants) to stabilize the dispersed phases during the reaction represents a drawback with respect to the product purity. Their presence can have detrimental impact on some of the final properties already at low percentages. Moreover, the careful control of the particle size distribution is a key aspect in many cases, given the relevance of this feature on the final properties of the material. Such control is frequently difficult, given the impact of complex phenomena like particle breakage and coagulation, in turn affected by the reactor fluid dynamics (Tobita and Hamielec, 2015).

Continuous STRs are used in both emulsion and suspension polymerizations. Many times, in emulsion polymerizations, the process scheme is designed with the aim of moving the classical semi-batch configuration toward a continuous one using a pre-emulsifier vessel followed by the reactor and then by a post-reaction section. In the former vessel, monomer droplets are formed due to the mixing with the emulsifier. Then the mixture is fed to the second container where the initiator is added in order to start the reaction and finally the products are purified to meet the market requirements (Pan et al., 2016). Sometimes the reaction tank is operated in a parallel configuration in order to optimize the dead times, thus the continuous way is effective only for the other two mentioned sections. Concerning emulsion polymerizations, a successful example is the use of tanks in series in order to satisfy the large market demand of styrene-butadiene rubbers (Agnely, 2006).

Concerning suspension polymerization, the requirement of a good mixing within the reaction mixture is often critical. In fact, mechanically stirred tanks are the favorite choice, even if backmixing may generate an oscillation in product and process features under specific operative conditions. To better understand the influence of mixing on the system dynamics, Alexopoulos et al. (2002) performed a CFD analysis of turbulence non-homogeneity in suspension polymerization carried out in well-mixed vessels. This computational technique is preferable than the experimental evaluations because of the accurate and detailed flow field reconstruction provided at different positions (Kunowa et al., 2007). They implemented a two-compartment model able to describe the mixing zone near the impeller as well as the circulation zone. In fact, the vortexes generated by mechanical agitation influence the final particle size, mainly driven by coagulation and breakage of the initial monomer droplets. In conclusion, their results described the time-evolution of the particle size distribution within the mixture and were in a good agreement with the experimental data. The results demonstrated that the impeller diameter and rector-scale up have a consistent impact on the creation of different turbulence conditions and hence micromixing within the reactive mixture.

The main advantages of TRs with respect to STRs are much narrower RTDs, resulting in narrower particle size distribution and more uniform polymer properties, high conversion, and superior heat transfer. On the other hand, clogging phenomena take place in heterogeneous polymerization carried out in tubular reactors. The main motivation is that dispersions are thermodynamically unstable and tend to precipitate or to form large aggregates. The additional lack of mixing make these systems hard to conduct (Asua, 2016). The loop reactor is the most industrially appealing tubular configuration for heterogeneous polymerizations. However, its use is limited by the coagulation phenomenon induced by the recycle pump that generates reactor fouling (Asua, 2016; González et al., 2007). Moreover, this configuration presents a backmixing behavior similar to CSTRs that could induce oscillation issues (Araújo et al., 2001).

An interesting technology used to overcome mixing hurdles in tubular reactors is pulsed flow. Here, the utilization of pulsations guarantee sufficient turbulence as to overcome the classic poor mixing associated to TRs. Thanks to this configuration, productivity and polymer quality remain particularly stable. Carvalho et al. (2010) studied a continuous emulsion copolymerization carried out in a tubular reactor equipped with oscillatory (pulsed) flow and fillers (sieve plate). The selected configuration was proved able to overcome fouling and mixing issues at remarkable extent (Sayer et al., 2002a,b). Reactors with superimposed secondary flow are another interesting solution proposed in the literature to solve the heat exchange, fouling and mixing issues typical of heterogeneous syntheses, which influence the size features and may also imply clogging (Asua, 2016; Pauer and Moritz, 2006). A well-known example is the Couette-Taylor reactor that comprises two concentric cylinders, with the inner one rotating on its axis (Fig. 3).

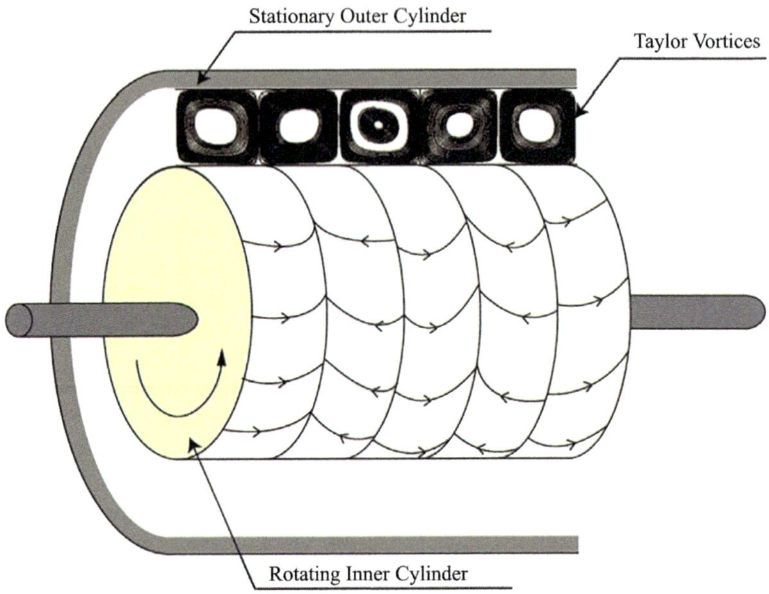

Fig. 3 Schematic representation of a Couette-Taylor reactor. *Figure reproduced from Jeon DH, Song JH, Hong JP, et al: Agglomeration of Li(NixMnyCoz)O_2 particles in Couette–Taylor flow reactor,* J Ind Eng Chem *76:524–531, 2019 with permission from Elsevier.*

The reaction environment is represented by the gap between these two structures and when a critical Taylor number (i.e., the ratio between centrifugal forces of a fluid that rotates around an axis and the viscous forces) is exceeded, a significant formation of eddies in the reaction mixture ensures an efficient mixing (Wei et al., 2001). This system permits the establishment of a narrow RTD, like the one typical of a train of CSTRs (Xue et al., 2002). The major drawback is represented by the huge viscosity changes as the conversion increases, with the corresponding large variation of Taylor number inside the reactor. However, since the Taylor number is directly proportional to the gap between the walls of the two cylinders, its reduction can be contrasted by a conical outer cylinder with increasing diameter. This interesting structure, known as Wicker reactor, can be conveniently employed to alleviate the coagulation issues (Apostel et al., 2001; González et al., 2009). Unfortunately, both of the last two solutions are not industrialized yet. González et al. (2009) investigated the use of a Couette-Taylor reactor in high solid content acrylic emulsion polymerizations. They evaluated the viscosity of the latex in a CSTR configuration and imposed a rotational speed in order to obtain a Taylor number higher than the critical one in the Couette-Taylor configuration. Final results demonstrated better performance, compared to the classic CSTR process, in terms of final latex conversion. Moreover, the PSD obtained with the superimposed secondary flow reactor is narrower and also allows the obtainment of higher yield (Liu et al., 2013b). These behaviors were achieved by means of an appropriate tuning of the residence time within the reactor. Finally, the authors proposed to actuate the start-up step filling the reactor with latex. The latter issue was investigated also by Liu et al. who studied the emulsion polymerization of methyl methacrylate in a Couette-Taylor reactor. They implemented a model and confirmed its reliability on the basis of several experimental results. Observing the obtained results, they asserted the hurdle in rapidly reaching homogeneous vertical conditions and this fact influences the conversion-time profile with lower values at the beginning. In a further elaborate, Liu and Kind (2013) examined the influence of fluid dynamics on the product properties (i.e., MWD). In particular, increasing the average residence time, they elucidated a decreasing trend of the weight-average molecular weight and higher polydispersity. Another example of reactor with superimposed secondary flow in suspension polymerization is Torus reactor (Pauer and Moritz, 2006). This structure works with mechanical stirrer that, together with its high symmetry, should guarantee narrow particle size distributions. Unfortunately, the polymerization of polystyrene beads produced a broad distribution due to the inhomogeneity induced by the presence of gravitational forces and therefore density gradients generated by the stirring within the system (Pauer and Moritz, 2006). Finally, Torus reactor was not suitable in terms of product quality with respect to the investigated suspension polymerization.

Another interesting tubular technology was investigated by Nogueira et al. (2017), who used vertically stirred tubular reactors closed at the extremities and hence conducted in batch to run the suspension polymerization of styrene. The system was equipped with axial tunable stirring disposed in vertical position so to promote the falling of the produced particles, by means of gravity, at certain reaction time. In particular, they implemented a model able to simulate the process and operated its validation through several experiments. Their results elucidated a good heat control also without the necessity of an external heat exchange system. The obtained particle size distribution resulted optimal for the typical final applications of such products even if the authors experienced an increasing average size when augmenting the number of axial impellers. Finally, it was also evaluated the pattern of the bead sedimentation which is a fundamental information concerning the switch toward continuous conduction.

Overall, the transition from batch to continuous in heterogeneous polymerizations is mainly constrained by the stability of the produced latex. An important prerequisite for the transition toward continuous processes is of course the dispersion stability in batch syntheses. Then, the fluid dynamics of the system must be carefully studied in order to reduce the clogging, which represents the most important issue in continuous reactors.

4. Microreactors

Microreactors have gained a huge interest in recent years as an appealing solution in the transition toward continuous processes (Buchholz, 2010; Tonhauser et al., 2012; Wilms et al., 2008). Microreactors are intended as three-dimensional structures with inner dimensions usually between 10 and 100 μm and in any case in the submillimeter scale (Jähnisch et al., 2004). Industry and academy proposed a wide range of different construction materials, i.e., polymers, glass, silicon, ceramics, and steel, trying to find out the best compromise between process chemical compatibility and microstructure production hurdles. Nowadays, poly(dimethylsiloxane) (PDMS) is the material of choice for the production of such devices, mainly due to its chemical inertness and flexibility, which reduces the shaping complexities. PDMS microreactors (Fig. 4) can be readily fabricated but with the disadvantage of incurring in swelling phenomena when in contact with common organic solvents.

Given the very small reaction volumes, all of the safety issues related to the toxic, exothermic, and even explosive nature of the reacting mixture become negligible (Tonhauser et al., 2012; Wilms et al., 2008). The result is a potential further improvement respect to the continuous configurations presented above. The most evident upgrade is the heat management thanks to the very high S/V ratio (Wilms et al., 2008) (i.e., between 10,000 and 50,000 m^2/m^3), more than two orders of magnitude higher than traditional reactors. Furthermore, since the reaction environment is in the size of the micro- or milliliters, mixing is ideally instantaneous, thus preventing any concentration gradient and hotspot that may lead to runaway phenomena. This is particularly true in the case of homogeneous FRPs where the diffusion limitations have a great impact upon the reaction proceeding. However, the management of clogging is particularly critical in microreactors, both on laboratory as well as on industrial scales, when dealing with

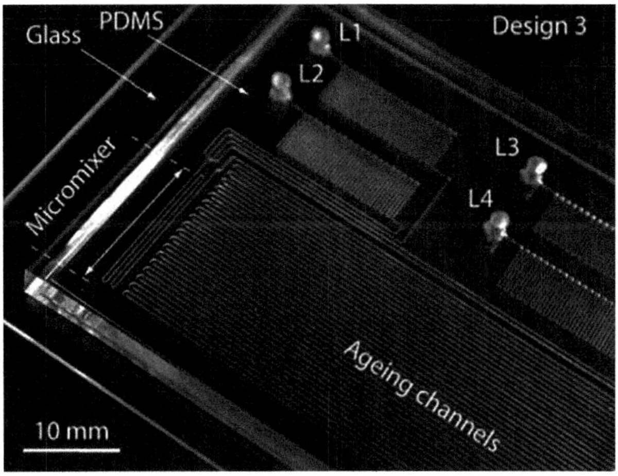

Fig. 4 Examples of poly(dimethylsiloxane) (PDMS) microreactor structures. *Figure reproduced with permission from Khan SA, Günther A, Schmidt MA, et al: Microfluidic synthesis of colloidal silica,* Langmuir *20:8604–8611, 2004. Copyright 2004, American Chemical Society.*

polymers (Tonhauser et al., 2012; Wilms et al., 2008). As a consequence, Wilms et al. (2008) underlined the importance to pay special attention to the judicious selection of temperature and dilution before implementing a polymerization process into a microscale reactor.

In the case of microreactors, all of the abovementioned characteristics significantly improve the product features in terms of selectivity, yield and product quality (Iwasaki and Yoshida, 2005; Tonhauser et al., 2012; Wilms et al., 2008; Yoshida et al., 2005). Indeed, Iwasaki and Yoshida (2005) made a comparison between three polymerizations (i.e., benzyl methacrylate and methyl methacrylate, butyl acrylate, vinyl benzoate and styrene) conducted in macro batch and continuous microreactors (see Table 1 for the comparison in the case of polymethyl methacrylate). They demonstrated that, using microreactors in these highly exothermic syntheses, a high impact improvement on MWD is obtained. Furthermore, they stated that higher exothermicity means higher polydispersity since the traditional reactor generates local hot spots which in turn provoke structure inhomogeneity. Instead, microstructures work with a very efficient heat transfer capacity.

Another key advantage in the application of this technology is represented by the kinetic behavior of the reaction system. Wilms et al. (2008) affirmed that the fast kinetics typical of FRPs is particularly suitable for these systems thanks to their characteristic homogeneities and the limited

Table 1 Comparison between poly(methyl methacrylate) MWDs in batch macro and microreactors.

Reactor	Residence time or reaction time (min)	Yield (%)	M_n ($\times 10^3$)	PDI (M_w/M_n)
Microreactor	0.5	8.6	8.6	1.84
	1	14.1	8.6	1.80
	2	27.1	8.4	1.82
	5	51.1	8.5	1.83
	10	75	9.1	1.84
Macroscale batch reactor	3	8.1	18	2.51
	5	21.9	13.1	2.21
	7.5	50.5	7.8	2.26
	10	68.5	7.0	2.26
	15	77.1	7.4	2.14

Table reproduced with permission from Iwasaki T, Yoshida J: Free radical polymerization in microreactors. Significant improvement in molecular weight distribution control, *Macromolecules* 38:1159–1163, 2005. Copyright 2005, American Chemical Society.

residence time of the continuous processes. As a consequence, the conduction of this kind of reactions in microreactors is achievable and their industrialization feasible. Following their desire of producing high amounts of polymer per unit time with reproducible chemical-physical features, Wolff et al. (2018) reported the study of a precipitation polymerization of two different microgel systems by means of a continuous tubular microreactor. In order to study the feasibility to move toward the interesting configuration, they compared the products in terms of quality (particle size distribution and temperature responsive behavior) to the ones obtained through a conventional batch process. The possibility of precisely tuning the residence time, reactant concentrations and temperature profile within the continuous reactor allows to optimally tailor and produce reproducible microgel features.

Rossow et al. (2016) reported some interesting results concerning the transfer of styrene and *n*-butyl acrylate emulsion polymerization from a semi-batch macro reactor to a continuous tubular smart-scale reactor. They concluded that it is impossible to directly transfer the process principally because of the strongly different concentration profiles which in turn affect the chain composition distribution. Therefore it is requested to suitably manage the parameters in the continuous mode.

The exploitation of microreactors to solve the safety and technical concerns related to continuous configurations is still at an explorative level but this reactor technology may become a valuable option from an economic viewpoint. Its translation to the industrial scale appears to be extremely challenging, mainly due to the large number of microreactors, operated in parallel, that would be required to satisfy the polymer market demand (Junkers, 2017; Parida et al., 2014; Tonhauser et al., 2012; Wilms et al., 2008). Nevertheless, many industries and research teams are working on the microreactor commercialization. On the other hand, since in these systems the polymerization reactions generate low heat and the heat exchange is optimal, the scale-up step in terms of quality is more trustable than in macro reactors. Also, microreactors are particularly suitable to conduct sensitivity lab-scale analyses without waste of time and materials. The continuous way allows an online modification of the operative conditions and, thanks to the small sizes, it is possible to work with small amounts of reactants and solvents (Newman and Jensen, 2013; Tonhauser et al., 2012; Wilms et al., 2008). As a consequence, first the tuning of the operating conditions to reach the desired product quality and then the scale-up step results effortless. Furthermore, the use of very small reactors represents an important contribution toward the implementation of continuous polymerization processes, in particular because of the modularity of the resulting reaction system (Iwasaki and Yoshida, 2005; Roberge et al., 2009; Tonhauser et al., 2012; Wilms et al., 2008; Yoshida et al., 2005).

5. Successful case studies

Once the challenges and the technological solutions currently available for the translation toward continuous FRPs have been clarified, it is worth analyzing the successful implementations of continuous reactors in the polymer production on an industrial scale.

With this regard, one of the most vivid example of the great interest toward continuous polymer production is represented by the European project F^3 Factory, which aims at making continuous plants more flexible, versatile and then suitable for multiproduct productions (Asteasuain et al., 2006; Bieringer et al., 2013; Padhiyar et al., 2006).

In this framework, Chevrel et al. (2016a) studied the non-ionized acrylic acid solution polymerization in tubular reactors equipped with static mixers. They proposed a methodology to switch from lab- to pilot-scale confirming their model with chemical-physical experimental data (i.e., kinetic parameters, molar mass distributions, viscosity variations) acquired by means of a rheo-Raman device (Chevrel et al., 2016b). The installation of static mixers enables optimal mass and heat transport phenomena, and consequently RTDs quite similar to ideal plug-flow reactors. In terms of pressure drops, it was necessary to pay attention to the solid content. In fact, fouling issues were observed after only 6–8 h running with high viscosity. The good agreement between experimental data and model outputs, in both scales without any adjustment, made the scale-up implementation successful. Once more, it is evident that the close relationship between modeling activity and validation is fundamental for both transferring a batch protocol to a continuous configuration and to scale-up the production. For instance Gao et al. (2010) built up a three-dimensional CFD model of a pilot-scale tubular loop reactor for the biphasic (i.e., liquid–solid) propylene polymerization. First of all, they confirmed its predictivity for the steady state thanks to the mixture pressure gradient data. After that, the model was used to point out the influence of reactor areas, flow velocities, slurry concentrations, particle diameters, external coolings on the temperature profile, which is a fundamental parameter used to optimize the conduction of this kind of processes. In this sense, another interesting elaborate on poly(acrylic acid) production in TR pointed out the importance to control product quality. Hashemi et al. (2016) implemented a model control strategy constrained by weight average molecular weight and residual monomer amount. The main result obtained with the presented pilot-scale process optimization is the possibility of a higher productivity than discontinuous technologies always preserving the polymer features.

An interesting concept arising from the F^3 Factory project is the modularity of chemical plants. Following this concept, the plant is fragmented in different (more or less) autonomous modules, each one corresponding to a specific operative section. Each module can be continuously moved and/or changed in order to design the optimal configuration to produce new products, to tune their properties, and to scale up their production. In this way, it is possible to combine the flexibility of batch reactors with the intrinsic safety and economic advantages of the continuous ones (Buchholz, 2010; Centi and Perathoner, 2008; Schoppmeyer et al., 2016). The hurdle is related to the recurrent unsteady state conditions. Some modeling work may help to foresee the system behavior, more specifically to guide the start-up, the grade transition products, and the shutdown steps (Asteasuain et al., 2006; Penlidis et al., 1985; Roudsari et al., 2013; Schoppmeyer et al., 2016; Schork and Ray, 1987). Since the product quality is not appropriate in these phases, their shortening means minimizing the waste disposal (Abou Hweij and Azizi, 2015; Sheoran et al., 2018). As an example, Kohlmann et al. (2016) run a pilot modular plant using INVITE center (BASF, Leverkusen) utilities. Important results were elucidated, such as the ability of the equipment to support both homopolymerization of AA with chain transfer agents as well as the copolymerization of AA. In addition, several operating conditions were examined in these processes. In this sense, it was possible to improve the flexibility of the continuous conduction and make it comparable to the discontinuous one. The major limitation concerns the maximum acceptable viscosity value which induces important depositions.

The traditional macroreactors are not the only ones that can be successfully scaled-up. Indeed, the microreactor team (Kockmann et al., 2011) in Lonza (Switzerland) transferred a laboratory scale microreactor to a multi-ton one in 2009 (Fig. 5). The fruitful achievement of this commercial structure was driven by considerations about the influence of the hydraulic diameter toward mixing and heat transfer behaviors at different production scales.

Finally, they were able using their technology to conduct in 2009 a long-term robust process producing more than to 2 tons in the same year.

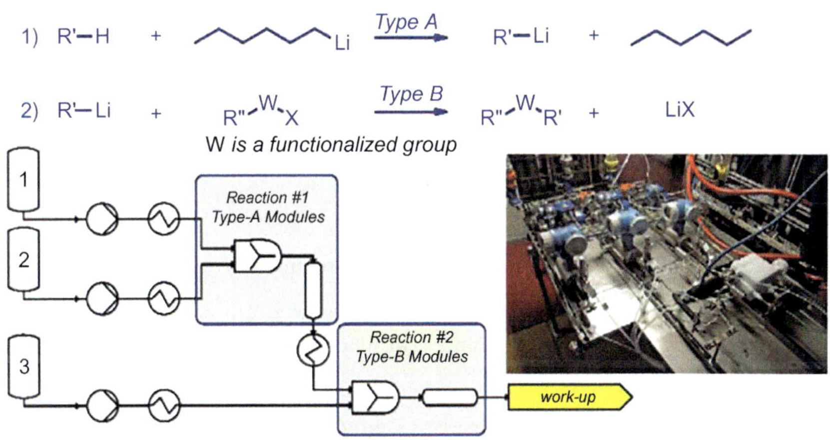

Fig. 5 Reaction scheme, process scheme and pilot-scale reactor operated in Lonza. *Figure reproduced from Kockmann N, Gottsponer M, Roberge, DM: Scale-up concept of single-channel microreactors from process development to industrial production.* Chem Eng J *167:718–726, 2011 with permission from Elsevier.*

6. Conclusions

Nowadays, one of the most interesting challenge in the polymer field is the transition from batch to continuous production. This trend is justified by the opportunity to improve first of all safety and environmental aspects, principally in free-radical polymerization. In addition, dealing with product quality and cost management perspectives, also the process yield is strongly improved.

In this chapter, a specific overview on the challenges for transferring a polymer production process from discontinuous to continuous is presented together with the reactor of choice for actuating such transition.

The main variables, which have to be considered to make this transition possible, are referred to the transport phenomena during the proceeding of the polymerization reactions. In this sense, it is worth highlighting the great impact of micro- and nano-scale technologies in this field. Moreover, modeling is a helpful instrument that can be used in order to minimize the experimental efforts. This mathematical tool allows overcoming thermodynamic, kinetic, and fluid dynamic hurdles. Finally, it is mandatory to design a smart on-line control system to implement performing continuous polymerization processes. In spite of an increase in the capital costs, the addition of this on-line control improves the product quality. This school of thought is perfectly in agreement with the industry 4.0 directives toward a continuous improvement of the polymer technologies.

Industry and research are working side by side first to manage in an optimal way the design of the entire process and, in terms of future perspectives, to make continuous processes more flexible overcoming the current profit margin reduction.

References

Abad C, de la Cal JC, Asua JM: Emulsion copolymerization in continuous loop reactors, *Chem Eng Sci* 49:5025–5037, 1995.

Abel O, Helbig A, Marquardt W, Zwick H, Daszkowski T: Productivity optimization of an industrial semi-batch polymerization reactor under safety constraints, *J Process Control* 10:351–362, 2000.

Abou Hweij K, Azizi F: Hydrodynamics and residence time distribution of liquid flow in tubular reactors equipped with screen-type static mixers, *Chem Eng J* 279:948–963, 2015.

Agnely M: Les latex de styrène-butadiène. In *Les latex synthétiques*, Paris, 2006, Lavoisier, p 417.

Albuquerque PBS, Malafaia CB: Perspectives on the production, structural characteristics and potential applications of bioplastics derived from polyhydroxyalkanoates, *Int J Biol Macromol* 107:615–625, 2018.

Alexopoulos AH, Maggioris D, Kiparissides C: CFD analysis of turbulence non-homogeneity in mixing vessels a two-compartment model, *Chem Eng Sci* 57:1735–1752, 2002.

Al-Tayyar NA, Youssef AM, Al-hindi R: Antimicrobial food packaging based on sustainable bio-based materials for reducing foodborne pathogens: a review, *Food Chem* 310: 2020, 125915.

Apostel M, Pauer W, Moritz H-U, Kremeskötter J, Hungenberg K-D: The Wicker tube reactor—characterization and modeling, *Chim Polym React Eng* 55:229–230, 2001.

Araújo PHH, De La Cal JC, Asua JM, Pinto JC: Modeling particle size distribution (PSD) in emulsion copolymerization reactions in a continuous loop reactor, *Macromol Theory Simul* 10:769–779, 2001.

Asteasuain M, Bandoni A, Sarmoria C, Brandolin A: Simultaneous process and control system design for grade transition in styrene polymerization, *Chem Eng Sci* 61:3362–3378, 2006.

Asua JM: Challenges and opportunities in continuous production of emulsion polymers: a review, *Macromol React Eng* 10:311–323, 2016.

Bieringer T, Buchholz S, Kockmann N: Future production concepts in the chemical industry: modular—small-scale—continuous, *Chem Eng Technol* 36:900–910, 2013.

Buchholz S: Future manufacturing approaches in the chemical and pharmaceutical industry, *Chem Eng Process Process Intensif* 49:993–995, 2010.

Capasso U, Ilare J, Romani C, Moscatelli D, Sponchioni M: Colloids and surfaces B: biointerfaces surfactant-free and rinsing-resistant biodegradable nanoparticles with high adsorption on natural fi bers for the long-lasting release of fragrances, *Colloids Surf B Biointerfaces* 190: 2020, 110926.

Carvalho ACSM, Chicoma DL, Sayer C, Giudici R: Development of a continuous emulsion copolymerization process in a tubular reactor, *Ind Eng Chem Res* 49:10262–10273, 2010.

Centi G, Perathoner S: Catalysis, a driver for sustainability and societal challenges, *Catal Today* 138:69–76, 2008.

Cheng KC: Effect of feed rate on structure of hyperbranched polymers formed by stepwise addition of AB2 monomers into multifunctional cores, *Polymer* 44:1259–1266, 2003.

Cheng KC, Chuang TH, Chang JS, Guo W, Su WF: Effect of feed rate on structure of hyperbranched polymers formed by self-condensing vinyl polymerization in semibatch reactor, *Macromolecules* 38:8252–8257, 2005.

Chern CS: Emulsion polymerization mechanisms and kinetics, *Prog Polym Sci* 31:443–486, 2006.

Chevrel M-C, Hoppe S, Meimaroglou D, Falk L, Durand A: Continuous pilot-scale tubular reactor for acrylic acid polymerization in solution designed using lab-scale Rheo-Raman data, *Macromol React Eng* 10:354–363, 2016a.

Chevrel MC, et al: Application of Raman spectroscopy to characterization of residence time distribution and online monitoring of a pilot-scale tubular reactor for acrylic acid solution polymerization, *Macromol React Eng* 10:406–414, 2016b.

Copelli S, et al: Emulsion polymerization of butyl acrylate: safe optimization using topological criteria, *Ind Eng Chem Res* 52:8625–8634, 2013.

Diani J, Gall K: Finite strain 3D thermoviscoelastic constitutive model, *Society*, 2006. https://doi.org/10.1002/pen.

Durand A, Engell S: Batch to conti transfer of polymer production processes, *Macromol React Eng* 10:308–310, 2016.

Ferrari R, Sponchioni M, Morbidelli M, Moscatelli D: Polymer nanoparticles for the intravenous delivery of anticancer drugs: the checkpoints on the road from the synthesis to clinical translation, *Nanoscale* 10:22701–22719, 2018.

Fleury PA: Bulk polymerisation or copolymerisation in a novel continuous kneader reactor, *Macromol Symp* 243:287–298, 2006.

Gao X, Shi DP, Chen XZ, Luo ZH: Three-dimensional CFD model of the temperature field for a pilot-plant tubular loop polymerization reactor, *Powder Technol* 203:574–590, 2010.

Ghadipasha N, Geraili A, Romagnoli JA, Castor C, Reed WF: First steps towards online optimal control of molecular weight in batch and semi-batch free-radical polymerization reactors, *Comput Aided Chem Eng* 38:1129–1134, 2016. Elsevier Masson SAS.

Gharaghani M, Abedini H, Parvazinia M: Dynamic simulation and control of auto-refrigerated CSTR and tubular reactor for bulk styrene polymerization, *Chem Eng Res Des* 90:1540–1552, 2012.

Goerke T, Kohlmann D, Engell S: Transfer of semibatch processes to continuous processes with side injections—opportunities and limitations, *Macromol React Eng* 10:364–388, 2016.

González I, Paulis M, de la Cal JC, Asua JM: (Mini)Emulsion polymerization: effect of the segregation degree on polymer architecture, *Macromol React Eng* 1:635–642, 2007.

González G, et al: Production of widely different dispersed polymers in a continuous Taylor-Couette reactor, *Macromol React Eng* 3:233–240, 2009.

Hamer JW, Ray WH: Continuous tubular polymerization reactors-II. Experimental studies of vinyl acetate polymerization, *Chem Eng Sci* 41:3095–3100, 1986.

Hamert JW, Engineering C, Because I: Continuous tubular polymerization reactors-I. A detailed model, *Chem Eng Sci* 41:3083–3093, 1986.

Harmain H, Chan CH: Batch-to-batch reproducibility studies of pilot-scale emulsion polymerization of poly(styrene-co-butyl acrylate), *Macromol Symp* 382:1–11, 2018.

Hashemi R, Kohlmann D, Engell S: Optimizing control and state estimation of a continuous polymerization process in a tubular reactor with multiple side-streams, *Macromol React Eng* 10:415–434, 2016.

Iwasaki T, Yoshida J: Free radical polymerization in microreactors. Significant improvement in molecular weight distribution control, *Macromolecules* 38:1159–1163, 2005.

Jähnisch K, Hessel V, Löwe H, Baerns M: Chemistry in microstructured reactors, *Angew Chem Int Ed* 43:406–446, 2004.

Jas G, Kirschning A: Continuous flow techniques in organic synthesis, *Chem A Eur J* 9:5708–5723, 2003.

Jiang B, Tsavalas JG, Sundberg DC: Morphology control in surfactant free polyurethane/acrylic hybrid latices—the special role of hydrogen bonding, *Polymer (Guildf)* 139:107–122, 2018.

Junkers T: Precision polymer design in microstructured flow reactors: improved control and first upscale at once, *Macromol Chem Phys* 218:1–9, 2017.

Kamesh R, Anand P, Rani KY: Evaluation of control strategies for a semi batch copolymerization reactor. In *2016 Indian control conference, ICC 2016—proceedings*, 2016, pp 474–480. https://doi.org/10.1109/INDIANCC.2016.7441177.

Kaur R, Kukkar D, Bhardwaj SK, Kim K, Deep A: Potential use of polymers and their complexes as media for storage and delivery of fragrances, *J Control Release* 285:81–95, 2018.

Kockmann N, Gottsponer M, Roberge DM: Scale-up concept of single-channel microreactors from process development to industrial production, *Chem Eng J* 167:718–726, 2011.

Kohlmann D, et al: Modular, flexible, and continuous plant for radical polymerization in aqueous solution, *Macromol React Eng* 10:339–353, 2016.

Kozub DJ, MacGregor JF: Feedback control of polymer quality in semi-batch copolymerization reactors, *Chem Eng Sci* 47:929–942, 1992.

Kunowa K, Schmidt-Lehr S, Pauer W, Moritz HU, Schwede C: Characterization of mixing efficiency in polymerization reactors using competitive-parallel reactions, *Macromol Symp* 259:32–41, 2007.

Li D, Grady MC, Hutchinson RA: High-temperature semibatch free radical copolymerization of butyl methacrylate and butyl acrylate, *Ind Eng Chem Res* 44:2506–2517, 2005.

Liu Z, Kind M: Continuous polymerization of methyl methacrylate in a Taylor-Couette reactor. II. Influence of fluid dynamics on polymer properties, *Polym Eng Sci* 53:950–955, 2013. https://doi.org/10.1002/pen.23342.

Liu H, Gao L, Shang Q, Xiao G: Preparation and characterization of polyurethane clearcoats and investigation into their antigraffiti property, *J Coat Technol Res* 10:775–784, 2013a.

Liu Z, Jin T, Kind M: Continuous polymerization of methyl methacrylate in a Taylor-Couette reactor. I. Influence of fluid dynamics on monomer conversion, *Polym Eng Sci* 53:96–104, 2013b. https://doi.org/10.1002/pen.23245.

Liu SH, Shu CM, Hou HY: Applications of thermal hazard analyses on process safety assessments, *J Loss Prev Process Ind* 33:59–69, 2015.

Manfredini N, Scibona E, Morbidelli M, Moscatelli D, Sponchioni M: 110th anniversary: fast and easy-to-use method for coating tissue culture polystyrene surfaces with nonfouling copolymers to prevent cell adhesion, *Ind Eng Chem Res* 58:22290–22298, 2019.

Meimaroglou D, et al: On the modeling of acrylic acid copolymerization in an aqueous solution: a modular, integrated approach, *Macromol React Eng* 10:389–405, 2016.

Meira GR, Oliva HM: Molecular weight distributions in ideal polymerization reactors. An introductory review, *Lat Am Appl Res* 41:389–401, 2011.

Mekonnen T, Mussone P, Khalil H, Bressler D: Progress in bio-based plastics and plasticizing modifications, *J Mater Chem A* 1:13379–13398, 2013.

Melo PA, Biscaia EC, Pinto JC: The bifurcation behavior of continuous free-radical solution loop polymerization reactors, *Chem Eng Sci* 58:2805–2821, 2003.

Monchatre B, Raveyre C, Carrot C: Residence time distributions in a co-kneader: a chemical engineering approach, *Polym Eng Sci* 55:1237–1245, 2015.

Newman SG, Jensen KF: The role of flow in green chemistry and engineering, *Green Chem* 15:1456–1472, 2013.

Nogueira AL, Quadri MB, Machado RAF, Claumann CA: Experimental and CFD study of a vertically stirred tubular reactor designed for suspension polymerization reactions, *Macromol React Eng* 11:1–18, 2017.

Oechsler BF, Poblete IBS, Melo PA, Pinto JC: Mixing effects in continuous free-radical solution copolymerization tank reactors: I—characterization of residence time distributions, *Macromol React Eng* 12:1–25, 2018.

Padhiyar N, Bhartiya S, Gudi RD: Optimal grade transition in polymerization reactors: a comparative case study, *Ind Eng Chem Res* 45:3583–3592, 2006.

Palmiero UC, et al: RAFT macro-surfmers and their use in the ab Initio RAFT emulsion polymerization to decouple nanoparticle size and polymer molecular weight, *Macromolecules* 49:8387–8396, 2016.

Pan Y, et al: Preparation and adsorption behaviour of cationic nanoparticles for sugarcane fibre modification, *RSC Adv* 6:33554–33560, 2016.

Parida D, et al: Flow inversion: an effective means to scale-up controlled radical polymerization tubular microreactors, *Macromol React Eng* 8:597–603, 2014.

Pauer W, Moritz HU: Continuous reactor concepts with superimposed secondary flow—Polymerization process intensification, *Macromol Symp* 243:299–308, 2006.

Penlidis A, MacGregor JF, Hamielec AE: Dynamic modeling of emulsion polymerization reactors, *AICHE J* 31:881–889, 1985.

Perale G, Hilborn J: *Bioresorbable polymers for biomedical applications*, 2017. https://doi.org/10.1016/c2014-0-02798-0.

Petty, J. D., Huckins, J. N. & David, A. (12) Patent Application Publication (10) Pub. No: US 2002/0187020 A1. 1, 12–16 (2002).

PlasticsEurope: Available at: https://www.plasticseurope.org/it, 2020.

Plessis C, et al: Seeded semibatch emulsion polymerization of n-butyl acrylate. Kinetics and structural properties, *Macromolecules* 33:5041–5047, 2000.

Plessis C, et al: Kinetics and polymer microstructure of the seeded semibatch emulsion copolymerization of n-butyl acrylate and styrene, *Macromolecules* 34:5147–5157, 2001a.
Plessis C, et al: Modeling of seeded semibatch emulsion polymerization of n-BA, *Ind Eng Chem Res* 40:3883–3894, 2001b.
Richards JR, Congalidis JP: Measurement and control of polymerization reactors, *Comput Chem Eng* 30:1447–1463, 2006.
Roberge DM, Gottsponer M, Eyholzer M, Kockmann N: Industrial design, scale-up, and use of microreactors, *Chim Oggi* 27:8–11, 2009.
Rossetti I, Compagnoni M: Chemical reaction engineering, process design and scale-up issues at the frontier of synthesis: flow chemistry, *Chem Eng J* 296:56–70, 2016.
Rossow K, et al: Transfer of emulsion polymerization of styrene and n-butyl acrylate from semi-batch to a continuous tubular reactor, *Macromol React Eng* 10:324–338, 2016.
Roudsari SF, Ein-mozaffari F, Dhib R: Use of CFD in modeling MMA solution polymerization in a CSTR, *Chem Eng J* 219:429–442, 2013.
Sayer C, Palma M, Giudici R: Modeling continuous vinyl acetate emulsion polymerization reactions in a pulsed sieve plate column, *Ind Eng Chem Res* 41:1733–1744, 2002a.
Sayer C, Palma M, Giudici R: Kinetics of vinyl acetate emulsion polymerization in a pulsed tubular reactor. Comparison between experimental and simulation results, *Braz J Chem Eng* 19:425–431, 2002b.
Schoppmeyer C, et al: Operation of flexible multiproduct modular continuous polymerization plants, *Macromol React Eng* 10:435–457, 2016.
Schork FJ, Ray WH: The dynamics of the continuous emulsion polymerization of methylmethacrylate, *J Appl Polym Sci* 34:1259–1276, 1987.
Shen M, Sun Y, Xu J, Guo X, Prud'Homme RK: Rheology and adhesion of poly(acrylic acid)/Laponite nanocomposite hydrogels as biocompatible adhesives, *Langmuir* 30:1636–1642, 2014.
Sheoran M, Chandra A, Bhunia H, Bajpai PK, Pant HJ: Residence time distribution studies using radiotracers in chemical industry—a review, *Chem Eng Commun* 205:739–758, 2018.
Sponchioni M, Palmiero UC, Moscatelli D: HPMA-PEG surfmers and their use in stabilizing fully biodegradable polymer nanoparticles, *Macromol Chem Phys* 218:1–12, 2017.
Sponchioni, et al: RAFT copolymerization of oppositely charged monomers and its use to tailor the composition of nonfouling polyampholytes with an UCST behaviour, *React Chem Eng* 4:436–446, 2019.
Szkudlarek M, Beginn U, Keul H, Möller M: Synthesis of terpolymers with homogeneous composition by free radical copolymerization of maleic anhydride, perfluorooctyl and butyl or dodecyl methacrylates: application of the continuous flow monomer addition technique, *Polymers (Basel)* 9:610, 2017.
Tobita H, Hamielec AE: Polymerization processes, 2. Modeling of processes and reactors, *Ullmann's Encycl Ind Chem* 1:50, 2015. https://doi.org/10.1002/14356007.o21_o01.pub2.
Tonhauser C, Natalello A, Löwe H, Frey H: Microflow technology in polymer synthesis, *Macromolecules* 45:9551–9570, 2012.
Vega JR, Gugliotta LM, Bielsa RO, Brandolini MC, Meira GR: Emulsion copolymerization of acrylonitrile and butadiene. Mathematical model of an industrial reactor, *Ind Eng Chem Res* 36:1238–1246, 1997.
Vladimir H, Puszynski JA, Viljoen HJ, Gatica JE: Model reactors and their design equations. In *Ullmann's encyclopedia of industrial chemistry*, vols. 1–39, 2000.
Wang C, et al: Novel synthesis of carboxy-functional soybean acrylic-alkyd resins for water-reducible coatings, *J Coatings Technol* 72:55–61, 2000.
Wang C, Ma C, Mu C, Lin W: Tailor-made zwitterionic polyurethane coatings: microstructure, mechanical property and their antimicrobial performance, *RSC Adv* 7:27522–27529, 2017a.

Wang X, Tian B, Wang C, Wu J: Mathematical modelling of residence time distribution in tubular loop reactors, *Can J Chem Eng* 95:1101–1108, 2017b.

Wei XUE, Takahashi H, Sato S, Nomura M: Continuous emulsion polymerization of styrene in a single Couette–Taylor vortex flow reactor, *J Appl Polym Sci* 80:1931–1942, 2001.

Weng S, Huo T, Liu K, Zhang J, Li W: In-situ polymerization of hydroquinone-formaldehyde resin to construct 3D porous composite LiFePO4/carbon for remarkable performance of lithium-ion batteries, *J Alloys Compd* 818: 2020, 152858.

Wessels, P. United States Patent (19). (1977).

Williams PA: *Handbook of industrial water soluble polymers,* 2007. https://doi.org/10.1002/9780470988701.

Wilms D, Klos J, Frey H: Microstructured reactors for polymer synthesis: a renaissance of continuous flow processes for tailor-made macromolecules? *Macromol Chem Phys* 209:343–356, 2008.

Wolff HJM, et al: From batch to continuous precipitation polymerization of thermo-responsive microgels, *ACS Appl Mater Interfaces* 10:24799–24806, 2018.

Xue W, Yoshikawa K, Oshima A, Sato S, Nomura M: Continuous emulsion polymerization of vinyl acetate. II. Operation in a single Couette-Taylor vortex flow reactor using sodium lauryl sulfate as emulsifier, *J Appl Polym Sci* 86:2755–2762, 2002.

Yong Q, et al: Preparation and characterization of low gloss aqueous coating via forming self-roughed surface based on waterborne polyurethane acrylate hybrid emulsion, *Prog Org Coat* 115:18–26, 2018.

Yoshida JI, Nagaki A, Iwasaki T, Suga S: Enhancement of chemical selectivity by micro-reactors, *Chem Eng Technol* 28:259–266, 2005.

Zeaiter J, et al: Operation of semi-batch emulsion polymerisation reactors: modelling, validation and effect of operating conditions, *Chem Eng Sci* 57:2955–2969, 2002.

CHAPTER EIGHT

Perspectives from industry

Gianmarco Polotti*
Department of Chemistry, Materials and Chemical Engineering "Giulio Natta", Politecnico di Milano, Milano, Italy
*Corresponding author: e-mail address: gianmarco.polotti@gmail.com

Contents

Advances in Chemical Engineering, Volume 56
ISSN 0065-2377
https://doi.org/10.1016/bs.ache.2020.07.003

Abstract

Polymer production is a mature sector but several inventions are still introduced in chemistry, process and plant activities. An extensive review of the most recent patents is presented. A selection and a clustering of different topics was done. The aim of the selection is to balance the traditional activities and the most and radical novelties. The aim of clustering is to facilitate the reader search. For each topic, some patents are presented by summarizing their conclusions after a short presentation of the problems involved. We strongly suggest to refer to the original documents by the bibliography. Almost all documents are freely available on line.

Polymer production is a worldwide activity and its market is global from the very beginning. Historically, new developments in polymer science are protected by a complete and sophisticated patent strategies. So, observing the patent production is a reliable way to understand trends and tendency in industry. In the first 10 months of 2019 patents issued about polymer reaction engineering, processes and formulations, are approximately 400 accordingly with Espacenet database (EPO, 2019). The research is limited to Worldwide (WO), European (EU) and American (US) patents issued in English.

Topics involve the whole polymer production and they have been ordered herein in homogeneous paragraphs. A selection of the patents that present a higher level of novelty has been also highlighted. Few general observations are:

- Traditional areas of polymer production, such as polyolefins and polymer compounding, still issue many patents. Approximately, more than 40% of inventions are in these two areas. New developments deal with plant efficiency and tailor-made polymer production.
- Actual trends such as biomass refining (or biorefining) has become more prevalent in industry. We are now observing the commercialization of integrated biorefineries that are capable of processing incoming biomass much the same as petroleum refineries process crude oil. Polymers are clearly influenced by using these new raw materials.
- Polymers with recycling capacity have attracted great interests because the traditional polymer networks crosslinked by covalent bonds are unrecyclable and are generally treated by landfilling or burning after their

lifetime. Associative exchangeable covalent bonds or reactions, such as imine bond and trans alkylation are considered in many patents.

- A small niche of inventions deals with polymeric additives with a double function in production and application. This allows a reduction of chemicals and a more intensive use of each one.

1. News in polyolefin production

1.1 Molecular weight control

There is a surprisingly high interest in producing ethylene-based polymers with multimodal molecular weight distributions, because they may combine the processability of resin with a balance of desirable properties such as strength, barrier properties, etc. (Lyondell-Basell, 2020). Each of the following patents provide multimodal distributions in different kind of plants, processes conditions and formulations and many of them claim specific advantages. The general idea of making multimodal distributions is changing the reaction conditions in some way. Low density (LDPE), high density (HDPE) and linear low density (LLDPE) are involved in this trend.

Lammens et al. (2019) deals with high pressure gas polymerization of LDPE (Fig. 1). The molecular weight control is achieved providing more than one injection point for the initiator. A standard continuous polymerization process is modified by the addition of an injection pump with

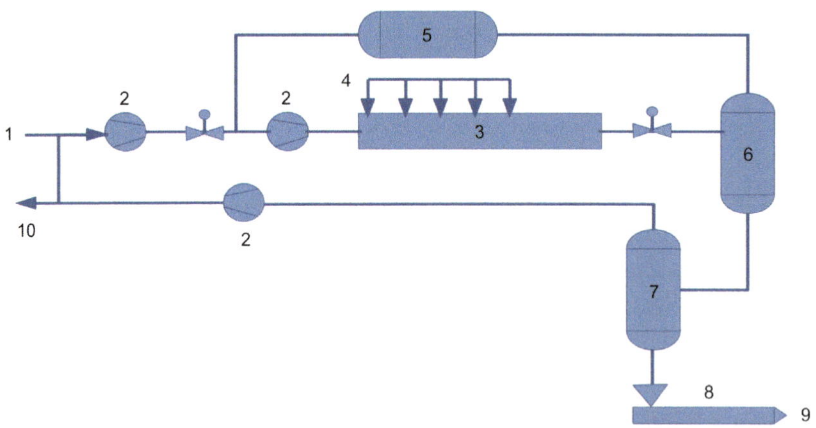

Fig. 1 High pressure gas polymerization of LDPE: 1-Monomer feed; 2-Gas compressor; 3-Plug flow reactor; 4-Initiator, chain transfer agent feeds; 5-Monomer purification; 6-Gas-polymer separation; 7-Polymer purification; 8-Polymer extrusion; 9-Polymer output; 10-Gas purge.

multiple entries along the tube reactor. But also, the reaction temperature is a key point for the molecular weight control. As the temperature rises, the chain transfer to polymer increases vs propagation and the polydispersity index increases. Unfortunately, polydispersity directly correlates with the haze of the polymer. Accordingly, when it is desired to manufacture a low haze grade polymer, it will be necessary to operate at low peak temperatures. Several injection points of the initiator, controlled by a temperature controller of the internal reaction zone, allows to modulate the temperature, so the polydispersity and the haze of the final product. Haze is an important parameter to define the optical properties of the polymer, fundamental for packaging applications. The lower the haze value the higher is the clarity of the polymer. Above 30%, the material is diffusing so it is considered opaque. Clymans (2019) extends the concept and the method by combining the initiator to chain transfer agent when necessary. Specifically, he describes a high pressure (100–350 MPa) continuous production in a plug flow reactor. A peroxide cocktail of up to five peroxide initiators active in a temperature range between 120 and 335 °C is used after dissolution in an organic inert solvent. Reactant feed (order: 160 t/h) is made by ethylene and eventually a comonomer (up to 10%) together with a fix amount of transfer agent (0.8% ca). The initiator solution is injected into the reactor not only at the beginning, as in the standard set up, but also in up to four points along the whole length of the tube reactor (diameter 65 mm ca, length 3000–4500 m). This causes different polymerization zones were different molecular weight polymers are formed. The peak temperature for each reaction zone will advantageously be in the range of 200–350 °C.

Gerhardus and Ulf (2019) deals with polymerization of ethylene, propylene and copolymers in gas phase at low pressure and with hydrogen as molecular weight regulator. The scope is always to achieve a polymer with a broad or even multimodal molecular weight distribution with advantages in blow-molding and extrusion applications. The inert gas is Nitrogen or propane and carbon monoxide or water are the inhibitors. Catalysts are Phillips chromium oxide or Ziegler-Natta types. The process is characterized by the presence of three or more polymerization zones which differ in the ratio of hydrogen to the sum of olefins in the gas mixture (Fig. 2). The ratio determines the molecular weight of the produced polyolefin: a larger ratio gives lower molecular weight and vice-versa. The amount of hydrogen varies among the zones up to 2 times or more, but it remains constant in a single zone. The growing polymer particles are made in a fluidized bed up to a minimal dimension. Then, they flow upward through a first

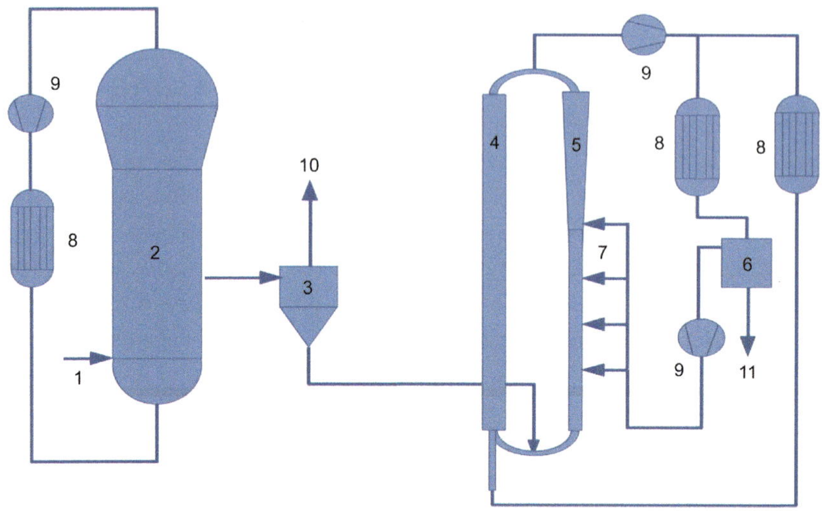

Fig. 2 Low pressure gas polymerization of polyolefins: 1-Monomer feed; 2-Fluidized bed; 3-Polymer particles separation and selection; 4-Riser; 5-Downcomer; 6-Monomer-Polymer separation; 7-Monomer recycle and Chain Transfer feeds; 8-Heat exchanger; 9-Gas compressor; 10-Gas purge; 11-Polymer output.

polymerization zone (riser) under fast fluidization conditions. Later, they enter a second polymerization zone (downcomer) through which they flow downward in a densified form. When particles leave the downcomer, they are reintroduced into the raiser, thereby establishing a circulation of polymer between the two units. The polymer passes a plurality of times through these two zones gradually changing its properties. In the downcomer the polymer particles are in a densified form (i.e. their bulk density is high compared to the one observed in the riser) and move in a plug-flow mode (i.e. with little or no back-mixing). Gaseous and liquid feed streams are introduced at one or more position to the downcomer to control the gas ratio composition. Streams contain one or more comonomers, inert gas or hydrogen. The riser is only needed to transport the growing polymer particles from the outlet to the inlet of the downcomer. So, it operates under fast fluidization conditions by feeding a reaction gas mixture at a velocity higher than the transport velocity of the polymer particles (0.8–5 m/s). In some cases, a solid/gas separator is arranged above the downcomer to separate the polyolefin and reaction gaseous mixture coming from the riser. In this way the gas composition is made uniform and ready to be modified by the new streams in the downcomer. To improve the difference of reacting conditions between riser and downcomer, a barrier liquid mixture is fed at the top of the downcomer.

This liquid may be the bottom phase of the gas mixture previously extracted and processed in a distillation column. The hydrogen rich gaseous fraction, withdrawn from the top of the distillation column, is fed in the ending part of the downcomer providing a significant boost of the hydrogen/olefin ratio. Bauch et al. (2019) suggests the possibility to regulate the hydrogen concentration in a recycle stream, where a catalytic reaction hydrogenates the unreacted monomer (e.g. propylene) to a correspondent alkane (e.g. propane). Rinaldi et al. (2019) instead claims the importance of controlling the bottom temperature of the riser. In this zone, they operate with temperature close to the condensation point of the reacting gas, in order to obtain a controlled amount of condensate. In this manner, the polymerization in the riser is not negatively affected and higher production rates are achieved in the downcomer.

The molecular weight control is also studied in solution polymerization using a standard loop reactor. Eswaran et al. (2019) and Wurnitsch et al. (2019) revise the ability to regulate reactor temperature and improve catalyst efficiency. In loop reactors, the reactor feed is introduced into the recirculation stream at a suitable point in the loop and the product is withdrawn at a different point from the recycle. The ratio by mass of the recirculating flow to fresh feed flow can be varied from 0 (to mimic the performance of a plug flow reactor) to 15 and above (to mimic the performance of stirred tank reactor). Authors suggest to feed Ziegler-Natta catalyst and activator solution by separate injection in at least two different streams. A more finely tuned control is achievable when feeds are more than two. Moreover, different feeds allow increased injection rates. The greater velocity (from 2.5 up to 20 ft/s) prevents the catalyst and the activator particles from settling in the feed lines and the increased number of injection points provides beneficial mixing.

Kuppuswamy et al. (2019) works on the composition of the catalyst instead than the number of feeders. The polymer structure is influenced by the catalyst: Ziegler-Natta catalyst and chromium-based catalysts often produce compositions with broad distributions, whereas metallocene catalysts typically produce compositions with narrow distributions. The same idea of working on the catalyst composition is adopted by Lee In et al. (2019) for the synthesis of HDPE in gas phase. However, he avoids multimodal molecular weight distribution, preferring a monomodal but extremely large distribution. This solution is optimal to obtain: high impact and transparency (provided by the low molecular weight fraction) together with hardness, heat and chemical resistance (given by the high molecular weight fraction). A single but broad molecular weight distribution (correlated with a high Melt Flow Index) is required for a good flowability in injection molding and extrusion. Since titanium-based Ziegler-Natta catalyst or chromium-based catalysts provide narrow distributions, the author provides combinations of catalysts to match the requirements.

Liu et al. (2019) produces a bimodal polyethylene distributed material, but it uses a quite simple methodology. He suggests to blend a fraction of polyethylene at low molecular weight (Mw = 20,000, Mn = 9000 g/mol) together with an equal or lower massive fraction of high molecular weight (Mw = 300,000, Mn = 16,000 g/mol). Two fractions may be produced by polymerizing ethylene in two cascade polymerization reactors that operate at different conditions. Alternatively, the two fractions may be produced in the same reactor by using two polymerization catalysts where the catalysts have different selectivity. The patent also releases the best application of this material in pipes to transport liquids and gas. Typically, pipes must be able to withstand pressure, because liquids or gas usually are pressurized. It is desirable to have pipes of good coilability facilitating easy installation and avoid fractures, if pipe is too stiff, fractures may occur. Therefore, the density of the material needs to be sufficiently low, but generally a low density corresponds to a poor pressure resistance. A good processability of the polyethylene composition is desirable and this is typical of low molecular weight materials. However high molecular weight is needed for meeting the requirements of good pressure resistance at elevated temperatures and low creep, even though processing of such high Mw resins alone is more difficult. A polymer with bimodal distribution is the easiest solution.

Dongming (2019) suggests that the blending of the two fractions of polyethylene should be done on melt state and gradually to preserve the molecular size. Dou et al. (2019) provides multimodal polyolefins distribution extending the number of blending components to three, among which one is an ultra-high molecular weight (UHMW) polymer. Typically, a UHMW has a molecular weight above 3 million and a density less than 935 kg/m^3, and its blending with lower size materials improves dramatically the mechanical properties. There are serious compatibility problems to incorporate high molecular weight to low molecular weight and the Patent seems to solve this problem adding an intermediate molecular weight component.

Although the blending of polymer with different molecular weights to broad the distribution seems to be an easy and efficient way, mechanical methods are limited by the physical mixing of the original polymers.

So, the mixture results different from a polymer having a similarly broad or multimodal molecular weight distribution produced in a single process. Pannell and Impelman (2019) suggests a further easy way to meet the target. Broad molecular weight distribution can be produced in a single reactor using at least two independently controlled catalysts feeders each supplying different proportions of at least two active catalysts. By controlling the catalyst feed rate, polymers having targeted molecular weight distribution can be readily produced. The composition of the catalyst system is detailed in the following section.

1.2 Composition control

Many polyolefins are copolymers of ethylene with other monomers (e.g. hexene). With copolymers, not only the molecular weight distribution affects properties but also the composition. The different composition influences: extractable content, environmental stress crack resistance, heat sealing, impact, tear and strength resistance. Much effort has been placed in understanding how the comonomer is distributed along the polymer carbon chain. The composition distribution of an ethylene alpha-olefin copolymer refers to the distribution of comonomer as a distribution of short chain branches. When the distribution of short chain branches varies among the backbone, the polymer is said to have a Broad Composition Distribution (BCD) and if the amount of comonomer is similar among the molecules of different molecular weights, the composition distribution is said to be narrow (NCD). Recently a third distribution has been documented as Broad Orthogonal Composition Distribution (BOCD) in which the comonomer is incorporated predominantly in the high molecular weight chains. Catalysts influence the polymer composition. Polymers made with Ziegler-Natta catalysts give broad distribution, but the high molecular weight fractions are higher density (i.e. less comonomer) than the lower molecular weight (high comonomer). In contrast, metallocene catalysts typically produce polyolefin with an NCD. Holtcamp et al. (2019) and the following patents focus on the production of polymer with BOCD distribution. BOCD behavior has been associated with a good balance of mechanical and optical properties. For example, LLDPE films strive for a good balance of stiffness, toughness, optical properties (haze and gloss) and processability. Sealing performance, key property in films, is affected mainly by density, it improves as density gets lower, but density has the opposite effect of stiffness. There is an interdependence between molecular weight and comonomer distribution on sealing performance and BOCD is the optimal solution. Ching-Tai et al. (2019) proposes a solution based on a combination of two specific metallocene catalysts and the method of their use. Boller et al. (2019) includes the possibility to use an activator (alumoxane compound) to convert the neutral metal compound to a catalytically active metal cation. Crowther et al. (2019) defines the comonomer types: C_3-C_{10} alpha-olefins and the temperature range: 10–75 °C.

Composition control can be achieved by a physical blend of different polymers as suggested by Brown et al. (2019) and Karjala et al. (2019). Blending LLDPE with LDPE increases bubble stability in part due to the high melt strength (MS) of the LDPE and the increase of MS helps provide for an increase of film output. There is a need to control the composition monitored by gel permeation chromatography (GPC) and find an optimized balance. Patents report equations useful to find a correct correlation between MS and blow film properties. Van Hoyweghen et al. (2019) blends ethylene α-olefin elastomer (65–90 wt%) having no crystallinity with isotactic polypropylene (10–35 wt%) to reach the optimal MS for film and other application.

Composition control is key when polyolefins are used for special applications such as the coating of low voltage electrical cable, object of (Macdonald and Werjefelt, 2019) work. Low Voltage Power cable operates at voltage below 6 kV. The conductor, typically copper, is coated with an insulation polyolefin layer by extrusion. Cables must resist to prolonged expose to wind and weather and the Oxygen Induction Time (OIT) is a good indicator of the resistance. OIT is a standardized measured of the time between melting and the onset decomposition in isothermal conditions (190–220 °C) achieved by Differential Scanning Calorimetry (DSC). The patent proposes a copolymer of ethylene and monomer with silane groups (VTES—vinyl trimethoxy silane, VTMS—vinyl trimethoxy silane) in the range of 0.5–3 wt%. Moreover, the copolymer must have a high molecular weight (express as MFR2 $<$ 10 g/10 min) and limit the number of short oligomers unable to crosslink in the extrusion. The copolymer is made in a standard process of high-pressure polymerization in a tube reactor with multiple injection feeds. The high pressure (up to 4000 bar) is achieved by piston compressors that require the use of a lubricant. As known, lubricants (typically PAG-poly alkylene glycols) leak through the seals, mix with the monomers and participate to the reaction as CTA. The Patent observes the influence of lubricant on the hot set elongation of the polymer, i.e. the permanent elongation after storage a 200 °C. Since an increase of elongation is detrimental for cable application, the inventor uses paraffinic hydrocarbons with MW lower than 480 as compressor lubricant, since their effect on the polymer is minimum.

Composition control is also a solution for LDPE onto paperboard extrusion laminations, where the advantage would be improved adhesion in the multi-layer structures, such as aseptic drink boxes, snack packaging, cheese packaging, etc. A good "fiber tear" adhesion is achieved when an adhesion between the polymer to the paper greater than the adhesion of the paper fiber to themselves. Strandberg and Palm (2019) claims that the optimal copolymer is made by ethylene with carbon monoxide (CO) having a content greater than 0 to <10 wt% CO on the whole polymer. Moreover, a low melt flow index (between 3 and 30 g/10 min) guarantees the fastest speed in extrusion. Such kind of copolymers are made by a high-pressure polymerization of ethylene in a plug flow reactor with several injection points of CO along all the tube length.

1.3 Optimal amount and type of condensing agents

There is a need and desire to increase production rates while maintaining the continuity of the microstructure properties (molecular weight, branching, composition, etc.) in a continuous process of gas polyolefin production. Brandl et al. (2019) and Pannell (2019) face this problem in a fluidized bed process and focalizes on the use of condensing agents. The condensing mode of operation in a fluidized bed reactor significantly increases the production rate by providing extra heat removal capacity through the evaporation of condensates in the cycle gas. Additional condensation is achieved by adding an inert condensing agent (ICA) into the reactor. ICA is a hydrocarbon with such as propane, isobutane or isopentane and their mixtures. Having a lower boiling point than the monomer, the ICA maintain the reactant close to the catalysts in the condensed phase. The amount of ICA must be kept below the "stickiness limit" beyond which the bed material becomes too sticky to discharge or to maintain a normal fluidization status. By modulating the amount of ICA into the reactor is possible to control the amount of the condensed phase and, consequently, the quality of the produced polymer (molecular weight, density, etc.). The patent provides a formula for the optimum vapor pressure for two or more condensing agents for increased production rate and increased polymer grade flexibility. The patent reports several data regarding the effect ICA blends on Melt index and density of the polyethylene produced. Impelman et al. (2019)

anticipated these conclusions suggesting to use a hydrocarbon coming from a countercurrent column placed on the gaseous recycle stream. The advantage of this column is to provide an efficient recovery of unreacted monomers solubilized in the liquid hydrocarbon and reinjected in the reactor together with a clean purge of inert components that do not accumulate in the plant. Finally, Drabish et al. (2019) and Mei et al. (2019) propose a under vacuum gas separator instead of the column for the pressure control and hydrocarbon recovery from polymer particles.

1.4 Silica quenching agent

A wide range of application to almost all the polyolefin synthesis is the improvement presented by Reimers et al. (2019). It describes a new method for quenching polymerization reactions by solid particulate such as amorphous silica. Polymerization processes typically require quenching agents to prevent further polymerization at the end. Unreacted monomers and solvents need to be subsequently recycled back into the reactor. Traditionally, small, polar, protic molecules such as water and methanol are used as quenching agents. However, their use causes both liquid-liquid and liquid vapor separation on the recycling stream. Such processing steps are undesirable as they increase the complexity of the process, often resulting in increased capital and operating costs. Patent proposes amorphous silica (with specific surface area from 30 to 800 m^2/g) as new type of quenching agent that does not partition into the recycle in a significant amount along with solvent and unreacted monomer. In fact, any reactive particulate having –OH groups on the surface may be useful as a quenching agent. The choice of silica is most likely done since its compatibility with the further applications of the polyolefin.

1.5 Antistatic for deposit reduction

Antistatic agents are used in process for the low-pressure polymerization of olefins to avoid electrostatic charging of the forming polymer particles. It is believed that the antistatic agents reduce wall sheeting and formation of polymer agglomerates in the polymerization reactor and downstream equipment. This is especially true in fluidized bed reactors and cycle reactors. Antistatic agents are made from or contain polar functional groups such as acid, ester, amine and hydroxyl groups. In spite of their low addition (500–1000 ppm), it is believed that some antistatic agents may negatively impact the activity of olefin polymerization catalysts. They may also preclude the use of polyolefins in food, beverage and medical packaging for the possibility to migrate outside the polymer. Mazzucco et al. (2019) proposes an antistatic agent without these negative effects, made by a water solution of low viscosity poly ethers (or polyglycol, poly alcohols). Notably, the water itself acts indistinctly on growing polymer particles independently from their size, while hydroxylated oligomers interact preferentially with the smaller particles that have higher capability to agglomerate and react with the surface. The antistatic agent is introduced directly in the reactor or in the recycle loop.

1.6 Highly reactive poly isobutene

Highly reactive poly isobutene (HR-PIB) is a request polyolefin, especially at low molecular weight, produced by cationic polymerization using Friedel-Craft type catalysts. Daire et al. (2019) shows a continuous polymerization process made by a cascade of two reactors: a first homogeneous pre-polymerization unit (HPPU) and a second polymerization in series (HERS). Both units share a heat exchanges and a coolant to control temperature. The HPPU unit is small and has a short residence time. It receives and mixes the reactants made by: isobutene (up to 30%), alkane solvent and catalyst (*t*-butyl chloride and initiator). The HERS unit can be a heat exchanger reactor with a temperature difference between coolant and internal reactants lower than 3 °C with a high resident time achieved by looping the fluid in the tubes. The tight temperature control is made using an evaporative coolant, i.e. a coolant that evaporates at least partially in the HERS. As typical of cation polymerizations, the process temperature is low and the coolant must be able to keep the temperature of the HPPU mixer below 0 °C and the HERS maximum temperature always below 100 °C. The pressure in the HERS is between 6 and 10 bar.

2. News in rubber production

2.1 Functionalized polymers for tire production

The increasing cost of solvent recovery pushes to reconsider the whole polymerization process. Polydienes may also be produced by bulk (or mass) polymerization, wherein conjugated diene monomer is polymerized in the absence of any solvent, and, in effect, the monomer itself acts as a diluent. This offers a number of economic advantages including lower capital cost for new plant capacity, lower energy cost and few people to operate.

The solvent-less feature also provides environmental advantages with emissions and waste water pollution being reduced. Bulk polymerization requires very careful temperature control, and there is also the need for strong and elaborate stirring equipment since the viscosity of the polymerization mixture can become very high. Tartamella et al. (2019) proposes a solution to these problems by a three-step continuous process. In first step diene monomers are polymerized by a lanthanide-based catalyst with low activity. The moderate reaction kinetic rate allows an efficient heat removal in the temperature range (20–100 °C) and a good control of the conversion up to 20%. Due to the lower amount of polymer the stream has a moderate viscosity and is easily moved to a second vessel. Here, a heterocyclic nitrile compound is added and acts as a radical terminator in mild condition (20–30 °C for 10–60 min). After a quenching with alcohols or water, the stream is sent to a devolatilization unit. A screw or paddle apparatus allows the complete residual monomer removal and the final processing of the polymer with a twin-screw extruder. The functionalized poly-dienes are employed in the manufacture of tires by standard rubber shaping, molding and curing techniques. They are employed to reduced hysteresis, i.e. the loss of mechanical energy to heat, and provide a low rolling resistance.

Padolewski and Tartamella (2019) proposes an intermediate solution. The polymerization starts as a bulk polymerization till the conversion reaches 30% ca. Then the reacting mixture including a polydiene and unreacted monomer is diluted in an organic solvent (e.g. aliphatic hydrocarbon). The polymerization is now a solution polymerization and advantageously provides effective temperature control, reduces the viscosity and increases the monomer conversion. This second part of the process can be carried out in a transfer pipe, in a dynamic or static mixer. Preferred set up of this process are: horizontal reactors are suitable for the bulk polymerization and a heated screw apparatus, such as a desolventizing extruder, for evaporating the solvent under atmospheric or sub-atmospheric pressure.

Mccauley et al. (2019) further refines the process focalizing on the quenching agent used to terminate the polymerization prior a significant monomer conversion. The ratio of water in the quenching agent to the lanthanide atoms in the catalyst must be <1500 to 1. A limited amount of water as quenching agent exhibits advantageous cold flow resistance. An excessive amount, as commonly employed, leads to decoupling of polymers that are believed to be coupled by the heterocyclic nitrile functionality. This results in a decrease of cold flow resistance which is problematic during tire storage.

2.2 Process improvements for rubber production

Rubbers are produced from multi-olefins and isoolefins by carbo cationic polymerization. The chemical mechanism is complex and involve an initiator and a Lewis acid such as aluminum trichloride. The type of monomers, the solvent and its polarity, the polymerization temperature and the specific combination of initiator and Lewis acid affect the propagation step. Industry has generally accepted a slurry polymerization in methyl chloride as a solvent, and the reaction is carried out at low temperature, generally lower than −90 °C. An acceptable low viscosity of the solution (40% polymer) has to be maintained enabling the heat of polymerization to be removed via heat exchange across the surface of the reactor. Several patents propose the use of condensed volatile solvent to improve temperature control. The evaporation requires a defined enthalpy and thus auto-control the temperature above the boiling point of the solvent. The evaporated refrigerant is typically recycled and used again. Hanns-Ingolf et al. (2019) proposes a process in ethane or carbon dioxide and gives all the details of the whole production in batch or continuous layout. Feller et al. (2019) defines optimal catalysts and process conditions.

When natural rubber is epoxidized under carefully controlled conditions, it can be converted to a totally new polymer with some properties more akin to specialty rubbers and some properties that appear advantageous for tire tread such as lower rolling resistance and better wet traction, even in combination. Maybe this is the undefined goal of (Sauty and Colvin, 2019) focused on a process for producing epoxidized polymers. The process involves dissolving a rubber in a solvent and epoxidizing the rubber in the presence of an epoxidation reagent having an aqueous phase pH in the range of 1.5–7.0. The solvent and the reaction conditions are selected such that the rubber is soluble therein and the epoxidized rubber self-precipitates. Solvents are liquid alkanes, e.g. n-pentane, the temperature is about −20 to 35 °C, the peracid is peracetic acid mainly. Several kinds of natural rubbers precipitate in 2.5 h after epoxidation and the material is achieved after steam stripping and drying.

A simple process for rubber epoxidation without the use of solvent or catalyst is presented by Binder et al. (2019). He found that if a unsaturated polymer and a peroxo acid are dry mixed in a mill or extruder or blade mixer the epoxidation may run up to 100% of substitution. The rubber could be any rubber with preference to halogenate polymer, the peracid could be peracetic acid (solid or in solution), the mixing temperature is self-maintained around

50 °C and the residence time 0.5 h approximately. In spite of the simplicity of the manufacturing process the epoxidized polymer may be further cures by any suitable method (e.g. sulfur-base cure) as the competitor materials.

Brominated isobutylene rubber (BIIR) is an elastomeric isobutylene-isoprene copolymer containing reactive bromine (Exxon-Mobil, 2020). It displays better co-vulcanization properties with unsaturated general purpose elastomers while maintain the physical strength and resistance of polyolefins. Bromo butyl compounds combine low permeability to the passage of air and moisture with high level of co-curing compatibility to elastomers. Now, BIIRs are very important ingredients in tire formulations and the interest is still growing. Nguyen et al. (2019) releases an interesting solvent free process to increase the impact resistance and thermostability of BIIRs. It functionalizes the rubber backbone by grafting pendant groups via an atom transfer radical polymerization reaction. In a 6-blade solid mixer (Brabender), the rubber is functionalized by a nucleophile such as thioacetate (e.g. potassium thioacetate) in presence of a catalyst (e.g. TBAB tetrabutyl ammonium bromide). The temperature is limited (up to 150 max) and the reaction time is quite limited (20 min).

2.3 Foam injection molding

Foam blow molding consists of using chemical or physical foaming additives with thermoplastic polymer pellets in an extrusion process. Chemical blowing agents are substances that releases gas upon high temperate conditions. Physical additives are inert gas such as Nitrogen or Carbon Dioxide. The extruder creates a single-phase fluid that is homogeneous while under the high temperature and pressure with the extruder. Upon exiting the ambient environment at the die tip, the sudden drop in fluid pressure results in the gas coming out from the melt and forming gas-filled cells. The cells ideally growth spherically and remain individual, yet numerous, as the extruded material is delivered in the mold and formed into its ultimate shape. The tendency of switch from chemical to physical blowing agents seems confirmed by Kishbaugh et al. (2019). They propose to feed the blowing agent (Nitrogen or Carbon Dioxide) in a new extruder hopper. The hopper feeds the polymeric material and, through a pressurized vessel, the gaseous blowing agent. In this way a better control, a complete mixing and a precise dosage are all achieved at the same time. The pressure level of the vessel is integrated in the electronic controller on board of the extruder. Processing techniques of injection or blow molding are discontinuous.

The screw may stop rotating after a charge of a new polymer or during the injection into a mold. Conventional blowing agent delivery systems inject instead blowing agent continuously and may lead to an uneven distribution of the blowing agent in the polymer melt. The production of microcellular foams is adversely affected if the blowing agent is not precisely controlled. A trial solution is positioning a valve close to the blowing agent port in the extruder and discontinuously introducing the gas. The frequent opening and closing of the valve, however, may limit the lifetime and precision of the delivery system. Burnham et al. (2019) has found an efficient solution to this problem. The gas is delivered from the source to the extruder by a flow restrictor and the pressure is measured and regulated at the entrance of the restrictor and only measured at the end. Blowing agent is introduced into the polymer when the pressure upstream to the restrictor is greater than the pressure downstream. If the downstream pressure is less, as it happens when the screw is stopped, the flow of blowing agent is null.

Rojek et al. (2019) relates to an extrusion process for the production of expandable vinyl aromatic polymer (e.g. styrene) adopted in thermal isolation. It is generally known that addition of athermanous fillers of heat absorbers (e.g. carbon black), heat scatterers (e.g. silica and titanium oxides) and heat reflectors (e.g. aluminum pigment and graphite) decreases the thermal conductivity of polymer. However, the presence of athermanous fillers often leads to a deterioration of the fire self-extinguishing and mechanical properties of the polymer foam. This problem is solved by an extrusion process for the production of expandable vinyl aromatic polymer granulate, comprising: a previous separate pre-mixing of each components with a quote of the vinyl aromatic polymer, followed by a granulation by extrusion. A preferred type of each re-mixing is a co-rotating twin-screw extruder (32D/40 mm with 2- to 6 mixing elements). Because of the separate addition each component, including the propellant, the process is highly flexible and allows for the processing of additives that have very different processing requirements. Kondratowicz (2019) increased the compatibility of polymer and filler proposing a calcinated anthracite pretreated with silane and siloxane on the surface.

Separate ingredient feeds although make the process more complex eliminate unwanted interactions, e.g. those between catalysis and additives. Drew (2019) exactly deals with this problem in the polyurethane foam reactions. Polyurethane foams have advantageous physical and mechanical properties but they are highly flammable and a complete pyrolysis can occur. Expandable graphite is formed from crystalline graphite flakes, which are intercalated with an expanding agent, such as sulfuric acid. When heated suddenly the sulfuric acid reacts with the carbon to form a blowing agent, which force the crystalline graphite apart expanding the structure a hundred times over. The expandable graphite is low density and non-flammable and acts as thermal heat shield insulating the underlying polyurethane foam and smothering any flame inside the foam. Therefore, conventional polyurethane foams incorporate expandable graphite as a flame retardant and premix graphite in the polyols feed. Unfortunately, the sulfuric acid included with the graphite is highly reactive with polyols and catalysts also dissolved. The effect is a decreased foam rise time and a reduced heat of polymerization. The patent suggests to split the polyol feed in at least two parts: one with dissolved graphite and the other with catalyst and other additives. All feeds are combined, in a controlled dispensing head, just before the extruder entry.

Polyurethane foams are also matter of (De Smet et al., 2019), that focus on the possibility of increasing "airflow." In some applications such as seating and bedding one often wants air to flow easily out of the foam when the foam is compressed and to re-enter the foam easily when the compressive force is released. The air bubbles made in the foam production are not enough big to provide this properties and large open cells in the foam are introduced via polymer polyols particles. These are round particles having multiple hydroxyl groups preformed and reticulated in the flexible polyurethane foam by the diisocyanates. Another main reason to incorporate polymer polyols particle is to increase load-bearing in resilient foam, i.e. the ability of the foam to support a compressive load. The patent synthetizes a styrene acrylonitrile copolymer in a continuous phase that includes one polyol like polypropylene oxide or polyethylene oxide polyols. The copolymer form particles with small controllable diameters that entangles polyols molecules.

It has long been desired to use reinforced high temperature plastic materials to replace metal in many articles especially in automotive industry. To be successful, the reinforced plastic material must have good mechanical properties and excellent dimensional stability, even at high temperatures. Weldlines (also known as knit lines see Fellahi et al., 1995) are intrinsically weak parts is submitted to load in most injection molded articles of even moderate complexity. They are generated when two separate melt streams join either in multi-gated molds or a consequence of flow around obstacles. The development of many interesting productions has been hampered by poor weldline strength. Zhujuan (2019) releases a polymer composition having an improved balance in mechanical properties at elevated temperature and mechanic strength at weldline. The invention consists: (1) a semi-crystalline semi-aromatic polyamide (SSPA); (2) a reinforcing agent, such as a glass or carbon fiber; (3) one or more additives (flame retardant, plasticizer, dye). The molded part with improved high temperature resistance may be an automotive part, for example a load bearing part or an engine part, or a part of electronic device, for example a part of housing or a frame.

Mueller-Cristadoro and Prissok (2019) releases composition and process to make rigid polymer foams suitable for high temperatures in the presence of moisture and high pressure such as in automotive engines. The foam is made by adding a standard polyol to a poly isocyanate component with polyimide groups and at least one Lewis base. The polyisocyanates is made by condensation of a polyisocyanate (e.g. MDI) with anhydrides of poly carboxylic acids such as the trimellitic anhydride or pyromellitic dianhydride after carbon dioxide release. Further gas is generated by the reaction of water with the isocyanate to form a carbamic acid that eliminates CO_2 to generate an amine. The CO_2 elimination is catalyzed by the Lewis base and provides the polymer foams rapidly and without addition of blowing agent. The amine-functional component present in the mixture and the isocyanate groups combine to form urea groups, while the hydroxyl of polyol form urethane groups with isocyanates. The polyurethane-polyurea polymer network reach a compressive stress at 10% relative deformation not less than 180 kPa.

Poly meth acryl imide (PMI) foams are widely used in composite material for light weight design in the aerospace, automotive, sport and medical device industries due to their light weight, high specific strength/modulus and high distortion temperature (80–240 °C). PMI foam is the optimal core material of the sandwich structure but its formation and moldability is under study (Tang and Chen, 2010). Kong et al. (2019) suggests a method by combining PMI particles with a thermoplastic resin and melt together by extrusion before molding covered by Willemann et al. (2019). The combination forms a flowable melt at temperatures at least 120 °C higher than the glass transition temperature of the thermoplastic polymer. The helpful effect of the thermoplastic that acts as a solvent, allows the extrusion and molding of articles with complex geometries impossible to obtain in alternative ways.

Polyethylene foams are utilized in footwear components as sole and midsole applications. Crosslinked ethylene-based polymers including ethylene-vinyl-acetate (EVA) copolymers have traditionally dominated this market since they can easily be foamed with a chemical blowing agent.

But cross-linked ethylene-based polymers are not recyclable and cannot be splintered because they are thermoset and not thermoplastic. Hu et al. (2019) proposes a mixture of ethylene-based copolymers that are not cross-linked and consequently fully recyclable. The components: ethylene/alpha-olefin multiblock copolymer made by hard (ethylene) and soft (e.g. octene) segments and a standard LDPE, are pre-modified by melting together with a peroxide (e.g. *t*-butyl peroxide) in an extruder. After the addition of a physical blowing agent (e.g. carbon dioxide in supercritical conditions) foam bids are produced with a very low density ($0.3\,g/cm^3$), but high strength (30 cN). The footwear soles are later modeled by hot sintering.

3. News in bio polymer and process

3.1 Biobased polyesters

Polyethylene terephthalate (PET) is the most common thermoplastic polyester used for its excellent combination of clarity, mechanical and gas barrier properties. Depending of its processing and thermal history, PET may exist both as amorphous (high transparency) and as a semi crystalline (low gas permeation) polymer. The polymer is synthetized by terephthalic acid and ethylene glycol or by transesterification between ethylene glycol and dimethyl terephthalate. To modulate transparency and workability some further ingredients (such as cyclo hexanedimethanol or isophthalic acid) can partially replace the main ingredients. This interferes with crystallization and lowers the polymer melting temperature and these comonomers are usually named "crystallization retarders." As a result, a comfortable blow molding provides articles clear and crystalline enough to be adequate barriers to gas as carbon dioxide. Due to the general trend of replacing petrochemically-derived raw materials by raw material coming from biologically based materials (or biomaterials) and thanks to the huge PET market, a vast research is running (Borgemenke et al., 2019). Briois (2019) grants the rights to cover many of these results. It relates to a process for producing PET from at least one ingredient (among terephthalate compounds or glycols) obtained from a bio-based material including compositions with the most common crystallization retarders. According to the invention, the process is carried out in industrial facilities already used for processing PET from fossil origin. The only difference is that all the starting materials are bio-based compounds, i.e. obtained from vegetal source. To make the definition clear, the patent specifies several path-ways already know in literature to obtain terephthalic acid and ethylene glycol by natural source. Sookray (2019) limits the origin of PET to the polymer chains having terephthalate containing carbons sourced from bio-based ethanol. He specifies how to detect the origin of the carbon used in the polymer by the method ASTM D6866. Isotope ^{14}C, oxidized into carbon dioxide, represents a small but measurable fraction of atmospheric carbon. Atmospheric carbon is cycled by green plants to make organic molecules during photosynthesis. Therefore, the ^{14}C that exists in the atmosphere becomes part of all life forms and their biological products, such as renewable based organic molecules. Petroleum-based carbon does not have the signature radiocarbon ratio of atmospheric carbon dioxide (or $<0.1\%$ of atmospheric amount) since the isotope already decomposed.

The only alternative left for the PET replacement is to find a polyester with different chemical composition. As PET alternatives, polyesters made by furanic monomers are highly considered (Poulopoulou et al., 2018). They arise from acid-catalyzed thermal dehydration of sugars. Also, the standard production of polyesters from furan dicarboxylic acids and diols by polycondensation reaction is revised. Nising et al. (2014) disclosed the possibility of making furanic polyester oligomers in a ring closing oligomerization step. Costa et al. (2019) discloses the process of making a polyester by using the previous polyester oligomers having furanic units together with a plasticizer. By this route, the process does not produce large quantities of water or alcoholic side products nor it requires high capacity devolatilization units. The plasticizer plays a key role not only reducing the viscosity and the melting point of the reactants, but improving the intimate contact with the catalysts. Plasticizers are optionally -substituted phenyl ether or xylene, a polyether (e.g. tetra ethylene glycol dimethyl ether) and mixtures. In some cases, the plasticizer could be a low molecular fraction (<1500 g/mol) of the resulting polyester itself. Finally, Sipsos (2019) and Van Berkel (2019) observe that semi-crystalline furanic polyesters undergo to solid state polymerization. Gradually increasing the temperature of polymer pellets from 90 to 200 °C (1–8 °C/h), the molecular weight of the polymer proportionally increases. Useful observation especially in polyester reprocessing.

3.2 Physically cross-linked lactide polymers

Lactide polymers are interesting since they are biodegradable and biocompatible. However, Lendlein et al. (2019) found another peculiar property i.e. the physical cross-linking in semi crystalline form. Polymer networks are made by covalent net points such as in rubber or carbomer, but non-covalent net points are possible. In semi-crystalline polymers for instance, crystallites may serve as temporary net points. Networks provide an elastic behavior useful in application even after molding (extrusion or 3D printing). The patent proposes a blend of two stereo complexes: poly-D-lactide (PDLA) and poly-L-lactide (PLLA) with a melting point crystallinity in the range of 200–230 °C. Both polymers are prepared by a solvent-based process. After the preparation of the blends a thermal post processing treatment is required (tempering) in order to obtain crystallites in sufficient number. Typically, the tempering is performed in a temperature range of 60–120 °C and required a long time.

3.3 Cellulosic fibers via ionic liquid solutions

Fibers made from renewable resources such as cellulosic fibers are very well know but their market will certainly improve in the future. Abels et al. (2019) focalizes on a new dry-wet process that allow s a wide performance property optimization (high strength and good elasticity). In a dry-wet spinning process, the fibers (obtained by the kraft process from wood) are made by extrusion, they pass through an air gap for partial hardening and finally they run into a coagulation bath which contains a precipitant for the polymer. The novelty of the process is mainly in the use of ionic liquid salt which has a melting point of <100 °C (e.g. ENIM: 1-ethyl-3-methylimidazolium octanoate) to dissolve the cellulose (degree of polymerization 400–800) prior the extrusion. The spinning solution can be produced by conventional processes, e.g. cellulose may be mixed with ionic liquid and dissolved at elevated temperature (110 °C). Cellulose is 10 to 14 parts by weight per 100 parts of ionic liquid. A single core-progressive screw extruder makes the fibers from the solution. The pressure reaches 200 bars at the die having a hole diameter of 60–90 μm. Two rotating rollers, which are arranged one above the other (godet duo) drive the fiber to the air trap and later to the coagulation batch. When immersing in the batch with a non-solvent (typically water), the fiber still contains ionic liquid and release it there. In this way ionic liquid accumulates over time in the coagulation batch and when it reaches the 30 wt% is recovered by distillation.

3.4 Ionic polymers for biomass degradation

Ionic liquids (IL) used to dissolve cellulose must be separated from the products after decomposition. Due to their toxicity the separation could be costly if food grade and human consumption sugars are expected. In order to facilitate the recovery of ILs, polymer of the component ILs have been proposed under the name of Ionic Polymers (IP). Dyson et al. (2019) shows a series of IPs functionalized with sulfonic acid (–SO_3H), that are very efficient catalysts in acidic cellulose hydrolysis. The polymers are made by radical solution polymerization on modified IL salts; a further polymerization with acrylic acid can bring to a film forming material and finally to a membrane. The biomass made by cellulosic residuals is heated in the presence of the IP in a water solution of a mineral acid. The decomposition brings lipids, sugars or furanic compounds depending of the initial vegetal source. The preferred application drives to sweetening or flavoring agent, bulking agents or substrates for fermentation process.

Siankevich et al. (2019) moves in the same area but avoids any separation because it incorporates IPs in membranes for the biomass process. Several IP structures are modified in order to add some allyl or vinyl groups (e.g. acrylic acid) together with the sulfonic groups. Then, either the IP itself may form a membrane or IP may be incorporated in a preformed membrane. In some cases, IPs are adsorbed on a solid support that is used a inorganic filler.

3.5 New melanin compound

Pimpalkar et al. (2019) shows how to provide a process for preparing a homogeneous compound of a synthetic polymer and melanins. In recent years, there has been interest in extracting, using and compounding biopolymers: melanin, cellulose, chitin and collagen. Melanins constitute a fascinating class of biopolymers that is found in most organisms including plants, fungi, animals and even bacteria. They have been attributed to several different functions that include adsorption of ultraviolet radiation and dissipation into harmless heat, metal ion chelation, thermoregulation, antibiotic activity and free-radical quenching. Natural melanins are amorphous pigments that are insoluble in almost all organic solvents and this has caused few attempts to create functional polymeric materials and composites which incorporate this biopolymer. The patent suggests the dissolution of polymer and melamine in tetrabutylammonium hydroxide solution under continuous stirring followed by incorporation in film, fibers, coating and shaped forms. Tetrabutylammonium hydroxide (TBAH) is a common strong bases used in water and alcohol solution in organic synthesis. The polymer is cellulose, silk, natural polysaccharides and proteins but (Shanmuganathan et al., 2011) extend to the synthetic class. Both natural melanin (from *Sepia officinalis*) and synthetic melanin derived by auto-oxidation of 3,4-dihydroxy-L-phenylalanine (DOPA melanin) are used.

3.6 Polymerization of β-butyrolactone

Poly 3-hydroxybutyrate (PHB) is a biopolymer produced by microorganisms as a strictly isotactic polyester of the monomer 3-(*R*)-hydroxybutyric acid. Interest comes as biodegradable and renewable source with thermoplastic properties similar to polypropylene. Up to now the fermentation of glucose is the main production way to produce PHB but (Adams and Rieger, 2019) proposes a synthetic route. β-Butyrolactone is a monomer obtainable by addition of CO on propylene oxide, i.e. largely available species. He found a catalyst/initiator system comprising a rare earth metal, a chiral ligand and optionally an alkali-based *co*-catalyst that allows to polymerize β-butyrolactone to obtain isotactic PHB with a value grade of crystallinity (above 60% ca). The catalyst/initiator system can be prepared in situ, i.e. the components of the system can be added to the monomer composition and the system provides the catalytic activity directly.

3.7 Catalyst in biobased polycarbonates production

Carbon dioxide (CO_2) is a highly attractive feedstock when copolymerized with epoxides to yield aliphatic polycarbonates (Cui et al., 2019). The synthesis requires a metal complex catalyst and the process result in a crude product containing the catalyst, which can degrade and discolor the polymer over time. Leeland et al. (2019) proposes a new bimetallic catalyst and (Kember et al., 2019) found a method to deactivate and remove the same catalyst. The catalyst can be quenched with an acid containing an anion that is effective to act as an initiator for the polymerization but is also effective to deactivate the catalyst. The molar ratio between the acid and the catalyst must be less or equal than 20:1. It is thought that a complex is formed between the catalyst and the anion of the acid (for example acetate, if acetic acid is used as the acid). Once, CO_2 is removed from the reaction, the presence of the catalyst polymer complex can lead to degradation of the polymer chain. Instead, the formation of the complex between the catalyst and the anion of the acid does not affect the polymer degradation. Most preferably, the acid may be a multifunctional carboxylic acid, such as dicarboxylic, tricarboxylic and salicylic acids.

3.8 New super adsorbent polymers

Being products, which adsorb aqueous solutions, super-adsorbent polymers (SAP) are used to produce diapers, tampons, sanitary napkins and other hygiene articles, but also as water retaining agents in market gardening.

SAP can adsorb and retain (becoming hydrogel) extremely large amounts of a liquid relative to their own mass measured in term of g of water/g of polymer. Synthetic super-adsorbents are based mainly on the bulk polymerization of melted maleic anhydride with comonomers like: methacrylic acid and styrene together with monomeric or polymeric cross-linkers. Suitable reactors are kneading reactors with contra-rotatory stirrer shafts or belt reactors.

The degree of crosslinking affects the adsorbent capacity and gel strength of SAP material. Polymers exhibiting inadequate gel strength will form a hydrogel that deforms inhibiting the fluid distribution throughout the article. Polymers having a low adsorbent capacity are incapable of adsorbing a sufficient amount of fluid especially in thin articles. Moreover, a high amount of low molecular weight polymers can leach out from the gel resulting in lower quality. The Absorbency Under Load (AUL) is the typical physical parameter that characterize the quality of SAP together with the centrifuge retention capacity (CRC).

Daniel et al. (2019) releases a new process to achieve materials with superior properties: CRC from 35 to 75 g/g and AUL from 20 to 50 g/g difficult to obtain in combination. The process requires a thermal post-crosslinking treatment on spherical polymer particles. Although the surface crosslinking is efficient since it is made at low temperatures (100–180 °C), the requirement of spherical particles makes the polymerization more difficult. Spherical particles are produced by polymerizing droplets of the monomer solution in a surrounding heated gas phase. Droplets are generated by an oscillating plate with a multitude of bores. A hot gas flows cocurrent to the free-following droplets from the top downward. This leaves a 15% unreacted monomer that need to be converted in a following fluidized bed. As a particular advantage of the round shape particles is the lack of edges that can be easily be broken by processing stress in diapers production and during swelling. There are no breakpoints on the surface that could lead to loss of mechanical strength.

The environmental impact of baby diapers is becoming an impressive problem due to rapid growth. So, Bucevschi et al. (2019) proposes a new biodegradable super-adsorbent compound. It adsorbs aqueous solution through hydrogen bonding up to 100 g/g. The water-soluble composite is made by:

- A standard linear poly (styrene-maleic acid) copolymer (SMAC)
- A natural biopolymer such as gelatin (10–15 wt%)
- Repeated thermal cross-linking operations after coating with PEG-200 or other hydroxyl based biodegradable compounds.

The process uses the standard reactors and does not requires plant upgrades. The material respects the standard technological requirements, i.e.:

- Absorbency under load (AUL) > 20 g/g
- Granulometry >100 μm

and add a correspondent biodegradation property up to the 20–22%.

Hygiene articles such as diapers and incontinent pads are designed to adsorb bodily fluids such as urine and consequently malodor is a common issue. Numerous attempts have been made through appropriate additions of additives for odor-inhibition or odor-reduction. The use of lower pH additives, enzyme inhibitors and chelants are only effective in delaying the formation of ammonia from urine and have little effect on organic molecule odor. Malodor is caused by large variety of chemicals including ketones, aldehydes, amines and mercaptans. One way could be a radical oxidation of the whole species. Loick and Smith (2019) releases a post treatment of the super-adsorbent polymers by the addition (up to 1.7%) of a peroxo compound. Both inorganic (e.g. peroxo mono sulfate triple salt) and organic (e.g. peroxo carboxylic acids) substances are considered. The additive may be added at the end of production process or just before the drying step.

4. News in polymer coating

4.1 Opaque coating compositions

Powder coating compositions is a class of paints that are substantially dry, finely divided, free flowing, solid materials have gained considerable popularity in recent years over liquid coating for a number of reasons. Powder coatings are environmentally friendly materials since they are free of harmful volatile organic solvents and eliminates emission problems associated with liquid coating. Moreover, powder coating compositions are fully recyclable and reusable since the sprayed powders can be reclaimed and recombined with fresh powder feed. Powder coating compositions are finely divided particles of a polymer and a crosslinker in case of a heat radiation curable powder coating composition that also contains pigments, fillers and other additives. In a variety of applications such as white goods (e.g. fridges, washing machines, dishwashers) coating with matt finish are required or highly desirable. These formulations pose a series of challenges for the formulator such as poor physical storage stability of heat curable powder, inconsistency of the matting effect and too high dependency of the curing conditions. Koning et al. (2019) provides a matt finish formulation by using only one heat-curable powder mixture (said 1 K component) and not a physical blend of two different, separate and distinct heat-curable formulation (2K components). The use of just one component simplifies the coating application process, increases capacity of a line and reduces dramatically the costs for preparing coated articles with matt finish as liked by the end-customers. The author makes a choice of selected chemicals among a large number of ingredients, such as:

- primer polyester (e.g. tri mellitic anhydride and cyclohexene oxide)
- copolymerizable polyester (i.e. with an excess of carboxylic groups internal to the backbone)
- cross-linker with at least two 3-hydroxyalkylamide groups.
- Standard matting agent (e.g. titanium dioxide)

Very important is also provide a powder with controlled and uniform particle size ($<180\,nm$) to increase the storage stability of the final product maintain the cure temperature below 255 °C.

Opaque effect is also required in liquid paints, where the pigment is suspended in water together with a polymeric binder that converts in a solid film after application to a substrate in a thin layer. Titan dioxide (TiO_2) is frequently used to provide opacity besides whiteness in water-borne coating compositions. It was found that the opacifying capacity is a function of the spacing between the particles of the pigments in the dried applied coating. This capacity is maximized when the light scattering of the particle is maximum. This happens when the particles have a certain diameter and spacing, so that the light scattering capability of each particle does not interfere with the light scattering capability of its neighboring particles. This condition occurs in coating containing sufficiently low level of TiO_2 such that the individual particles are isolated from each other. But such low level of TiO_2, however, does not provide sufficient whiteness and hiding at typical dried coating thicknesses. Achieving the desired levels of whiteness requires high levels of TiO_2 and a consequent close proximity of the particles on the surface. In short, the efficiency of the TiO_2 as opacifying pigment is reduced when the particles are not homogeneously dispersed in the coating composition. Particles tend to agglomerate upon film formation and drying. It has been discovered that the spacing among particles can be improved by adsorbing a polymer latex on the surface of TiO_2 particles. Rost et al. (2019) designs a binder made by an emulsion polymerization of acrylic and methacrylic esters with styrene. The procedure to produce this binder is carefully detailed. It starts with a styrene seeded polymerization and alternate additions of acid monomer (e.g. methacrylic acid) are superimposed to a continuous feed of the other monomers (e.g. methacrylic esters). The binder is able to adsorb on the TiO_2 particles without the need of phosphorous containing monomers as reported in literature.

Lundgard and Mecca (2019) avoids agglomeration of TiO_2 particles coating them by a thermoplastic polymer. The polymer is melted in an extruder together with the inorganic particles. It is interesting to note that the water dispersion is made in the extruder itself by pumping water gradually. A concentrated dispersion is obtained for future incorporation in the water-based varnish.

A polymeric matte dispersion suitable as a clear top coat for leather is described in Bohling et al. (2019). The aqueous dispersion of polymeric microspheres (60% dry content—diameter between 1 and 25 μm) remains stable and flowable only if the particle size distribution is narrow. In fact, particles with reduce size (below 500 nm) promote the formation of gel. The dispersion is made by a seed polymerization of standard acrylic monomers and esters. The key finding of the Patent is the use of some reactive surfactants belong to the class of either a nonionic or anionic polyalkylene oxide salt of a distyryl or tristyryl phenol.

Finally, Syurik and Lscher (2019) does not attributes too much importance to the polymer composition (polystyrene, polycarbonate and polymethylmethacrylate can be equally used) vs the relevant aspect of the surface roughness. It proposes a diluted polymer solution (2 wt%) with water and a volatile organic solvent such as acetone, ethyl acetate or chloroform. Preferred molecular weights of the polymer are in the range of 120,000–200,000 g/mol with narrow distribution. A film is formed by spraying in cold condition (15 °C) and evaporating solvent between 20 and 40 °C. The white conditions are achieved if the porous film has a microstructure comprising a plurality of interconnected nodes which have and average diameter of 150–1000 nm and a plurality of rod-shaped elements (rod like) having an average diameter of at least 40 nm and length of at least 700 nm.

4.2 Varnish with high solvent stability and low tendency to soiling

A common effort is to find coating with high solvent stability and low tendency to soiling or elevated resistance to staining. Gashler et al. (2019) proposes a varnish made by a polymeric water dispersion. The dispersion is produced by a water polymerization in the presence of a carboxylic reach polymer particles, previously synthetized. The polymer particles increase their size by the addition of the new forming polymer on the surface.

High solid content and considerable particle size (ca 650 nm) are obtained. The varnish is then sprayed on the surface and tested. Limited tests made only on ethanol and coffee show a superior performance vs reference products.

Bobaru et al. (2019) moves to the same goal working on the physical properties of the polymer surface of plastic household appliances. Altering the surface structure of a polymer substrate allows variation of the surface properties. By creating micro-scale and nanoscale structures the hydrophobicity of a surface changes as such that it becomes highly water repellent. Since microorganisms live in the water, a water-repellant surface is safer. Inspired by the structure of lotus leaves, the patent proposes a surface covered by micro-pillars high between 5 and 15 μm and distant <1 μm. The surface is imprinted by a Thermal Nanoprint Lithography (Th-NIL). Th-NIL is a mechanical forming technique that physically molds deformable material into a desired shape. The method is applicable to the most common thermoplastics.

A more drastic approach is utilized by Hyberger and Scheltjens (2019). Fluoro-acrylate and silane monomers are sprayed and surface-polymerized by plasma radiation. Inventors claim not only the hydrophobic effect of the polymer but also physical repulsion due to the multi-stacked nanostructure of the polymeric layer.

4.3 Compact process coating for automotive

Traditional automotive paint process requires up to five layers, each followed by a heating cured period. Each layer provides a specific function such as chip resistance, UV opacity, substrate smoothing, coloring, etc. The tendency in the last decade is to reduce the number of intermediate bake and cured cycles and so the energy, cost and complexity while maintaining high level of optical quality and appearance. The general name given to such modified paint processes is "Compact Process." Lutz et al. (2019) proposes a two steps process based on water-based core-shell particles and a cross-linker. The polymer core of the particles is made by at least 20% of vinyl ester and amides of carboxylic acids and lactams. Such high level contributes to the smoothness and chip resistance of the resulting composite coating. The core may be internally crosslinked through the use of monomers (up to 10%) having multiple double bonds such as glycol di methacrylate. The shell is designed to be more polar using monomers having hydroxyl and acid groups able to undergo a later external reticulation. The second layer in the procedure is made by a more traditional polyester, polyacrylate or polyurethane. The crosslinking agent includes poly isocyanates, polyacids and anhydrides able to react with the high variety of hydroxyl groups on the surface when the temperature reaches 140 °C.

4.4 Low shrinkage polyisocyanurates

Processes known of polyisocyanurate plastics from monomeric aliphatic di isocyanates (e.g. HDI 1,6 -di isocyanato hexane) have the fundamental disadvantage that a considerable shrinkage in volume occurs in the course of a trimerization reaction, which can present problems particularly in the casting of bodies of defined geometry and in film protection. Laas et al. (2019) protects the idea of preforming the aliphatic polyisocyanurates and produce the film using the polyisocyanurate together with a small amount (20 wt%) of monomeric diisocyanates. The comparatively low monomeric isocyanate produces less heat of reaction and facilitates the production of large-volume. Preformed aliphatic polyisocyanurates are achieved as a modification of the isocyanate synthesis and are a mixture of several coupling structures (uretdione, isocyanurate, allophanate, biuret, iminooxadiazinedione and oxadiazinetrione).

4.5 Radiation curing

Liska et al. (2019) proposes new initiators and process conditions for the photo-polymerization of epoxides and polyols. It combines UV-induced cationic photo-initiators together with a thermal frontal polymerization. Important advantages of the cationic UV-cure systems are improved adhesion due to the low shrinkage of film (less than 2–5%), low viscosity reactants, solvent-free formulations and high production rate. However, photo-induced polymerizations are limited to thin layer articles since the light has a limited penetration. The addition of a selected thermal initiator allows to start a frontal polymerization that makes possible to cure thick samples of epoxy monomers (Mariani et al., 2004). The coating of optical fibers and fabrication of printed circuit boards may have a beneficial impact. When the reaction is carried out using the combination of a cationic photoinitiator and a thermal free radical initiator, the initial irradiation with UV-light triggers a reaction cascade. A front will develop at the interface between the polymers which have already been generated and the monomers which have not yet been reacted; as long as sufficient heat is related to decompose the thermal initiator activates the cationic initiator and new cationic polymerization reactions will continue. If the two initiators are available in sufficient amounts, this front will propagate through the entire mixture which is to be polymerized, until the polymerization tops due to the lack of further unreacted monomers. The patent achieves this aim by the use of standard cationic UV-curable photo initiators (as diaryliodonium salts) and a benzo pinacol (tetra phenyl ethylene diol—TPED) derivate as thermal initiator.

A new photocurable material is presented by Hagiwara (2019). Achieved by grafting an organo-siloxane on a polyvinyl alcohol backbone the curable material has good film forming properties, transparency and solubility in organic solvent. In short, the invention makes curable the polyvinyl alcohol already widely used in films, sheets and containers. When radically curable organo-siloxanes graft polyvinyl alcohol, the crystallinity of the polymer is lowered, the solubility in organic solvents increases and the reactivity with modifying agents becomes higher, so that the new material becomes inexpensive and excellent. The grafting is made possible preparing first an organo-poly siloxane containing an isocyanate group. The hydroxyl groups of the polyvinyl alcohol are then effectively reacted with the previously prepared siloxane.

Photocurable polymers have found new applications in the lithographic process for the photofabrication of three-dimensional (3D) printing, especially in biomedical applications such as "Lab-on-Chip" (Chee et al., 2015). In general, lithographic processes involve the use of UV light to locally cure a polymer at specific locations in superimposed layers to form 3-dimensional articles. Photocurable polymers are liquid at room temperature and easy to use in a printer, but the articles have moderate mechanical properties and thermal stability. Pollino and Jeol (2019) tends to fill this gap by a functionalized poly aryl ether sulfone (PAES). The polymer backbone such as a poly biphenyl ether sulfone is prepared by condensation in presence of a base (4,4′-dihydroxy biphenyl bi phenol and 4,4′-dichlorodiphenyl-sulfone). The polymer is then functionalized by addition of an alpha olefin (1–10 carbons) at least at one terminal of the molecule. The olefin comes from an halogenate compound (e.g. 4-bromo-1 butene) able to replace the hydroxyl ends, in organic solvent substitution. This application is probably part of a broader strategy of introducing PAES as block copolymer in many applications including reactive extrusion as proposed by Jeol and Pollino (2019).

4.6 Polymer and process for electronic

Electronic uses high-quality, high-purity polymers in almost all devices and production steps. Many applications require the assembly of polymers in thin films with different layers having different features (domain spacing and morphology). These are combined with either "bottom-up" (substrate modification) or "top-down" (lithography) techniques for locally controlling the morphology.

In the production of Integrated Circuits (IC), circuits are printed on a silicon wafer partially covered with a complementary metal-oxide-semiconductor (CMOS). The production starts etching holes by a reactive ion beam (DRIE) in a front side of the CMOS layer. Then, holes are filled with a sacrificial material (photoresist) to provide a planar frontside surface for further steps. North et al. (2019) suggests to make a preliminary layer of a thermoplastic polyimide and warm it above its glass transition temperature (ca 90°C) before making the photo imageable polyimides layer. Thermoplastic polymer, which has a lower viscosity than the photo resists material, can be reflowed when heated and fills the holes in a uniform way. With the same purpose, Urano et al. (2019) makes a uniform film of polyimide by UV curing from a low viscosity polyimide precursor (tetra carboxylic acid diester). As alternative, Dos Santos Oliveira et al. (2019) introduces a new polymer made by repeated aromatic unit (benzene or naphthalene rings) connected by Nitrogen as planarizing agent. Finally, Maruyama and Kondo (2019) simplifies the whole procedure with a single photosensitive polymer instead of the double layer of polyimides. The polymer has a silicone structure containing groups able to better cover the holes and smooth the surface before hardening under the effect of light.

The wide spreading flash memory market and the demand for increased storage capacities drive forward process improvement. Two or more circuit patterns by photoresist lithography are produced in the same line thanks to a continuous improvement of miniaturization technology going toward 10nm. Extreme ultraviolet light of wavelength 13.5nm is capable of forming pattern with a size of the order of 10nm via single exposure, but suffers from the problems of still low laser power and low productivity. The request of new polymeric materials required for the circuit design by negative lithography is very high since existing materials are insufficient for this purpose. Negative lithography requires etching a substrate by using a resist pattern as a mask. The resist pattern is obtained on the substrate by irradiating a thin-film with light to induce crosslinking thereby remarkably changing the solubility of the thin film. The following development treatment with a solvent remove the excess of the unreacted monomer. Fukushima et al. (2019), Hasegawa et al. (2019), and Sagehashi et al., (2019) propose new monomers (and related polymers) forming a negative pattern in alkaline developer and having the required etch resistance. Authors insert in the monomer a substituent group (fluorine based) capable of polarity switch under the action of acid. The polymer comprising the monomer has a high affinity and dissolution rate in the alkaline developer liquid. But the polymer eliminates the fluorinated group with the acid generated by an acid-generator activated by the exposure to light. After elimination of the fluorinated group, the polymer forms a negative pattern not more soluble in alkali.

The lithography using electron beam has a specific problem of electrification phenomenon (charge-up) during exposure. This happens when the substrate, irradiated with electron beam, is coated with an insulating resist film as happens in the negative lithography. The substrate accumulates electric charge and bent the new coming electron beam and the precision of the drawing is significantly reduced. The solution could be a photoresist material made by a conductive polymer with low resistivity and high ability to release the charge improving the antistatic property of the film. Nagasawa (2019) proposes a polyaniline composites using a polyaniline that has a high electric conductivity but poor dispersibility in water, together with a dopant polymer reach of sulfo-groups to improves water affinity.

Lassiter et al. (2019) deals with multi-layered thin film encapsulation of organic light-emitting diode (OLED) in order to protect the device from moisture and oxygen. The multilayer thin films are composed of stacks of alternating layers of the polymeric films and inorganic barrier layers (SiO_2, Al_2O_3, TiO_2 etc.). The inorganic layer is made by plasma enhanced chemical vapor deposition (PECVD), instead the polymeric layer is now deposed by inject printing. The curable ink typically includes one or more primary film forming monomers, one or more multifunctional cross-linking agents and one or more photo-initiators. The light source used to activate the photo-initiator is narrow spectrum (365–420 nm) UV light coming from a low power (200 mW/cm^2) emitting diode (LED). The low power and the fast vitrification of the surface, leave unreacted a lot of photo-initiator. It migrates in the polymeric film as volatile species and reacts with functional layers deteriorating the whole performance of the device. The Patent suggests a second exposure of at least 50 s to a plasma irradiation. The high energy ions and radicals easily react with residual photo-initiator prior its migration. The process results convenient because it does not require further apparatus and the second plasma irradiation can be easily combined with PECVD.

Wang and Qiang (2019) finds a totally different solution to multiple layer deposition. Standard techniques lead to an orientation switch, an order-disorder transition and a limited size variation of the thickness. The new method exposes a polymer film comprising a first polymer to a vapor comprising a polymerizable monomer, wherein the polymerizable monomer diffuse into, and swells, the polymer film. Then, the polymerizable monomers are reacted into a second polymer on the first polymeric layer. This method enables the hierarchical manipulation of nanopatterns in polymer thin film by changing their underlying composition. Changes in the thin film that can be controlled by the methods include changes in the morphology (shape), size and spacing in the domains within the polymer film. The initiator, when used, can be embedded in a polymer film by dissolving the first polymer and the initiator in a solvent, applying the solution as a coating on the substrate surface, and allowing the volatile components of the solution to evaporate. After the second monomer deposition the initiator is activated via exposure to an external stimulus, such as radiation, heat, and electron beam. The new layer is a self-assembled block copolymer organized in a plurality of lamellae or cylinders, horizontally or vertically oriented with respect the substrate surface. The thickness of the new layer goes from 10 to 500 nm. Example of material used are: polystyrene-b-polyisoprene-b-polystyrene (SIS) as first layer, diphenyl (2,4,6-tri-methylbenzoyl) phosphine oxide (TPO) as photo-initiator and isoprene as second monomer.

Photovoltaic devices are devices able to convert the energy of light radiation into electric energy. Currently most of photovoltaic devices uses pure crystalline silicon with high production cost. Alternative organic materials having conjugated, oligomeric or polymeric structure, are under scrutiny for the production of solar cells. They are light and cheap and as organic material can be recycled at the end of its use. Bianchi et al. (2019) found that anthradithiophene derivatives can be used in the synthesis of electro donor polymers. Electrodonors adsorb a photon from the light and dissociate it in two charges: a negative one transferred to an acceptor substance (typically a fullerene derivative) and a positive one maintained in the donor. Donor and acceptor transfer the charges to connected anode and cathode starting the electrical flow. Many polymers are already identified as donors and the Patent list them, however, they are not optimal because only a small part of solar radiation is converted to energy. Anthradithiophene derivatives can be functionalized to achieve electron rich and electron-poor monomers. Alternating monomers in the same backbone brings to an efficient polymer donor. The polymerization is done by a sequence Stille coupling reactions.

Many of the previous processes use polyimide as insulating material for semiconductor wafers because its outstanding physical properties. Unfortunately, polyimide typically requires a long time to cure when conventional heating techniques are used. A cure cycle of several hours is typical and often becomes the rate determining step in semiconductor fabrication. In addition, when polyimide is cured in a conventional surface the outer surface of the resin cures faster than the center portions. This causes various physical defects, such as the formation of voids, reduced modulus, enhanced swelling, increased solvent uptake and high coefficient of thermal expansion. A polyimide precursor may be applied to the substrate, as said by Urano et al. (2019) and then dried to prepare for imidization of the polymer. A goal of the drying process is to remove the solvent (NMP—*N*-methyl-2-pyrrolidone) and it can be also be important to remove oxygen without forming any bubbles in the layer. Moffat (2019) provides a process and (Moffat and Mccoy, 2019) a specific tool (vapor trap) for polyimide drying. The process uses a sequential set of descending pressure operations that allow for time efficient processing of wafers. The set of descending pressure operations are interspersed with evacuation steps using heated gases, which combine heating and solvent evacuation. The process results in layers with reduced or eliminated voiding, discoloration and solvent retention.

Liquid crystal display elements (LCD) have excellent visible light permeability and properties such as heat, chemical, crack resistance. Silicone polymers have been gaining attention because they have many of required properties. A silicone polymer can be synthesized by conducting a hydrolysis-condensation reaction of hydrolysable silicone compound in the presence of a base (or acid) and then removing a solvent for concentration (Smith, 1987). Mizuta et al. (2019) teaches how to make a silicone polymer with a reduced number of silanol end groups. When many of such silanol groups remain in the polymer, they keep reacting on the storage after production and the viscosity increases. The reaction is a hydrolysis/condensation in a water/alcohol solution using a base (e.g. tetramethylammonium hydroxide) as catalyst. The reactant is a commonly used silane (methacryl oxypropyl trimethoxy silane, methyltrimethoxy silane) and it is added dropwise keeping the temperature around 40 °C. After the polymer formation, the polymeric growth is stopped by the addition of a short chain organic acid (e.g. citric, acetic acid).

5. News in polymer composites

5.1 Cold-sintered ceramic polymer composite (CSP)

Schwartz et al. (2019) releases a quite broad methodology (composition and process) to produce cold sintered ceramic polymer composites. The cold process (<200 °C) and a wide variety of monomers permit the incorporation of diverse inorganic compounds in a polymeric matrix. The actual sintering process involves the application of high temperatures, typically above 1000 °C, to densify and to enhance properties of materials (strength, electrical/thermal conductivity, translucency). Although some advances where made reducing the production temperature (between 450 and 750 °C), the author claims incredibly low temperatures. The polymer composite is made by:

- An inorganic compound (mainly metal oxides) in the form of particle of 30 μm disperse in a solvent
- A monomer or oligomer or reactive polymer chosen in a broad variety of materials involving ring-opening, radical, thermal polymerization
- Additives for polymerization (initiator or catalysts) and fillers

In a stainless-steel die, exposing the mixture to a temperature <200 °C where polymerization starts and the solvent evaporates under a pressure control (<5000 MPa). Bolvari et al. (2019) enlarges the production process assuming that the inorganic compound is mixed with a preformed polymer solubilized in a solvent. In this case, the required sintering temperature must be higher than the melting point of the polymer and the boiling point of the solvent.

Ferranda (2019) covers the use of volcanic rock as polymeric additives. Volcanic rock of igneous or eruptive composition resulting from the cooling of the magma are porous and have high capability of adsorbing liquid and organic compound and can be used as deodorizing agent. Chemically, the structure of volcanic rock is composed on average by 90 wt% amorphous silico-aluminate and other metallic oxides.

5.2 Solid polymer electrolyte for batteries

Solid polymer electrolytes provide promising opportunity to tackle the growing market of rechargeable (secondary) batteries. They do not suffer of leakage but also, they do not promote lithium dendrites, i.e. microscopic damaging metal fibers. Hamon et al. (2019) proposes a compound of silica, fluorinated polymer and lithium salt, usable as solid electrolyte with improved electrical conductivity. In the form of a film between the two battery electrodes, the compound releases lithium ions when discharging and absorbs lithium ions when the battery is being charged. Keys finding of the patent are: the polymer is a polyvinylidenefluoride-*co*-hexafluoropropylene (PVdF-HFP, 98–2 mol%); the nanoparticles of silicon dioxide are dispersed in water having a pH higher than its isoelectric point; the mixture of the two ingredients is finely dispersed by a high shear mixer (e.g. microchannel homogenizer) to form a gel before drying; the lithium salt is an ionic liquid with pyrrolidium rings; the compound film is made by hot pressing.

A fuel cell is a cell for converting a reaction energy of a gas as raw material directly into an electric energy and hydrogen-oxygen fuel cells are studied for their little adverse effect on the global environmental since they produce water. They use a proton conductive ion exchange membrane as a polymer electrolyte and particularly a membrane of perfluorocarbon polymer having sulfonic acid groups., In polymer electrolyte fuel cells, gas diffusible electrode layers are disposed on the both side of the ion exchange membrane and a gas containing hydrogen as a fuel and a gas containing oxygen (e.g. air) as oxidizing agent are supplied to an anode and to a cathode respectively to generate electric energy. The reduction reaction of oxygen on the cathode proceeds via hydrogen peroxide generation especially when the humidity of the air is low. Only perfluoro polymers with sufonic groups have a greater stability to radicals than hydrocarbon polymers. Kawazoe et al. (2019) discovers an improved stability when cerium or manganese ions are present in the membrane, probably proving a rapid decomposition of the adsorbed peroxide. The process proposes a membrane produced from a liquid solution of a perfluorpolymer with a sulfonic terminal and carbonate of cesium (or manganese) in water (or alcohol).

5.3 Plastic metallization

Lin et al. (2019) deals with metallizing plastic components. Metallized plastic components have the same high-quality look and feel as chromium plated metal part but are lighter, corrosion-proof and less costly to produce. For these reasons, they are used in auto manufacturing for radiator grilles or mirror caps, as well as for electronic appliances and bathroom fittings. Plastic is gradually replacing metal as the basic material for such applications for good reason and their production method, injection molding, is a much cheap and environmentally compatible form of manufacturing. However, the patent highlights two major drawbacks of the industrial production:

- At the first step, the polymer (acrylonitrile-butadiene-styrene, polycarbonate or their blends) surface is etched using chromo-sulfuric acid to form micro-caverns in which metal will deposit and adhere. Chromo-sulfuric acid-based solutions are unfortunately toxic.
- After the etching step, the polymer surface is activated by means of a noble metal (i.e. palladium) and then metallized using electroless plating. Unfortunately, noble metals are rare and expensive.

The patent suggests to avoid both these problems making the polymer surface conductive by the superposition of a single or multiple layer of a pristine graphene material with null or limited non-carbon elements. Then a metal layer comprising a plated metal layer is deposited on the electrically conducting graphene and polymer surface. The quality of metal coating, mediated by the graphene sheet, can take a luxurious chrome look and exhibit superior abrasion resistance, barrier properties, heat radiation reflective properties, corrosion resistance, strength and hardness. The patent also discloses an apparatus that can be used to produce the surface-metalized polymer films.

5.4 Plastic on metals

Much more common in use is the problem of deposing a polymer on a metallic surface. However, some applications e.g. metal dental prosthesis, bring additional constrains to make the procedure very challenging. Implanted material is seen as "foreign body" by the living system, trigging an inflammatory response and a rejection of the implant. Promising polymeric prosthetic materials have been developed to allow the osteointegration of implants. They are made by "bioactive polymers" intended to reduce fibro-inflammatory issues. Migonney et al. (2019) proposes a direct grafting process wherein the polymer is grafted directly onto the surface of metal, without any intermediate molecule as anchor point. The technique comprises the following steps:

- The metallic surface is activated by acid and oxidizing bath of H_2O_2, where hydroperoxide groups (–OOH) are generated in a control manner
- Active donor species of free radicals are generated by UV irradiation (power 162 mW/cm^2 for less 90 min) in presence of a biocompatible monomer (e.g. sodium styrene sulfonate)
- The biocompatible polymer is formed by addition of other acrylic/methacrylic monomers in an oxygen free environment.

5.5 Fiber composite using thermoplastic polymers

Fiber composites are commonly made by impregnating fibers with low viscous thermosetting monomers/oligomers. The following thermal crosslinking makes a three-dimensional structure that hardens the material permanently. Example of thermosetting prepolymers are unsaturated polyesters, vinyl esters, epoxy and phenolic ones. A major disadvantage is the matrix rigidity and the impossibility of recycling. Thermoplastic polymers do not present these disadvantages, but they are too viscous for fiber impregnation even at high temperature or solution. Gerard et al. (2019) (1) and Escale et al. (2019) (2) offer a solution made by the mixture:

- Methacrylic polymer (mw > 50,000 g/mol in (1) and < 100,000 in (2))
- Methacrylic monomer
- Peroxide initiator (only in (1))

Used in a correct dosage, the mixture is able to completely wet the fibrous substrate (preferably mineral glass fibers). The thermoplastic composite material has satisfying mechanical properties such as high stiffness and a young modulus of at least 15 GPa. Under recycling, the polymeric material is partially recovered. Any special manufacturing process is available for the use of the new composite material.

5.6 Membranes made by thermoplastic and water-soluble polymers

A membrane is a thin, semipermeable structure capable of separating two fluids or separating molecular and/or ionic components or particles from a liquid. A membrane acts as a selective barrier, allowing some particles, substances or chemicals to pass through, while reining others. Many separations use membranes (osmosis, filtration) and all require high water resistance in combination with vapor permeability. The traditional material is expanded PTFE with subsequent bidirectional stretching, but it is under investigation for environmental concerns. Prissok et al. (2019) proposes a thermoplastic polymeric membrane, having pores with an average diameter <2000 nm, and a new production process. In a new imprinting polymerization method, the process starts to produce a film by a thermoplastic and a water-soluble polymer. As soon as the film is shaped, the water-soluble polymer is removed obtaining a porous membrane. The patent covers a large variety of polymer combinations: preferred thermoplastic polymers are polyurethanes and preferred water-soluble polymers are polyvinylpyrrolidones and polyethylene oxides.

5.7 High performance polymer and carbon nanotube composite

Poly ether ketone (PEK) and poly-2,5 benzidazole (ABPBI) belong to the family of high performance semi-crystalline thermoplastic polymers used in injection molded production of articles for high quality applications. The blend has a reasonable processability and superior performance in chemical resistance, temperature stability and low moisture absorption. However, the blend has not the required high electrical conductivity and tends to deform at low temperature (HDT—heat deflection temperature). Kapadia et al. (2019) finds a solution by adding multi wallet carbon nanotubes (MWCNT) between 0.5 and 5 wt%. Polymers and nanotubes are mixed in a dry powder blend, the blend is extruded to obtain granules followed by injection molding for the preparation of useful articles. The HTD growths up to 170–240 °C and the storage modulus still remains very high (150–400 MPa at 300 °C). To avoid agglomeration MWCNT fibers are pre-treated by ultra-sonification and later blended with the polymeric granules (0.1 10 μm).

Barcala Dominguez et al. (2019) mechanically disperses MWCNT in epoxy resin precursors. The dispersion is enough stable to coat different kind of surfaces. He claims that only a special three roll mill, described in Giovannelli (2019), provides enough shear rate (above 1×10^7 1/s) to break the strong nanotube interactions.

5.8 Antimicrobial composites

Chlorhexidine gluconate (CHG) is a preferred skin antiseptic over tinctures of iodine, iophors and alcohol. It is a substituted diguanidine salt with a high degree of antimicrobial activity, low mammalian toxicity and the ability to bind to the stratum corneum layer of skin and to mucous membranes. Articles with antimicrobial properties are largely request in sanitary applications but unfortunately, CHG cannot be processed at high temperature and above 40 °C decomposes giving the toxic p-chloroaniline. Mcguire (2019) releases a process where CHG is chemically bounded with an acrylic adhesive precursor and later fixed on a polymeric film by radiation curing. All the process is done at ambient temperature and CHG can be probably entrapped in the film formation. The author does not explain if the CHG is chemically or physically bonded to the substrate.

5.9 Nanocellulose polymer composite

Nelson et al. (2019) releases composition and process for a composite made by a hydrophobic polymer together with a nanocellulose -lignin material.

Unutilized lignocellulosic biomass feedstocks have the potential to be much cheaper than petroleum as well much better from an environmental life-cycle assessment. Lignocellulosic biomass normally comprises primarily cellulose, hemicellulose and lignin. The first two are natural polymers of sugar and the last one is an aromatic/aliphatic hydrocarbon polymer reinforcing the entire biomass network. It is desirable to separate from the biomass the cellulose, as nano fibrils and crystals, and the lignin. Then, both components form a masterbatch by extrusion or pelletization. The patent describes a whole multistep process where:

- after separation by mineral acid, the hemicellulosic and lignin fraction is fermented and reduced to sugars and repolymerized in a whole polymer structure
- the cellulose-rich fraction is instead blended with a hydrophobic synthetic polymer
- both parts recombine in a single composite pellet, that is some examples is 100% renewable.

5.10 Repelling animal polymer composite

Cables, such as power transmission, telephone, optical fiber cables, are used to transmit electricity/signal/data over distance. Rodents have been known to chew on cables and cause their failure. Baker (2019) proposes a protective material based on a multistep procedure. The method of preparing the aversive additive involves infusing an aversive material into an encapsulant, depositing a protective material around the encapsulant and compounding the aversive additive with a polymer. More precisely:

- Encapsulants are formed from a cell (e.g. yeast cell) from which the cellular contents have been eluted.
- The aversive material will trigger a flavor, olfactory or tactile response in the animal, so suggested materials are: cinnamaldehyde, wintergreen oil, capsaicin, peppermint oil, etc. It is fed to encapsulant by infusion.
- The protective material is a negatively charged component including: polyacrylic acid, ialuronic acid, sodium alginate and polystyrene sulphonate. Sometimes a layer-by layer deposition is performed using positively charged polydiallyldimethyl ammonium chloride (PDADMAC) and negatively charged polyacryliacid (PAA) sodium salt.
- The polymer is taken from a broad variety of materials including thermoplastic, and thermoset polymers. A preferred material is medium density polyethylene (MDPE) and the compound is made by extrusion

5.11 PVA mixture as support in 3D printing

3D printing process requires the deposition of drops of melted thermoplastic polymer. Certain structures like hollow sections cannot be built directly and first need a support structure printed on or around the final structure from the thermoplastic polymer is build. The support structure is prepared from a printable polymer which can be removed by thermal or solvent treatment after the final structure is made. Several thermoplastic polymers are already discovered as support polymer such as waxes, polyethylene glycol, high impact polystyrene, polylactic acid or hydroxy propyl methyl cellulose. Kumaki et al. (2019) covers the use of poly vinyl alcohol (PVA) because is a thermoplastic material with a broad melting range and can be easily removed from the final printed structure by dissolving in alkaline solution. PVA is mixed with polar polymers to improve the adhesion to the thermoplastic persistent material. Polar polymers are selected in a wide list including polylactic acid, polyvinyl butyral and polycaprolactone in a range between 5 and 50 wt%.

6. News in polymer plants

6.1 Reactor for PET recycling

The manufacture of bottles, tires or any kind of gum-like polymers is now promoting the recycling of articles, which are increasingly consumed by people, to reduce the volume of solid waste. An interesting method of Poly ethylene terephthalate (PET) recycling, called "chemical recycling," is based on the concept of depolymerizing the condensation polymer through solvolytic chain cleavage into low molecular-weight products which can be purified and reused as raw materials for the production of high-quality chemical products. The chemical pathway is detailed by Carta et al. (2003). Cortes Vergara (2019) provides a device, system and procedure for recycling with a high level of efficiency and simple and low-cost construction. The reactor is formed by a vertical structure composed of two ferrous concentric tanks that form an air jacket between them. The recycling material, broken in small granules, enters by a rocket valve from the top. In the lid of the reactor a chimney is also arranged for the discharge of flow of air from inside the reactor. Inside the inner tank, a stirring unit composed of a vertical axis with a plurality of blades of different positions and sizes, stir the reacting material. A hot air inlet at the bottom enters in the external jacket and from it to the inside tank gradually, allowing the suspension of the material, emulating a fluidized bed reactor. The reactor design allows different zones where different level of pressures and temperatures. In this way, in the upper part of the reactor temperatures in the range of 200–250 °C are obtained, while in the intermediate part the temperatures range between 300 and 400 °C. This difference of temperatures is also achieved thanks to other factors such as the great pressure inside the reactor, the exothermic reactions produced by the material processing and to the presence of the thermal jacket wrapping up the reactor with different grade of insulation.

6.2 Mixer-free reactor for solution polyolefin polymerization

Catalyst based solution polymerization of polyolefins requires a good temperature control achievable with a good mixing. The mixer-equipped reactor has a disadvantage of requiring an additional expensive motor for driving the mixer. Therefore, mixer-equipped reactor is disadvantages in terms of high facility investment cost, operation and maintenance cost attributable to the expensive motor. Kim (2019) projects a new reactor structure capable of preparing polymers having homogeneous physical properties by improving flowability of a reaction without using a stirrer. The scope is reached controlling and direct the flow inside a cylindrical vessel. A feed inlet installed at the lower portion of the reactor vessel allows the feed monomer, solvent and catalyst into the reactor. A guide pipe having a cylindrical shape being open at the respective ends is installed along the central axis of the vessel and divides an internal space into an up-flow region where the reaction mixture flows downward. A swirling flow-inducing blade attached to the exterior surface of the guide pipe, causes the reaction mixture to rise along the exterior surface of the guide pipe while forming a swirling flow. Patent shows qualitative reactor drawings.

6.3 Reactor for expandable microspheres

In cosmetic, hollow, air filled microspheres are used to provide volume and texture to hair without weighting it down over time. A microsphere consists of a thermoplastic polymer shell initially filled with a fluid (liquid or gas) such that, when heated, the shell softens and the fluid inside expands, resulting in a sphere the expands like a balloon up to as much as four times its initial size. After the heat source is removed, the shell stays in its deformed state. One common method for expanding polymeric microspheres is by slurring the spheres in a solvent, typically water, and heating in and extruder or spray dryer. Trahan and Mclaughlin (2018) provides a new solvent-free process that relays on a new type of reactor. The reactor is a jacketed vessel fitted with a single motion agitator comprising an anchor-type frame and shaft, with horizontal, pitched blades welded to the frame and shaft, and scraper blades affixed to the frame, sufficient to provide adequate turnover of the microspheres along the sidewalls of the tank, such that the resultant product is a free-flowing powder of specific, larger particle size. Patent shows drawings of reactor and stirrer from 40 to 200 gal volume. The particles size is between 10 and 120 μm and the chemical composition could be any thermoplastic polymer.

6.4 Aqueous polymer solutions device

Bernd et al. (2019) and Loesh et al. (2019) release an apparatus for the production of an aqueous polymer solution from polymer dry powders. Hydrated polymers are widely used in industry as thickeners and rheology modifiers. In solution they are in the form which is typically referred to as a gel. They may be synthetic (i.e. carbomers, polyacrylamides) or natural (i.e. CMC, guar, xanthan gums) and they are supplied in the form of dry powders. Such polymers are normally made up into aqueous solution at the site where they are intended to be used. If the powder is added to water incorrectly the hydrating particles can stick to the equipment and to each other, resulting in lumps or agglomerates in solution. Unfortunately, lumps or agglomerates tend not to dissolve once they are formed. Most commercially available powder make-up systems employ a screw feed to meet the powder into the water. The feeder consists of an Archimedes Screw type augur or scroll connected to a drive motor. Quite often the screw is blocked by the particulate materials, made sticky by the adsorption of water vapor.

The apparatus and the procedure suggested by the patents provide a uniform and lump free polymer solution in a continuous way. Core of the invention is a device for cutting the hydrate polymer particles into small pieces. Polymer particles drop in a tubular container. At the bottom of the container, a centrally mounted nozzle forms a water ascending stream that impacts the incoming particles. A high-pressure stream of water is ejected perpendicular to the axis of the container and rotates and forms a circular disc pattern around the central nozzle. A sieve tray is located in the bottom, beneath the nozzle and prevents oversized polymer lumps from passing. Several possible variations of the spraying elements are presented depending of the polymer and plant requirements.

6.5 Separation vessel for polyolefins

In high pressure polyethylene production (Fig. 1), after the tube reactor only 40% of the ethylene monomers react and a separation is required to recirculate the gas phase. This is done in two steps at decreasing pressure. Specifically, the second step at 300 bars is required to recover the last 10% residual gaseous ethylene and recover the solid polymer for admitting it into an extruder. The outer walls of the low-pressure separator vessel are continuously heated by means of steam jacket in order to maintain the polymeric material flowable. Unfortunately, the inner surface of the vessel, due to its closeness to the steam jacket, is covered by cross-linked degraded polymers. If the degraded polymers are not periodically removed, they can contaminate the final polyethylene and degrade its appearance. Cross-linked residue has a high molecular weight and increase the polydispersity of the melt polymer, and this increases the haze and reduces the transparency of the blown film products. To solve this problem, Cornelissen et al. (2019) suggest a removable cover of the vessel together with an internal 0.05 to 0.15 mm layer of polytetrafluoroethylene lining. The cleaning procedure suggests a cooling water in the jacket in the ending part of vessel unload. In this way a thin polyethylene film is formed on the internal surface but it has a very poor adhesion to the wall. The film is easily removed when the vessel cover is open. The cleaning process takes only 2 h compared with the one-day routine maintenance.

6.6 Spiral heat exchange for devolatilization of polyolefins

Also, in solution polyolefin production monomers and solvents need to be recycled back into the polymerization process. Usually, removal of monomers and solvents is achieved through devolatilization of the polymer solution stream. Devolatilization is achieved by raising the temperature of the polymer solution above the boiling point of the polymer solution and removing the volatile components. Typically, shell and tube heat exchangers (STHE) and plate heat exchangers (PHE) have been used as the preheater during devolatilization. None of them is fully satisfactory due to the high viscosity of the stream. Hong and Reimers (2019) introduces a new heat exchanger with increased heat transfer, smaller footprints, lower temperature gradients and lower pressure drops. The polymer solution stream enters to a spiral heat exchange from the top and produces a second effluent stream. When the first effluent stream flows through the spiral heat exchanger in a cross-flow direction relative to spirals performs a separation of the second effluent reverted back the exchanger. The separation provides a polymer free of solvent and a gaseous stream made by solvent and unreacted monomer. Patent provides drawing and schema of the unit.

Van Egmond et al. (2019) suggests to passivate the catalyst before devolatilization. The new unit is added in the recycle loop where the polymerized solids are contacted with catalyst poison compounds (selected from carbon monoxide or dioxide, oxygen, water alcohols or amines), thereby forming a passivated stream. A further gas stripping reduces the VOC content of at least 10% of the initial content.

6.7 Screen classification for polymer particles

Superabsorbent polymer particles must have a size between 150 and 850 μm for the use in diapers. Unfortunately, standard classification methods over screen of different pore size are not efficient. The most frequent problem is the blockage of the screen surface and the deterioration in the classifying efficiency. A further problem is the caking tendency of the product particles which leads to undesired agglomerates before, after and during the screening. A higher screening quality is typically achieved by adding surfactants to the product which serve to increases the free flow and reduce the stickiness of the particles. Funk et al. (2019) eliminates the use of surfactants by inventing a new screening process. A screening machine with 4 screens of different size operates under moderate vacuum conditions. A key condition is: the pressure above the uppermost sieve is 4.0 mbar per sieve higher than the pressure below the undermost sieve. The pressure profile inside the machine is the control parameter. High pressure drops over the sieves reduces the separation efficiency because particles that could not pass increase their agglomeration with the pressure drop. These findings combine with Pfeiffer et al. (2019), where the oversize fraction of polymer particles is recycled after deagglomeration with a roll crusher.

6.8 Multi-cascade reactor flow sheet for polyolefins

During certain intervals stirred tank reactors of reactor cascade used for slurry polymerization require maintenance work, for example because of failure or reactor equipment such reactor slurry pumps. The process has to be stopped and one or all reactors need to be emptied. Only the last polymerization rector can be isolated and its material recovered since fits commercial product specification. Polymers in other reactors must be disposed and treated as waste. Therefore, plant flow sheets should be revised to minimize the amount of off-spec materials. Rohrbach and Perzlmeier (2019) proposes a plant with two or more reactors equipped with and external loop able to recirculate on themselves but, if required, to transfer the whole content in one of each of the other reactors. On this set up, all the reacting streams can be easily processed up to the final conversion in the reactors do not involved in maintenance.

Siraux and Dams (2019) proposes the same three reactor cascade with another purpose: reduce the consumption of catalyst. Catalyst and co-catalyst are injected only in the first reactor, the co-catalyst is added in the second and nothing in the third. The consumption of catalyst is minimized without modifying the quantity and quality of the product.

6.9 Methods to avoid water contamination in polyolefin production

Water is generally a poison to polymerization catalysts and therefore its presence in the reactor system is undesirable. Startup is enabled by charging a seed bed of polymer to the empty reactor prior the introduction of the reactants. This seed bed is likely to have been exposed to the atmosphere and contains adsorbed water. Moisture trapped in polymer granules can generate electrostatic charges and modify the reaction course. Utilizing nitrogen purge to remove water from the reactor system prior startup can take days, with additional delays encountered as any residual water reacts with the first catalyst that is introduced. Hendrickson et al. (2019) proposes the use of a desiccant and a sacrificial amount catalyst in the polymer seed bed at startup. The desiccant can comprise a molecular sieve (dehydrated silica or alumina) because water does not have to chemically react with a scavenger and a simple physical sequestration can be sufficient to preserve the catalyst.

Singh (2019a) suggests a different solution that does not modify the composition of the bed. After the reactor is filled with the seed bed, an alkyl solution is injected; the bed is then soaked for at least 1 h; the concentration of ethane in the reactor is continuously monitored; the purge ends and the monomer can be fed when the ethane concentration reaches an equilibrium. The equilibrium of ethane concentration indicates that moisture has been removed and stripped out from the polymer bed. The alkyl solution comprises triethylammonium (TEAL) able to react with water and free ethane as byproduct.

6.10 Fluid bed reactor with improved cooling capability

One of the biggest challenges of exothermic polymerization reactions in gas solid olefin polymerization reactors is the cooling of the reactor. The fluidization gas moving upward through the dense phase where the polymerization takes place removes the most part of the heat. In order to prevent excessive powder entrainment into the fluidization gas stream, the superficial gas velocity has to be limited which results in inefficient mixing of the polyolefin and low heat exchange. Since the flow rate is limited two possible options are left: the entry temperature of the gas stream and the efficiency of the contact between gas and solid phase. Nyfors et al. (2019) and Weickert et al. (2019) basically redraw the fluid dynamic inside the reactor to maximize the heat exchange. An external loop moves back to bottom the gas stream and its temperature is efficiently reduced by external heat exchanges. The re-injection of the gas is split in several points from the bottom to the top in order to maintain a uniform movement of the powder inside the reactor. Patents detail the physical conditions of each entries.

7. News in polymer compositions

7.1 Dynamic cross-linked polymers

Thermoset polymers have superior dimensional stability, mechanical, thermal and environmental resistance and are irreplaceable in many demanding, but cannot be reshaped and reprocesses by heat or with solvent. In contrast thermoplastics, when heated can flow which permits extrusion, injection and molding of articles. The key is to design the chemistry so that at high temperature, exchange reactions enable stress relaxation and malleability and upon cooling, the exchanges become so slow that the topology of the network is essentially fixed (Montarnal et al., 2011). Transesterification as an exchangeable reaction is an equilibrium process, where the exchange between ester and alcohol creates a new pair of ester and alcohol through the alkoxy moiety interchange. Alidedeoglu et al. (2019), Chellamuthu et al. (2019), Groote et al. (2019), and Kumar et al. (2019) disclose methods of forming dynamic cross-linked polymers in a batch process using functional chain extenders. The aim is to dynamically cross-linked polybutylene terephthalate (PBT) for electronic and automotive applications improving process yield and reducing cost on an industrial scale. Butylene terephthalate oligomers are proposed in a two steps process with a straight control of the carboxylic acid end groups (10–60 mmol/kg) and the intrinsic viscosity

(0.7–1.2 dL/g). The compound is then used in a conventional polymer forming process, such as compression molding, profile extrusion, injection molding or blow molding to form the articles.

7.2 Biodegradable polymer for therapeutic applications

Synthetic peptidic drugs are efficient in cancer therapy since show high specificity, but they require high dosage if they are not delivered exactly into the cytosol of cancerous cells. Singh (2019b) provides a biodegradable block copolymer in a form of nanoparticles used as carriers for various agents, including therapeutic and targeting agents. The patent covers several polymers essentially made by a polylactic acid (PLA) and hydrophilic-hydrophobic block copolymers from synthetic origin. Among them, authors focus on the block copolymer of PLA with a polyethylene glycol-polypropylene glycol-polyethylene glycol (PEG-PPG-PEG). Limits are given on the molecular weight of the polymer to make particles with a size in the range of 30–120 nm. The therapeutic agent can be surface-conjugated or adsorbed on the biodegradable nanoparticles. The preparation starts dissolving PLA and PEG-PPG-PEG in an organic solvent together with a carbodiimide (DCC—*N,N*-dicyclohexylcabodiimide) and leaving the coupling reaction takes place (20 – 30 h). Then, the mixture is dissolved in water without the need of any emulsifiers, where particles of the required size are formed. The nanoparticles are amphiphilic in nature and may load hydrophobic and hydrophilic drugs in relevant amount due to the absence of emulsifiers.

7.3 Fracture resistant polymers

A material with high fracture toughness is well suited for use in ticker walled parts e.g. machine components, pipes and casings. Polyetheretherketone (PEEK) is a common material for many commercial applications because it is high crystalline and has outstanding chemical resistant properties. Both properties are needed in the electronic industry for components of mobile phones, tablets, etc. together with a good fracture toughness. Fracture toughness measures the energy required to propagate a crack through a test bar until the bar breaks. The propagation of a crack requires less energy in brittle than in ductile materials. Allcock et al. (2019) proposed a detailed procedure for a nucleophilic polycondensation of one or more bisphenols with one or more organic dihalide compounds, in the presence of sodium or potassium carbonates solubilized in an aromatic sulfone. The presence of organic dihalide compounds as residual impurities in the resulting polymer is minimized by a long termination step at high temperature (up to 320 °C) gradually reached in the course of the reaction. The molecular weight is monitored by measuring the torque of a stirrer motor driving paddle in the reaction mixture which has been previously calibrated.

7.4 Laurolactam polyamide with low yellowing

Laurolactam is a macrocyclic lactam used as a monomer in engineering plastics, such as nylon 12 and copolymers. Its use is mainly for covering cables and insulating material, while in automotive is used in oil and gasoline resistant tubes. However, it is used also as matting agent in cosmetic (body powder and skin creams) where a white color is required. Micoine et al. (2019) proposes a new process for monomer and polymer production which show lower yellowing compared to the prior art. It uses a standard method for the monomer synthesis (a Beckmann rearrangement of cyclododecanone oxime to give lauro-lactam in the presence of a catalyst), but introduces a purification to remove impurities before making the polymerization. It has been found that the presence of cyclic substance having 24 carbon atoms and at least one heteroatom (oxygen or nitrogen) and having a molar mass between 300 and 380 g/mol affects the color of the polymer. Impurities are generated as secondary reaction in the monomer preparation, can be easily detected by spectrometry (GC–MS) and removed by crystallization or distillation with at least five theoretical plates.

7.5 Polyamides with high melting point or high molecular weight

Semi aromatic polyamides with high melting points are commercially requested but are difficult to produce. It is known in the art that as the molar concentration of terephthalic acid increases in a polyamide, the melting point of the polyamide increases as well. However, as the melting point of the polyamide increases in solution polymerization, the temperature must be increased in order to keep the viscosity low. This causes several problems including a decomposition of the polymer itself before a valuable molecular weight is reached. Mutel and Doshi (2019) finds a solution splitting the production in two parts: a polymerization of a polyamides with a standard composition of aromatic acids; a second polymerization of short oligomers amine terminated, made almost exclusively by terephthalic acids. The two polymers are later joined in a reactive extrusion at 350–380 °C that last no more than 4 min. The process supplies a product with a required molecular weight and a tunable melting point above 300 °C.

Khare et al. (2019) proposes high molecular polyamide 6/66 copolymers. Interestingly, he utilizes cyclic oligomers analog of caprolactam, such as cyclic dimers and higher analogs (C3–C9). The compounds are unwanted by-products that are formed during the production of PA6 and PA6/66 copolymers. In fact, after the usual polymerization the cyclic oligomers are removed by a leaching process. The inventor found that cyclic oligomers react with diamine at elevated temperature (260 °C) to ring-open the cycle oligomers to produce amide-prepolymers that are end capped with amine groups. The prepolymers are then reacted with a diacid to form the 6/66 copolymers with a very high molecular weight (2–4.5 relative viscosity by ISO307:2007).

7.6 Polymer with urethane and thiourethane groups

Since polyurethanes are important industrial polymers, there is a demand to find process without the need of toxic isocyanates. Porta and Weis (2013) discloses radically polymerizable compounds with a cyclic five membered carbonate ring system (alkylidene-1,3-dioxolan-2-one). Urethanes groups are formed by reacting these compounds with diamine species. Rudolf (2019) broads the concept including mono thiocarbonate (MTC) instead of a simple carbonate. An MTC is a compound with one five-membered cyclic MTC group, i.e. a ring system with 5 atoms three of them are from MTC, –CO(=O)–S–, and the remaining two members are carbons, closing the cycle. They may have a molecular weight from 104 up to 100,000 g/mol. The polymerization conditions are similar to those used for the known carbonates and the polymers seem to have different properties. Patents do not cover any specific application but claim a better thermal resistance than traditional polyurethanes.

7.7 Bioactive polymer: Imidazolium macromolecules

Design of novel imidazolium-based polymers (Anderson and Long, 2010) remains as an exciting field since their ability to bond to metal as a ligand. The polymeric nature inherits almost all the properties of its structural unit, well known from ionic liquids. Such properties are tunable via varying anions without changing the polycation main chain and can serve a broad range of applications (cosmetic, pharma and electronic). It is not surprising though that (Lindner, 2019) releases a method of preparing polymeric imidazolium compounds without monoaldehyde and the related problems of an expensive purification. The process starts from an alpha-dicarboxylic compound, an oligo amine, a protic acid and less than 1 mol of a monoaldehyde, reacted with one another. Polymers of high molecular weight (ca 13,000 g/mol) are produced but with broad distribution (polydispersity >5).

7.8 Urea-urethane thickeners

Urea-urethane polymers are generally prepared by a two steps procedure where in the first step a monohydroxyl compound is reacted with a diisocyanate to give a monoisocyanate adduct; and in the second step the adduct is combined with a diamine in the presence of a lithium salt and solvent. Kou et al. (2019) proposes one of these polymers as thixotropic rheology modifier in paint and coating formulations. He observes that if the monohydroxyl compound (preferred are: butyltriglycol, 2-ethylhexanol and methoxy polyethylene glycol) is in molar excess (up to 1:1.5%) to the isocyanate (preferred is toluene isocyanate) the isocyanate still remains monosubstituted. This allows the further combination of adducts with diamines and reduces the viscosity of the adduct compound since the excess of monohydroxyl compound acts as solvent. This simple observation avoids unnecessary isocyanate distillation and the addition of further organic solvents.

7.9 Polymeric suspending agent for nanoparticles in lubricants

The effect of lubricant to reduce the friction between surfaces is sometime achieved to the use of solid inorganic nanoparticles (such silica and titration dioxide). The challenging problem is to stably suspend such particles in an organic solvent without agglomeration and sedimentation even when heat and shear forces stress the lubricant in use. Literature shows that polymers can be added to the surface via surface-initiated polymerization. A small molecule is first reacted with the particle surface and then a polymer is built from it. One problem of this method is that cross-linking is likely to occur at high monomer conversions because of the high density of reactive sites on the particle surface. Wieber et al. (2019) solves the problem attaching on the surface a preformed comb polymer, i.e. ester of methacrylic acid with a hydroxylated hydrogenated polybutadiene (Mw = 500–1000 g/mol). The backbone of the polymer is formed by the interlinked unsaturated groups of the methacrylic esters. The alkyl groups and the hydrogenated polybutadiene chains form the side chains. The macromonomers can be prepared by transesterification of short chain methacrylates with hydroxylated hydrogenated polybutadienes (i.e. a macro-alcohol). Inorganic nanoparticles used in the example are hexagonal boron nitride (hBN) nanoparticles with diameter between 5 and 100 nm.

7.10 Amphoteric and anionic polymeric flocculants

To concentrate an aqueous solid suspension which are mineral in nature such china clay, gold, copper or nickel tailing, is widely used a polymeric flocculant. Aqueous suspensions are concentrated in a gravity thickener vessel in a continuous manner. Suspension and a water solution of flocculant are met in the middle of a vessel, the flocculated solid settle to the bottom to form a bed of solids, which tend to undergo consolidation at the lower end of the vessel. The water accumulates to the top of the vessel where increasing the clarity. The solid is withdrawn continuously by a mechanical device at the bottom and a clear solution is pump away from the top. Polymeric flocculants are evaluated by respect the combination of dosage, settling rate and overflow clarity. It is known that anionic polymers are efficient in clarity and rate and non-ionic polymers excel in low dosage. Chipfunhu et al. (2019) finds the best compromise between two categories suggesting and amphoteric polymer made by: a backbone of acrylamide and alternate groups of anionic monomers (AMPS—2-acrylamide-2-methylpropane sulphonic acid) and cationic monomers (e.g. quaternary ammonium salts of dimethyl amino acrylate). The water solution polymerization runs till the formation of a solid gel shown to be very efficient in the flocculation of china clays.

Anionic flocculants are instead requested in paper and paperboard manufacturing process. They help the agglomeration of bleached kraft hardwood and softwood when combined with a cationic flocculant. Jehn-Rendu et al. (2019) rebalances the previous formulation to achieve an anionic polymer. The cationic monomer is then replaced by acrylic acid, but AMPS and acrylamide still remain in almost the same proportions. In this case, an inverse emulsion polymerization could be the preferred production method. Luly and Pointer (2018) protects the application of using anionic flocculants of acrylamide/potassium acrylate copolymers (Mw from 5 k to 5 M g/mol) to recover hydrogen fluoride (HF) in the manufacture of fluorides and separation of uranium isotopes.

7.11 Polymeric flame retardants

Brominated compounds are known to be highly effective as flame retardants and in many cases, they constitute the only viable option for reducing the fire risk of synthetic materials. It is postulated that the higher the molecular weight of the brominated compound, the lower is its volatility and its migration outside the plastic. So, the bio-accumulation in living tissue is reduced. Mark et al. (2019) found that traditional brominated polystyrenes (BPS) can be used as flame retardants in impact polystyrene (HIPS) and acrylonitrile-butadiene-styrene (ABS), but the idea was completed in Gelmont et al. (2019). In fact, it discovers a novel class of high molecular weight polymers having pendent penta-bromo-benzyl groups ($-CH_2C_6Br_5$), which are suitable retardants in many flammable materials (e.g. polyamide, polypropylene and HIPS and ABS). Low molecular weight compounds containing a penta-bromo-benzyl moiety are known in the art as pendant in acrylate polymers, but the patent largely expands the polymeric substrate. It found that the $-CH_2C_6Br_5$ group can be incorporated into aromatic ring containing polymers via a Friedel-Craft alkylation. The composition includes polymers where the brominated group is in the form of the side chains (like in polystyrene) or as part of its backbone chain (like in poly phenyl ether). The flame-retardant polymers have a molecular weight preferably less than 3500 g/mol with a Bromine content not less than 60%. The Friedel-Craft alkylation starts from a penta-bromo-benzyl halide (Chlorine or Bromine) in an organic solution of the base polymer. The reaction use $AlCl_3$ as a catalyst, runs between 60 and 90 °C and generates hydrogen bromide. The end of the reaction is determined by gas chromatography on the reacting halide or by cessation of hydrogen bromide evolution.

Ponticello et al. (2019) ideally completes the application of Bromine-polymeric retardant when they are used in expanded or extruded foam articles for insulation of buildings. For many years, hexabromo-cyclodecane has represented the main flame retardant, used industrially by all the main self-extinguishing expandable polystyrene manufactures. Recently, it has been observed that carbon fillers used as athermanous agents significantly accelerates the brominate flame retardant degradation process, with the consequent release of significant quantities of hydrobromic acid. This is responsible for corrosion on the production plants with the consequent deterioration of the metal parts and the transformation characteristic of the expanded polystyrene. The patented solution is the use of a flame-retardant mixture instead of a single species. The mixture is: a non-polymeric brominated flame retardant, a polymeric flame retardant containing at least 50% bromine and a synergistic additive. The synergistic agent is defined as a substance able to generate radicals that induce the degradation of the halogenated flame retardant with the formation of halogen radicals at the flame temperature. Said additives may be peroxide (di-cumyl peroxide) or an unstable hydrocarbon (2,3 dimethyl 2,3 diphenyl butane).

7.12 High refractive polymers for optical applications

A number of organic polymers have been developed as replacement for glass application such as optical lens, fiber optics, windows in nautical and aviation. They can provide advantages relative to glass, including shatter resistance, lighter weight, ease of molding and demolding and ease of dyeing. In ophthalmic applications, the use of polymeric materials with high refractive index require thin lens which is generally desirable. Moreover, preferred materials have a low adhesion to the interior of the glass molds during the casting process, leading articles with less defects. Kryger et al. (2019) releases a composition and process to produce polymers with high refractive index and ease demolding. Over a common although optimized polyurethane composition the invention focalizes on the addition of ionic fluoride and/or ionic fluoride precursors. The ionic fluoride is in the form of salt: potassium fluoride, cesium fluoride, tetrabutylammonium fluoride, etc. The fluoride precursor is a material capable of forming ionic fluoride in situ such as hexafluorosilicic acid and tris dimethylamino sulfonium difluorotrimethyl silicate. The fluoride materials can be added to the final organic polymerizable composition prior forming the articles. Patent also reviews several thermosetting organic polymers and indicates as best solution for transparent high refractive polymer a mixture of cycloaliphatic diisocyanates (e.g. H12MDI) and a polyol with $MW < 500\,g/mol$ (e.g. ethoxylated bisphenol A) and a polyol/polythiol containing both amine and hydroxyl functional groups (e.g. diethylene glycolamine).

7.13 Film forming and wash active polymers

Portion wise packaged detergents and cleaners (mono-dose units) are manifold for easy handling and correct dosage. Polyvinyl alcohol (PVA) film is used as packaging material since it dissolves at the start of washing process and releases ingredients that develop their effect. The biggest disadvantage of PVA is the inability to contribute to the washing performance. Some patents remedies to this introducing a copolymer made by a polyacrylic portion joined with an etho (Detering et al., 2019; Salzinger et al., 2019a) or etho–propoxylated (Fuchs et al., 2019) part. This is a film forming polymer soluble in water and able to act as a nonionic surfactant when fully solubilized. Patents adapts a radical solution polymerization method taken from literature (Bromberg, 1998) and covers several combinations of monomers and additives. A broad range of molecular weight 2000–100,000 g/mol is considered.

7.14 Cationic thickener for detergency

There is a steady demand for new and cheap thickeners for home and personal care cationic formulations (fabric softeners, dishwashing gels). Traditional thickeners are cross-linked cationic polymers made by inverse emulsion polymerization. Leyrer et al. (2019) and Weiss and Leyrer (2019) introduce a novelty because the product is made by two cationic polymers with different molecular weight so that their distribution is bimodal. The process provides the direct mixture and avoid coagulum formed if two different productions are simply mixed. The small molecular weight polymer (approximately 25 wt%) is water soluble and is linear. It works as anti-resoling agent and leaves the fibers clean of soil particles, such that graying has been minimized. By contrast, the high molecular weight part has a rather high amount of crosslink. It is almost insoluble in water and swellable in water acting as thickener. The polymerization is carried out in two subsequent steps. In the second step the cross-linker is completely absent and it is carried out after the end of first step.

8. News in flow chemistry

8.1 Polyamides in plug flow reactor

Although continuous process for polyamides production is known, many of the existing processes relate to batch operation. In a batch process, a water solution of diacids and diamine is polymerized in a cooled vessel, requiring large volumes of heat transfer fluid in the jacket to remove the reaction enthalpy. The rising polymer melt viscosity complicates water bubble formation, migration to the melt surface and steam disengagement which contributes to excessive foaming. In case of inefficient heat removal undesired polymer branching (gelling) is formed. Sutton et al. (2019) discovers that by concentrating the aqueous monomer solution by evaporation and continuously flowing it on the shell side of shell-and-tube reactor surprisingly provides a significant drop in the overall heat flux. This use of a plug flow reactor advantageously provides an overall heat efficiency, high conversion and low level of skin temperatures, foaming and branching. The reactor may be operated at a temperature from 200 to 300 °C, pressure from 13.5 to 18 bar, conversion greater than 85% and residence time from 50 to 60 min. The patent also highlights the possibility for retrofitting an existing process. In this case, the existing reactor is in series between the evaporator and the new shell-tube reactor.

Fogle et al. (2019) couples a polyamide continuous production with a continuous compounding process. Since polyamides are frequently compounded with fillers, reinforcing agents and modifiers, the invention proposes of introducing the molten polyamide from the reactor directly to the extruder. No significant solidification is allowed to occur between the polymerization and the subsequent compounding. By introducing a polyamide feed already in molten form, additional extruder barrels, normally used for melting the resin, can alternatively be used for increasing vacuum exposure. This increased vacuum exposure in turn results in increased volatile impurity removal and a clearer product. Additional benefit resulting from applying vacuum in more zones is that the final pressure can be higher and a reduced cost of the vacuum apparatus. Locating the extruder <500 ft from the reactor output, the patent claims the compound at the lowest possible cost.

8.2 Conducting polymer in continuous flow reactors

Kinlen and Zweig (2019) and Kohl et al. (2019) disclose a continuous flow process for synthesizing polyaniline. Polyaniline is a conducting polymer that has attracted considerable attention for its environmental stability an because its conductivity can be reversibly controlled by oxidation/reduction or by protonation/deprotonation (Ravindrakumar, 2014). The key problem of its production is a narrow molecular weight distribution hardly achievable in batch reactors. The polymerization temperature has a strong effect on both the yield and the molecular weight of the polymer. They both decrease with increasing the reaction temperature. Batch synthesis problematically introduces an exothermic spike at start-up degrading the whole production. The patent proposes a tight temperature control (<3 °C of variation) to avoid formation of low molecular weight polymers (not less than 30 monomer units). An excellent temperature control can be achieved using:

- Small scale reactors: 1 L or less in case of stirred tank; 10 mm or less as diameter in case of plug flow reactor
- High mixing capability improved with the addition of static/dynamic mixers

- Diluted monomer solution (0.2–0.5 M)
- Short residence time (0.85–1.2 h)

Reaction starts with an organic stream made by aniline converted in salt by a protonic acid (e.g. di nonyl naphthalene sulfonic acid) solubilized in a solvent such as an aliphatic/aromatic hydrocarbon. An oxidant stream is made by dispersion of a free radical initiator (e.g. ammonium persulfate) in aqueous solution. Both streams are cooled down before joined in at least one mixing device and then send to a tube or a plate reactor.

8.3 Silylated polymers for coatings and sealants

Silylated polymers are produced reacting a polyurethane prepolymer with a silylating agent (e.g. 3-isocyanoto propyl trimethoxy silane) in a tubular reactor with minimum back-mixing. It is important to reach high conversion because the unreacted monomer continues to react during storage changing the product viscosity and color. Huang et al. (2019a) provides compositions of tin-free formulations stable when moisture may be present. Instead, Banevicius et al. (2019) introduces a stabilization step after silylation in the continuous process. The stabilization is managed with a scavenging package able to reduce all the residual NCO groups. Example of the substance used as terminators are caprolactam and vinyl trimethoxy silane. Patent provides plant details where the stabilization step is performed into static mixers, heat exchangers or in combination with other reaction steps in an effort of process intensification.

9. News in emulsion/suspension polymerization

9.1 Core-shell re-dispersible powder for construction and optics

Re-dispersible polymer powders are widely used in cement-based waterproofing systems, floor screeds, tile adhesives, etc. and provide flexibility and a controlled rheological profile. They are made compatible to the matrix by spray drying on the polymer latex some anti-caking additives (colloids and silica). These additives prevent the irreversible aggregation during storage and promote the redispersion in the final application. Dietzsch et al. (2019) proposes a brilliant solution with a core-shell particle having a polymer core and a continuous silica shell. The key steps of the process are:

- Providing an aqueous polymer dispersion comprising polymer particle made by standard emulsion polymerization. The patent covers almost all the existing polymers in civil applications but the choice is not limited since commercial latexes can be used.

- Adjusting the pH in the range 7–10 by mineral acids after the addition of a mediator agent, if required. The mediator agent is a non-ionic or cationic surfactant the make the surface of the particle more neutral.
- Adding sodium or potassium silicate (water glass) to the aqueous polymer dispersion in the range of 5–40%, based on total weight of polymers.
- Drying (by fluid bed or spray tower or drum dryers) and achieving a powder.

The silicate makes an external layer (1.5–10 nm thick) over particles with diameter between 100 and 1500 nm. Thus, the powder has a high storage stability, does not lead to any coloration and improves the wet strength of the final product. The polymer core cannot unfold its properties and is protected from outside influences. Only after the shell is at least partially removed, by dissolution or shear stress, the core can develop its properties and is accessible to influences of the surrounding medium. Putzien et al. (2019) further improves the invention including water soluble salt of multivalent metal and anions (aluminate, borate, carbonate).

Core polymer particles with silicate shell are reference materials in sensor technology, in medical diagnostic and optic as point out in Sarma and Rurak (2019). Light scattering particles with spherical shape, diameter about 0.5 to 10 μm, refractive index of 1.59 for excitation at 589 nm and density between 1.09 and 1.53 g/cm^3 are very requested in optical applications. Common silicon dioxide particles are used but their density close to 2 g/cm^3 drives a rapid sedimentation. The patent uses polystyrene or polyvinylpyrrolidone particles covered with 5–50 wt% of a silica layer. Depending of the composition of the core, the surface can be modulated from smooth to multi pores and the area growths of up to three times. The silicate shell is formed directly on the core particle mainly by electrostatic and acid base interactions using a sol-gel first and a functionalization with an organic chlorosilane or alkoxysilane derivate.

9.2 Phase change materials (PCM) via emulsion polymerization

"Pickering emulsions" are special type of emulsifier-free latexes. Colloidal solid particles (nanoparticles of silica, zinc oxide, etc.) are positioned in the interface of the two miscible phases of the emulsion preventing the particles from coalescing due to the fact that total free-energy is reduced (particle-water, particle-oil interactions). Strictly speaking, this polymerization is a dispersion polymerization more than an emulsion polymerization. Beside that (Bucchi Alencastre Moroz et al., 2019) associates this method with the interesting concept of PCM. PCMs are materials able to storing and realizing thermal energy during the melting and solidification process. When the material initially in the liquid state cools down, it releases energy in the form of solidification (or crystallization) latent heat. Inversely, when a material initially in solid state melts some energy is absorbed from the environmental. PCMs in the form of thermo-functional nanostructures may generate products in cosmetic, pharmaceutics and medical equipment. The patent proposes a process where:

- Emulsification of the PCM (triglycerides of vegetal oil with melting point of 32 °C) with the colloidal oxide nanoparticles in aqueous phase, making a pre-emulsion
- Reducing the drop size of the pre-emulsion by high pressure homogenizer (50 °C–900 bars)
- Adsorption of the monomers (methacrylic acid and hydroxy ethyl methacrylate)
- Polymerization by sodium persulfate radical initiator at 70 °C
- Drying the product in solid form by fluidized bed drying or spray driers

The thermal effect is given by the ability of the nanostructure to reflects and diffuses UV radiation. A different method, described in Iwamoto et al. (2019), produces similar materials via interfacial polymerization of cyanoacrylate type monomers.

9.3 Composition control in suspension polymerization

Ion-exchange resins are produced in a form of polymeric beads by suspension polymerization. Typical monomers are styrene (S) and divinylbenzene (DVB) cross-linked copolymers. After polymerization a standard sulfonation process, using sulfuric acid, introduces (up to 95 mol%) sulfonic acid groups making the final strong acid cation (SAC) exchange resin. SACs are valued mainly by measuring the amount of water adsorbed by the polymeric beads via the MHC (moisture hold capacity) method. The goal of Savo et al. (2019) is to increase the MHC value and reduce the amount of DVB, because multifunctional vinyl monomers are more expensive than others. It is well known that due to the different reactivity in the suspension monomer droplets the more reactive monomer (DVB) tends to polymerize first than the less reactive (S). This causes a not uniform structure in the polymer particles between internal and external layers. The Patent proposes a semi-batch process where the ratio of DVB/S in the feed during polymerization is lower than the ratio used in the initial load. The change of the composition counter reacts the difference of reactivity and the whole copolymer composition results more uniform. Patent claims a saving of 12% of DVB per equal MHC.

9.4 Dispersion aids in suspension polymerization

The suspension polymerization of PVC requires two kind of suspension aids:

(1) Primary dispersants added to stabilize the vinyl chloride monomer droplets and control their size. They are mainly poly vinyl alcohols (PVAs) with high solubility and supplied in water solution

(2) Secondary dispersants needed to increase the porosity of the polymer particles after reaction and avoid their adhesion to the reactor wall. They are PVAs with a low saponification degree (70% or less) water insoluble and supplied in alcohol solution.

PVAs are industrially produced by polymerizing a vinyl ester monomer in an alcohol solution at atmospheric pressure and saponifying the resultant vinyl ester polymer. More hydroxyl groups are generated when saponification is high and the polymer becomes more water soluble.

Kaneshima (2019) focalizes on the production of primary PVA dispersants with low molecular weight discovering three new kinds of water-soluble molecular weight regulators. It discovers the transferring agent effect of alcohols, aldehyde or thiols having an aliphatic hydrocarbon group with 6–12 carbon atoms.

Omori and Kozuka (2019) found a secondary PVA dispersant efficient in its role and stable in water solution. The hydrophobic tails in the PVA are reduced combining adjacent tails in the structure: –C1CC(*R*)OC(–)O1 where R is an alkyl group. The new PVA is made by a sequence of 6 atoms rings with two Oxygens and a short alkyl branch. The method of obtaining the constituent unit is not limited and the preferred is by acetalization of PVA (partially saponified) with aliphatic aldehydes (e.g. acetaldehyde). By acetalization the hydrophobicity of the dispersant can be modulated to optimize the solubility and the surface activity. The residual alcohols of acetalization are left in the dispersion as useful stabilizers.

10. News in polymer extrusion

10.1 Polymeric polyphosphate additives in extrusion

When the screw speed is increased, defects occur on the surface of the extruded material known as "orange peel" or "sharkskin." Processing aids are used to improve the melt processing of polyolefins by reducing melt fracture but it is necessary that they do not lead to a deterioration in the physical properties of the extruded resin (mechanical properties, yellowing index, melt flow index). Jakupca et al. (2019) finds a series of polymeric phosphites that avoid melt fracture and do not cause issues. Polyphosphites are achieved by a nucleophilic substitution in aromatic phosphite ethers (e.g. triphenyl phosphite) with polyglycols and alcohols (e.g. polypropylene glycols) (Ogata and Okano, 1949). Additives can be applied in many target applications such as: blow film, pipe, windows and door frame in limited amount (0.01–2%). Comparatives examples show superior performance compared with common monomeric and oligomeric phosphites.

10.2 Porous article preparation

Microporous membranes or hollow fibers are required in a wide spectrum of applications covering microfiltration and ultrafiltration, hemodialysis, batteries separators reverse osmosis pretreatment and food packaging. The standard preparation methods use precipitation of the polymer in a solvent (such as *N*-Methyl-2-pyrrolidone or *N*,*N*-dimethyl formamide) followed by evaporation of the solvent that leaves cavities in the polymeric matrix. The use of solvent generates substantial environmental and disposal problems. Roller et al. (2019) over passes these problems using a mixture of a film forming polymer and a water-soluble additive (not <28 wt%). The mixture is made in an extruder and mold into an article. In contacting the article with water, the additive dissolves creating a pore network within the shaped article. Polymers tested are semi crystalline polymers like poly(phenylene sulfide) (PPS) and similar. A broad class of water-soluble additives are presented with preference for those that form a salt of sulfonic acid or carboxylic acid (e.g. Sodium benzenesulfonate, Sodium acetate).

10.3 Optimized additives for poly vinyl chloride extrusion

Siphuma and Van Rensburg (2019) provides a new lubricant able to reduce the viscosity, fusion time and torque in the extrusion of poly vinyl chloride (PVC). Additives are used in the processing of PVC because it melts in a non-homogeneous mixture of fused powder with different micro-domains of un-fused particles of different size (from 1 to 150 μm) causing an extreme high friction in the extruder. Some fatty acid salts or esters are used as internal lubricants since they reduce the fusion temperature and then the viscosity. Other lubricants (external) such as paraffin or polyethylene waxes are instead able to migrate to the surface of the particles and reduce the friction on the metal and interface. The mixture of the two types of lubricants in a single masterbatch is not feasible since the high viscosity of the blend. The Patent overpass all these problems finetuning the composition of the mixture (balance between oxidized and non-oxidized hydrocarbons) and carefully selecting the molecular structure of each ingredients (carbon atom per molecules between 40 and 100 in a linear backbone of more 75 wt%).

Claessens et al. (2019) relates to uses of acrylic polymer additives in PVC extrusion to provide oriented polyvinyl chloride (o-PVC). The acrylic polymers are methyl methacrylate and C_2–C_6 alkyl methacrylate copolymers of at least 4 M g/mol weight. o-PVC offers significant improvement over regular PVC in pipe production by maintaining burst and impact strength of the pipe while reducing its weight. To provide the o-PVC pipe, a regular pipe is extruded at half the desired final diameter and then stretched in the hoop direction to twice the original diameter. In doing so, the thickness of the pipe wall is also reduced by half, and the PVC chains align and orient, which increases the strength of the PVC material. Therefore, given a desired pipe diameter, an o-PVC pipe is half the weight and requires half the raw material of a regular PVC pipe, yet it maintains the same or better strength.

10.4 Grafting process with a low gel content

Functionalized ethylene-based polymers are made from polyethylene (LDPE, LLDPE, HDPE) grafting on the backbone small amount (0.5–5 wt%) of maleic anhydride (MAH) by extrusion. The grafted functionality helps chemical interaction of the ethylene-based polymers to polar polymer like nylon, ethylene vinyl alcohol (EVOH) and inorganic fillers. The grafted functionality also increases the surface polarity of ethylene-based polymers to improve adhesion to paints and inks. Unfortunately, the methods used to make highly grafted ethylene-based polymers tend to have inherent process inefficiencies, resulting in low production rates and typically produce polymers with high gel levels. In a typical preparation, the two ingredients are premix with a radical initiator (peroxide or azo compound) for an extended period and fed to the extruder. Ideally, the mixture should be homogenously melted before the reaction take place. Often at least, some of the initiator is activated due to either the shear forces generated within the extruder or reaching the initiator activation temperature before a homogeneity and this, in turn, allows the polymer to cross-link as opposed to react with the MAH. Depending upon the nature and amount of gels in functionalized ethylene-based polymer, the polymer may be unacceptable for application. Demirors et al. (2019) finds a solution feeding the MAH and initiator just in the middle of the extruder (6th barrel over a total length 12 barrel for a twin-screw extruder L/D 1:48). Moreover, the use of the combination of a 3,4-diethyl-3,4 diphenyl hexane (DEDPH) and a 2,5-dimethyl-2,5-di (*t*-butyl peroxy) hexane (DBPH) is optimal because they decompose at different temperatures and with different kinetic. Walther et al. (2018) combines the extrusion production of grafted ethylene-based polymers together with a polymer melt of pure ethylene coming directly from the polymerization reactor. The mixture is efficiently made by a static mixer before the production of pellets.

10.5 Extrusion of ultra-high molecular weight polyolefins

Ultra-high molecular weight polymers (UHMW) are polymers with a molecular weight higher than 0.75 Mg/mol. They cannot be processed by common techniques because their poor fluidity. However, they have desirable properties including high tensile strength, a low coefficient of friction, high impact and abrasion resistance. Tapsak et al. (2019) proposes a method to make them melt processable and injection moldable. The method includes a combination of applying a shear force to a polymer and heating the polymer. Initially a shear rate of at least 0.2 kwh/kg is applied to the UHMW at a temperature lower than its melting point. The polymer is transformed in a sheared polymer by an extruder (single, twin or triple screw) or mixer (turbo blender or granulator). Depending of the device, the step is called solid-state melt extrusion (SSME) or solid-state shear pulverization (SSSP). It is believed that applying a shear force to a polymer to form a sheared polymer, selectively cleaves very large chains but leaves shorter chains intact. In addition, free radicals generated during the process may react with shorter length polymers and thereby raise the overall number average molecular weight. Thus, the initial step lowers the average molecular weight (Mw) and raises the number average molecular weight (Mn), improving processability of the sheared polymer while improving tensile strength, impact resistance and toughness. The second step includes extruding the sheared polymer to form a processed polymer by heating it above its melting point. Here, both Mn and Mw decreases allowing the processed polymer to be useful for injection molding. The whole production can be performed in a single machine or two or more separated machines.

References

Abels F, Cwik T, Beyer R, Hermanutz F: *Process for the preparation of polymer fibers from polymers dissolved in ionic liquids by means of an air gap spinning process,* US Patent 48490, 2019.

Adams F, Rieger B: *Process for polymerizing beta-butyrolactone,* WO Patent 120846, 2019.

Alidedeoglu HA, Chellamuthu M, Kumar P, Groote R: *Methods of forming dynamic cross-linked polymer compositions using functional chain extenders under batch process,* US Patent 270847, 2019.

Allcock N, Ainsworth R, Harrington G, Scherrington J, Slater N: *Polymers and process for their manufacture,* EP Patent 3515966, 2019.

Anderson EB, Long TE: Imidazole and imidazolium containing polymers for biology and material science applications, *Polymer* 51(12):2447–2454, 2010.

Baker LS: *Method for encapsulating hydrophobic materials in stabilized yeast cells suitable for processing with polymers,* WO Patent 75010, 2019.

Banevicius J, Huang M, O'Keefe B, Kumar V, Barbera B: *Process for the preparation of silylated polymers having low color and color stability,* US Patent 112415, 2019.

Barcala Dominguez P, Giovannelli A, Giovannelli D: *Process for preparing suspensions of carbon nanotubes in polymers, suspensions thus obtained, and their use in production of composites,* WO Patent 3095, 2019.

Bauch CG, Englehorn BE, Lawson KW, Edwards TS, Trapp KW: *Olefin polymerization processes,* US Patent 144573, 20195.

Bernd S, Jeck S, Soetje O, et al: *A process and apparatus for producing an aqueous polymer solution,* US Patent 292330, 2019.

Bianchi G, Pasini D, Nitti A: *Anthradithiophene derivatives, process for the preparation thereof and polymers that contain them,* WO Patent 175367, 2019.

Binder B, Elliott S, Davidson GJ, Guo S: *Process for epoxidation of unsaturated polymer,* US Patent 002597, 2019.

Bobaru SC, Calabia GN, Plou VDM, Villuendas EM: *Thermoplastic polymer body with a surface structure, process for its manufacture and household appliance comprising the thermoplastic polymer body,* WO Patent 186312, 2019.

Bohling JC, Eryazici I, Harsh PR, Majumdar PS, Nungesser E: *Process for preparing an aqueous dispersion of polymeric microspheres,* US Patent 71540, 2019.

Boller TM, Crowther DJ, Pannell RB, Diop JM, Li W: *Polymerization processes,* US Patent 144580, 2019.

Bolvari A, Hoeks T, Dash R, et al: *Ceramic-polymer composites obtained by a cold sintering process,* US Patent 185382, 2019.

Borgemenke J, Cui S, Qin Y, Li ZLeY: Chapter five—bio-based polycarbonates from renewable feedstocks and carbon dioxide, *Adv Bioenerg* 4:183–208, 2019.

Brandl CL, Ehrman FD, Lattner JR: *Polymerization process,* US Patent 218318, 2019.

Briois J-F: *Process for producing a bio-based polyethylene terephthalate (pet) polymer, entirely from bio-based materials,* US Patent 185617, 2019.

Bromberg L: Novel family of thermogelling materials via c-c bonding between polyacrylic acid and polyethylene oxide-b-polypropylene oxide-b-polyethylene oxide, *J Phys Chem B* 102(11):1956–1963, 1998.

Brown HA, Karjala TP, Kardos LL, Gillespie DT, Polk ZL, Ortega J: *Ethylene-based polymers and processes to make the same,* EP Patent 3458512, 2019.

Bucchi Alencastre Moroz J, Iwamoto H, Chu Yan Ling Okuda J, et al: *A process for producing a thermofunctional nanostructure obtained via polymerization in emulsion,* US Patent 31936, 2019.

Bucevschi MD, Meiri Z, Shahal S: *Process for the production of biodegradable superabsorbent polymer with high absorbency under load based on styrene maleic acid copolymers and biopolymer,* WO Patent 195272, 2019.

Burnham T, Cockell B, Dix S, Kishbaugh L: *Blowing agent introduction in polymer foam processing,* WO Patent 083938, 2019.
Carta D, Cao G, D'Angeli C: Chemical recycling of polyethylene terephthalate (pet) by hydrolysis and glycolysis, *Environ Sci Pollut Res* 10:390–394, 2003.
Chee MB, Sum HN, Ho HLK, Yong-Jin Y: 3D printed microfluidics for biological applications, *Lab Chip* 15:3627–3637, 2015.
Chellamuthu M, Groote R, Alidedeoglu HA, Kumar P: *Methods of forming dynamic cross-linked polymer compositions using functional chain extenders under continuous process,* US Patent 276590, 2019.
Ching-Tai L, Dongming L, Holtcamp MW, et al: *Catalyst systems and polymerization processes for using the same,* US Patent 161559, 2019.
Chipfunhu D, Misslitz H, Haskell TM: *Amphoteric polymer, process for production thereof, and use thereof, to treat aqueous dispersions,* US Patent 100448, 2019.
Claessens S, De Smet A, Pel G, Hunter D, Delk V: *Process making polymer polyol having monodisperse dispersed polymer particles,* EP Patent 3436496, 2019.
Clymans PJ: *Process for manufacturing ethylene polymers and using modifiers,* EP Patent 3483188, 20195.
Cornelissen PH, Lammens HA, Franco FC, Morgan CK, Reynders L: *Separation vessels for use in polymerization processes and methods for cleaning same,* 2019.
Cortes Vergara SG: *Reactor, system and method for processing polymers,* EP Patent 3517205, 2019.
Costa LI, Fleckenstein PJ, Rosenboom J-G, Storti G, Morbidelli M: *Process to prepare polyester polymer and polyester polymer obtainable thereby,* US Patent 233583, 2019.
Crowther DJ, Boller TM, Pannell RB, Diop JM, Li W: *Polymerization processes,* US Patent 144579, 2019.
Cui S, Borgemenke J, Qin Y, Liu Z, Li Y: Biobased polycarbonates from renewable feedstocks and carbon dioxide. In *Advances in bioenergy*, vol. 4, 2019, Elsevier Inc, pp 183–207.
Daire ED, Galewski J-M, Simoens A, Soulage A: *Polymerisation unit and polymerisation process,* US Patent 118158, 2019.
Daniel T, Herfert N, Bauer S, et al: *Process for producing surface-postcrosslinked water-absorbent polymer particles,* US Patent 135993, 2019.
De Smet A, Claessens S, Pel G, Hunter D, Delk V: *Process making polymer polyol having monodisperse dispersed polymer particles,* EP Patent 3436496, 2019.
Demirors M, Ewart S, Karjala TP: *Process for producing functionalized ethylene-based polymers with a low gel content,* US Patent 31806, 2019.
Detering J, De Moragas M, Fuchs Y, et al: *Washing- and cleaning-active polymer films, process for their production and their use,* WO Patent 48474, 2019.
Dietzsch M, Leidinger P, Putzien S, Albrecht G, Schlesinger M: *A process for preparing core-shell particles having a polymer core and a continuous silica shell, an aqueous polymer dispersion obtainable by said process, a redispersible polymer powder, and a composition comprising the redispersible polymer powder,* US Patent 119453, 2019.
Dongming L: *Process for producing polymer compositions having multimodal molecular weight distribution,* WO Patent 067055, 2019.
Dos Santos Oliveira MG, Martins de Carvalho Josa M, Fernandes RDS, Tiago MSR: *Polymer material and process of obtaining it,* EP Patent 3467021, 2019.
Dou Q, Liu Y, Sumerin V, et al: *Polymer composition and a process for production of the polymer composition,* US Patent 256695, 2019.
Drabish J, Van Egmond J, Gelzer T: *Remote pressure sensing for polymer reactor control,* WO Patent 152105, 2019.
Drew KA: *Improved system and process for the manufacture of polymer foam with additives,* US Patent 092919, 2019.

Dyson P, Fei Z, Siankevich S, Savoglidis G: *Ionic polymers and use thereof in processing of biomass,* US Patent 31797, 2019.

EPO: *Espacenet, 2019:* 2019, European Patent Organisation. 1 October 2019. [Online]. Available, https://www.epo.org/index.html.

Escale P, Gerard P, Couffin A, Inoubli R, Hajji P: *Liquid composition or syrup comprising a multistage polymer and a (meth)acrylic polymer, syrup for impregnation process, impregnation process for a fibrous substrate, a method of polymerization and structured article obtained thereof,* US Patent 233603, 2019.

Eswaran VR, Reimers JL, Gallow KC: *Metallocene catalyst feed system for solution polymerization process,* US Patent 247821, 2019.

Exxon-Mobil: Exxon™ bromobutyl rubber compounding and applications manua, 2020: [Online]. Available: https://www.exxonmobilchemical.com/en/library/library-detail/2252/bromobutyl_rubber_compounding_and_applications_manual_en, 2020.

Fellahi S, Meddad A, FIsa B, Favis BD: Weldlines in injection-molded parts: a review, *Adv Polym Technol* 14(3):169–195, 1995.

Feller R, Hanns-Ingolf P, Wiesner U, et al: *Process for the production of isoolefin polymers with improved initiator system preparation,* WO Patent 011813, 2019.

Ferranda C: *Dispersion in the form of a stable suspension of volcanic rock, masterization of an engineered polymer, a polymeric composition, a process for preparing a polymeric composition, a product and use of a polymeric composition,* EP Patent 3464450, 2019.

Fogle RE, Calvert TD, Carben ZJ, et al: *Polymerization coupled compounding process,* US Patent 177495, 2019.

Fuchs Y, Detering J, Meise M, Schmidt-Hansberg B, Esper C: *Washing- and cleaning-active polymer films, process for the production thereof and use thereof,* US Patent 77895, 2019.

Fukushima M, Sagehashi M, Hasegawa K, Adachi T, Katayama K: *Monomer, polymer, resist composition, and patterning process,* US Patent 64664, 2019.

Funk R, Weismantel M, Peterson M, Hamilton P, De Kaey R, Possemiers K: *Classification process for superabsorbent polymer particles,* WO Patent 025210, 2019.

Gashler W, Enck S, Berkhout L, Guitman L, Kaczum J, Hajnal D: *Process for preparing an aqueous polymer dispersion having high chemical resistance,* US Patent 26394, 2019.

Gelmont M, Yuzefovich M, Yoffe D, Yelizarov R, Frim R, Melamed L: *Process for preparing bromine- containing polymers,* US Patent 144754, 2019.

Gerard P, Cauchois J-P, Hochstetter G, Glotin M, Perrin C, Gilles FR: *Impregnation process for a fibrous substrate, a liquid (meth) acrylic syrup for the impregnation process, its method of polymerization and structured article obtained thereof,* US Patent 241728, 2019.

Gerhardus M, Ulf S: *Olefin polymerization process in a gas-phase reactor having three or more polymerization zones,* US Patent 309107, 201910.

Giovannelli A: *Method for the dispersion of nanoparticles in a fluid,* EP Patent 2868370, 2019.

Groote R, Chellamuthu M, Kumar P, Alidedeoglu HA: *Methods of forming dynamic cross-linked polymer compositions using functional monomeric chain extenders under batch process,* US Patent 127519, 2019.

Hagiwara M: *Radically curable organosiloxane graft polyvinyl alcohol polymer and process for producing same,* US Patent 256663, 2019.

Hamon C, Pieri RR, Schlegel B, Mobidelli M, Wu H, Caimi S: *A process for the preparation of a solid polymer electrolyte useful in batteries,* WO Patent 170694, 2019.

Hanns-Ingolf P, Lovegrove J, Leiberich R, et al: *Process for the production of isobutene polymers with improved temperature control,* WO Patent 011814, 2019.

Hasegawa K, Fukushima M, Hatakeyama J: *Monomer, polymer, resist composition, and patterning process,* US Patent 79399, 2019.

Hendrickson GG, Lowell JS, Schwerdtfeger ED: *Method for startup of a gas phase polymerization reactor,* US Patent 309098, 2019.

Holtcamp MW, Dongming L, Ching-Tai L, et al: *Catalyst systems and polymerization processes for using the same,* US Patent 119413, 2019.
Hong Y, Reimers JL: *Spiral heat exchanger as a preheater in polymer devolatilization processes,* EP Patent 3507310, 2019.
Hu X, Yu H, Chen H, Gong Y, Zhang Y: *Process for producing foam beads using a modified ethylene-based polymer,* WO Patent 41216, 2019.
Huang M, Kunar V, Lim T, Kondos C, O'Keefe B: *Process for the preparation of silylated polyurethane polymers using titanium-containing and zirconium-containing catalysts,* EP Patent 3476876, 2019.
Hyberger R, Scheltjens G: *Soft plasma polymerization process for a mechanically durable super-hydrophobic nanostructured coating,* WO Patent 038378, 2019.
Impelman R, Stolz R, Dechellis M, Corona G, Sandell D: *Gas-phase olefin polymerization process,* WO Patent 099134, 2019.
Iwamoto H, Bucchi Alencastre MJ, Ling OJCY, et al: *A process for producing nanoencapsulated temperature regulating agent (ntra) via inferfacial polymerization,* US Patent 31937, 2019.
Jakupca MR, Lance JM, Regula JT: *Use of a polymeric polyphosphite as a polymer processing aid,* WO Patent 160845, 2019.
Jehn-Rendu C, Thomas A, Corberan RR: *Composition comprising cross-linked anionic, organic polymeric microparticles, its preparation and use in paper and paperboard making processes,* WO Patent 048587, 2019.
Jeol S, Pollino J: *Process for preparing functionalized poly(aryl ether sulfones) polymers and block copolymers resulting therefrom,* WO Patent 145235, 2019.
Kaneshima T: *Process for producing vinyl-alcohol-based polymer,* EP Patent 3466989, 2019.
Kapadia AS, Trivedi PD, Patil AT, Bhattacharyya AR: *Polymeric blend composite and a process for preparing the same,* WO Patent 138361, 2019.
Karjala TP, Kardos LL, Brown HA, Gillespie DT, Polk ZL, Ortega J: *Ethylene-based polymers and processes to make the same,* EP Patent 3420026, 2019.
Kawazoe H, Endoh E, Nakagawa H, Terazono S: *Liquid composition, process for its production, and process for producing membrane-electrode assembly for polymer electrolyte fuel cells,* US Patent 67722, 2019.
Kember M, Muir D, Chartoire A, et al: *Method for quenching a polymerisation process,* WO Patent 81932, 2019.
Khare N, Germain D, Nelliappan V: *A process for polymerizing cyclic oligomers of polyamides,* WO Patent 27593, 2019.
Kim Y: *Reactor for metallocene-based solution polymerization process of polyolefins,* EP Patent 3483190, 2019.
Kinlen PJ, Zweig AM: *Flow reactor synthesis of polymers,* US Patent 62502, 2019.
Kishbaugh L, Dix S, Burnham T: *Blowing agent introduction into hopper of polymer foam processing system,* WO Patent 169252, 2019.
Kohl T, Tsanaktsidis J, Hornung C, et al: *Continuous flow process for preparing conducting polymers,* WO Patent 113640, 2019.
Kondratowicz FL: *Vinyl aromatic polymer granulate and foam containing treated particles of anthracite as an athermanous additive and process for the production thereof,* WO Patent 057891, 2019.
Kong Q, Willemann R, Holleyn D, Shi Z, Yan K: *Process for extrusion moulding polymer compounds comprising poly(meth)acrylimide foam particles,* WO Patent 062731, 2019.
Koning C, Veldhuis T, Molhoek LJ, Buijsen PF, Drijhourt JP: *Polymers, processes, compositions and uses,* US Patent 292313, 2019.
Kou H, Tritschler U, Peter W, Auschra C: *Process for preparing urea urethane polymer,* WO Patent 96611, 2019.
Kryger MJ, Zawacky SR, Hickenboth CR: *Process for producing a molded polymeric article,* US Patent 112449, 2019.

Kumaki Y, Baier M, Keable HB: *Use of mixtures of polyvinyl alcohol with less polar polymers as support structure in 3d printing process,* EP Patent 3529049, 2019.

Kumar P, Alidedeoglu HA, Chellamuthu M, Groote R: *Methods of forming dynamic cross-linked polymer compositions using functional, polymeric chain extenders under batch process,* US Patent 119455, 2019.

Kuppuswamy S, Holtcamp MW, Bedoya MS, Stevens KA, Lue C-T, Rangwalla H: *Mixed catalyst systems containing bridged metallocenes with a pendant group 13 element, processes for making a polymer product using same, and products made from same,* US Patent 284312, 2019.

Laas HJ, Mager D, Matner M, Achten D, Hocke H: *Polyisocyanurate polymers and process for the production of polyisocyanurate polymers,* EP Patent 3424975, 2019.

Lammens HA, Clymans PJ, Comelissen PH: *Apparatus and process for high pressure polymerization of ethylene,* US Patent 2019299182, 201910.

Lassiter B, Moro L, Ramos TA, Tai E, Serrato A, Steinmann V: *Two-step process for forming cured polymeric films for electronic device encapsulation,* US Patent 136075, 2019.

Lee In J, Kang Won J, Kim Dong O, et al: *High-processability high-density ethylene-based polymer using hybrid supported metallocene catalyst, and preparation method therefor,* US Patent 169323, 2019.

Leeland J, Kember M, Landel M, Smith R, Morton A, Drane S: *A polymerisation process,* WO Patent 48878, 2019.

Lendlein A, Kratz K, Zhang Q: *Method of preparing a physically cross-linked, lactide-based polymer system, polymer system prepared by the method, and process of manufacturing an object from the polymer system,* WO Patent 63609, 2019.

Leyrer RJ, Fonseca G, Flores-Figueroa A, Boyko V, Dykstra RR, Sivik MR: *Process for obtaining a cationic polymer with an at least bimodal molecular weight distribution,* US Patent 62478, 2019.

Lin Y, Lee S-Y, Jhong Y-D, Zhamu A, Jang BZ: *Process for graphene-mediated metallization of polymer films,* US Patent 292675, 2019.

Lindner J-PB: *Process for preparing polymeric imidazolium compounds without or with less monoaldehyde,* US Patent 16858, 2019.

Liska R, Bomze D, Kern W, Knaack P: *Process for frontal polymerization of cationically polymerizable monomers,* US Patent 309123, 2019.

Liu Y, Sumerin V, Tupe R: *Polymer composition and a process for production of the polymer composition,* US Patent 292283, 2019.

Loesh D, El-Toufaili FA, Meckelnburg D, Ostermayr M: *A process and apparatus for producing aqueous polymer solutions,* WO Patent 81324, 2019.

Loick C, Smith S: *Superabsorbent polymers with improved odor control capacity and process for the production thereof,* US Patent 248939, 2019.

Luly MH, Pointer BE: *Process to recover hydrogen fluoride from hydrogen fluoride-polymer compositions,* WO Patent 67511, 2018.

Lundgard R, Mecca J: *A coating composition comprising polymer encapsulated metal oxide opacifying pigments and a process of producing the same,* EP Patent 3421552, 2019.

Lutz KE, Swarup S, Olson K, Schwendeman JE, Humbert LA, Boggs CL: *Coating processes using waterborne curable film-forming compositions containing polymers derived from natural gas raw materials,* EP Patent 3517222, 2019.

Lyondell-Basell: *Technical-and-comparative-data-on-petrothene-broad-molecular-weight-lldpe-for-blow-molding-and-extrusion,* 2020. «docplayer,» 6 March 2020. [Online]. Available.

Macdonald PJ, Werjefelt C: *Under instrument panel emergency vision apparatus,* US Patent 106219, 2019.

Mariani A, Bidali S, Fiori S, et al: UV-ignited frontal polymerization of an epoxy resin, *J Polym Sci A Polym Chem* 42(9):2066–2072, 2004.

Mark AG, Layman WJ, Kolich CH, Kumar G: *Low molecular weight brominated polymers, processes for their manufacture and their use in thermoplastic formulations,* EP Patent 3470441, 2019.

Maruyama H, Kondo K: *Silicone structure-containing polymer, photosensitive resin composition, photosensitive resin coating, photosensitive dry film, laminate, and pattern forming process,* US Patent 49844, 2019.

Mazzucco A, Mei G, Caputo T, et al: *Olefin polymerization process in the presence of antistatic composition,* US Patent 023885, 2019.

Mccauley KM, Tartamella TL, Bush ZA, Baldwin SM: *Process for producing functionalized polymers,* US Patent 185598, 2019.

Mcguire JE: *Methods of processing chlorhexidine-containing polymerizable compositions and antimicrobial articles formed thereby,* US Patent 29258, 2019.

Mei G, Mei G, Penzo G: *Polymerization process including discharging polyolefin particles from a gas-phase polymerization reactor,* WO Patent 025560, 2019.

Micoine K, Zumbansen K, Baumann FE, et al: *Process for preparing polymers from monomers comprising laurolactam,* US Patent 177478, 2019.

Migonney V, Baumann J-S, Falentin-Daudre C, Chouirfa H: *Process for grafting bioactive polymers onto metallic materials,* US Patent 16093, 2019.

Mizuta M, Seki H, Kobayashi H, Kato H: *Production process for silicone polymer,* US Patent 23848, 2019.

Moffat W: *Method for the rapid processing of polymer layers in support of imidization processes and fan out wafer level packaging including effiecient drying of precursor layers,* US Patent 10319612, 2019.

Moffat W, Mccoy C: *Trap assembly and system for trapping polymer vapors in process oven vacuum systems,* WO Patent 99401, 2019.

Montarnal D, Capelot M, Tournilhac F, Leibler L: Silica-like malleable materials from permanent organic networks, *Science* 334(6058):965–968, 2011.

Mueller-Cristadoro AM, Prissok F: *Process for producing polymer foams comprising imide groups,* US Patent 185611, 2019.

Mutel AT, Doshi SR: *Polymerization process,* WO Patent 67517, 2019.

Nagasawa T: *Conductive polymer composition, coated article, and patterning process,* US Patent 80888, 2019.

Nelson K, O'Connor R, Retsina T: *Nanocellulose-polymer composites, and processes for producing them,* US Patent 218372, 2019.

Nguyen PT, Freas D, Pham T, Chen Y-J, Blok E, Dias A: *Solvent-free process for functionalizing isobutylene-based polymers,* WO Patent 195111, 2019.

Nising P, Costa LI, Tancini F, Pfister D, Storti G, Morbidelli M: *A process to prepare a cyclic oligomer and a cyclic oligomer obtainable thereby,* WO Patent 139603, 2014.

North A, O'Reilly R, Mcavoy G: *Process for filling etched holes using first and second polymers,* US Patent 263657, 2019.

Nyfors K, Weickert G, Elovainio E, Kanellopoulos V, Prisen EJ: *Method for improving the cooling capacity of a gas solids olefin polymerization reactor,* WO Patent 96557, 2019.

Ogata Y, Okano M: Nucleophilic substitution in aromatic ethers. ii. kinetics of the methanolysis of 2,4-dinitrodiphenyl ethers, *J Am Chem Soc* 71(9):3212–3213, 1949.

Omori T, Kozuka Y: *Dispersion aid for suspension polymerization, and process for producing vinyl-based polymer using same,* EP Patent 3536714, 2019.

Padolewski JP, Tartamella TL: *Hybrid process for improved productivity in bulk polymerization,* WO Patent 133338, 2019.

Pannell RB: *Polymerization process,* EP Patent 3519447, 2019.

Pannell RB, Impelman RW: *Catalyst supply system and process for producing polymers,* US Patent 284310, 2019.

Pfeiffer T, Funk R, Weismantel M, Peterson M, Possemiers K, De Kaey R: *Process for producing superabsorbent polymer particles,* WO Patent 11793, 2019.

Pimpalkar NK, Shete AP, Ram F, Shanmuganathan K: *Process for preparing a homogeneous solution of a polymer and melanin,* US Patent 297880, 2019.

Pollino J, Jeol S: *Photocurable polymers, photocurable polymer compositions and lithographic processes including the same,* US Patent 153226, 2019.

Ponticello A, Ghidoni D, Pasquali F: *Expandable compositions containing aromatic vinyl polymers having self-extinguishing properties and improved processability,* US Patent 144622, 2019.

Porta GM, Weis M: *Polymerizable alylidene -1-3-dioxolan-2-ones and use thereof,* WO Patent 144299, 2013.

Poulopoulou N, Bikiaris DN, Papageorgiou DG, Floudas G, Papageorgiou GZ: Sustainable polymers from renewable resources: polymer blends of furan-based polyesters, *Macromol Mater Eng* 303(8):153–161, 2018.

Prissok F, Neff RA, Ahlers J, Weber M, Gronwald O: *Process for preparation of a porous membrane from at least one thermoplastic polymer and at least one water soluble polymer,* WO Patent 72754, 2019.

Putzien S, Albrecht G, Gaedt T, et al: *A process for modifying polymer particles of an aqueous polymer dispersion, an aqueous polymer dispersion obtainable by said process, a redispersible polymer powder, and a composition comprising the redispersible polymer powder,* US Patent 71545, 2019.

Ravindrakumar BG: Synthesis of polyaniline (PANI). In *Thesis, Jalgaon, SOPS, NMU,* 2014.

Reimers JL, Hong Y, Hagadorn JR: *Silica quenching agents for use in polymerization process,* US Patent 256621, 2019.

Rinaldi R, Mazzucco A, Balestra E, Pantaleoni E, Tartari D: *Process for gas-phase polymerization of olefins,* WO Patent 057595, 20193.

Rohrbach P, Perzlmeier J: *Multi-reactor slurry polymerization process,* EP Patent 3532507, 2019.

Rojek P, Kondratowicz L, Mikoszek-Operchalska M, Utrada K: *Process for the production of expandable vinyl aromatic polymer granulate having decreased thermal conductivity,* US Patent 283286, 2019.

Roller DB, Branham KD, Di Nicolo E, Campanelli P, Jeol S: *Porous article comprising a polymer and an additive, processes for their preparation and use thereof,* US Patent 241712, 2019.

Rost T, Balk R, Willerich I, et al: *Process for preparing an aqueous polymer latex,* US Patent 144585, 2019.

Rudolf P: *Process for the manufacturing of a polymer with urethane groups,* WO Patent 34470, 2019.

Sagehashi M, Fukushima M, Hasegawa K, Hatakeyama J, Katayama K: *Monomer, polymer, resist composition, and patterning process,* US Patent 235381, 2019.

Salzinger S, Brodhagen A, Fuchs Y, Lazinger D, Witteler H: *A process for the preparation of a polymer composition,* EP Patent 3497138, 2019.

Sarma D, Rurak K: *Process for synthesizing hybrid core-shell microparticles comprising a polymer core and a silicon dioxide shell with controlled structure and surface,* US Patent 91647, 2019.

Sauty N, Colvin H: *Process for producing epoxidized polymers,* US Patent 127495, 2019.

Savo AM, Jones T, Ostrander M: *Polymerization process involving addition of monofunctional vinyl monomer,* WO Patent 94199, 2019.

Schwartz E, Evans TL, Hoeks T, Van De Grampel RD, Leenders CA, Armstrong MJ: *Ceramic-polymer composites obtained by cold sintering process using a reactive monomer approach,* US Patent 225549, 2019.

Shanmuganathan K, Hee Cho J, Iyer P, Baranowitz S, Ellison CJ: Thermooxidative stabilization of polymers using natural and synthetic melanins, *Macromolecules* 44(24):9499–9507, 2011.

Siankevich S, Savoglidis G, Dyson PJ, Fei Z: *Ionic polymers and use thereof in biomass processing,* WO Patent 58270, 2019.

Singh D: *Methods of preparing and monitoring a seed bed for polymerization reactor startup,* WO Patent 173030, 2019.

Singh H: *Polymeric nanoparticles and a process of preparation thereof,* US Patent 91280, 2019.

Siphuma L, Van Rensburg VJ: *Chlorine-containing polymer composition comprising a chlorine-containing polymer and a wax comprising a fraction consisting of oxidized hydrocarbons and a fraction consisting of non-oxidized hydrocarbons, method of processing the polymer composition and T,* US Patent 233631, 2019.

Sipsos L: *Process for the production of a solid-state polymerized poly (tetramethylene-2, 5-furan dicarboxylate) polymer and polymer thus produce,* EP Patent 3523349, 2019.

Siraux D, Dams M: *Process for the preparation of ethylene polymers using a number of reactors arranged in series,* US Patent 194371, 2019.

Smith KA: Polycondensation of methyltrimethoxysilane, *Macromolecules* 20(10):2514–2520, 1987.

Sookray SH: *Processes for the production of polymer compositions from bio-based ethanol and compositions thereof,* US Patent 106533, 2019.

Strandberg R, Palm M: *Pipelined analog to digital converter,* EP Patent 3479483, 2019.

Sutton JP, Zabcik JM, Schwier C, Uzunpinar C, Akbari S: *Process for preparing polymers,* US Patent 263967, 2019.

Syurik J, Lscher HH: *Process for producing a porous polymer composition, porous polymer composition and article including the porous polymer composition,* EP Patent 3511368, 2019.

Tang HY, Chen L: Preparation and characterization of polymethacrylimide foam, *Adv Mater Res* 160–162:1309–1313, 2010.

Tapsak MA, Janse M, Patel B, Brunner P: *Processable polymers and methods of making and using,* EP Patent 3439846, 2019.

Tartamella TL, Mccauley KM, Bush ZA: *Process for producing functionalized polymers,* US Patent 10370460, 2019.

Trahan L, Mclaughlin R: *Process for expandable polymeric microspheres,* WO Patent 89722, 2018.

Urano H, Takemura K, Iio M, Hasegawa K, Funatsu K: *Tetracarboxylic acid diester compound, polymer of polyimide precursor and method for producing same, negative photosensitive resin composition, patterning process, and method for forming cured film,* US Patent 18320, 2019.

Van Berkel JG: *Process for enhancing the molecular weight of a polyester by solid state polymerization,* US Patent 031826, 2019.

Van Egmond JW, Goad JD, Kaarto JK, Chrismar DJ: *Process for efficient polymer particle purging,* EP Patent 3512624, 2019.

Van Hoyweghen D, Mitchell CA, Van Loon AM, Dharmarajan N, Datta S: *Low crystalline polymer compositions prepared in a dual reactor,* EP Patent 3436521, 2019.

Walther BW, Bawiskar SS, Pan J: *Process to form a composition containing functionalized and un-functionalized ethylene based polymers,* WO Patent 98796, 2018.

Wang M, Qiang Z: *A process for hierarchical manipulation of self-assembled polymer thin film patterns through in-film polymerization,* WO Patent 89915, 2019.

Weickert G, Prisen EJ, Nyfors K, Elovainio E, Kanellopoulos V: *Method of splitting the return fluidization gas in a gas solids olefin polymerization reactor,* WO Patent 96556, 2019.

Weiss T, Leyrer RJ: *A process for obtaining a polymer based on anionic or nonionic monomers with an at least bimodal molecular weight distribution,* WO Patent 016334, 2019.

Wieber S, Sirak S, Ness D, et al: *Polymeric-inorganic nanoparticle compositions, manufacturing process thereof and their use as lubricant additives,* WO Patent 145307, 2019.

Willemann RL, Kong Q, Hu P, Lu H: *Process for injection moulding polymer compounds comprising poly (meth) acrylimide foam particles,* WO Patent 052360, 2019.

Wurnitsch C, Vijay S, Elovainio E: *Polymerization reactor system comprising at least one withdrawal valve,* WO Patent 086623, 2019.

Zhujuan W: *Polymer composition, molded part and processes for production thereof,* US Patent 225760, 2019.

Index

Note: Page numbers followed by "*f*" indicate figures, "*t*" indicate tables, and "*s*" indicate schemes.

D

E

L

M

N

O

P

Q

R

9780128206454